T0297674

# A HAMMER IN THEIR HANDS

# A HAMMER IN THEIR HANDS

## A Documentary History of Technology and the African-American Experience

**edited by Carroll Pursell**

Published in cooperation with the Lemelson Center for the Study of Invention and Innovation at the Smithsonian Institution

The MIT Press
Cambridge, Massachusetts
London, England

First MIT Press paperback edition, 2006


This book was set in Stone serif and Stone sans by Asco Typesetters, Hong Kong.

Library of Congress Cataloging-in-Publication Data

A hammer in their hands : documentary history of technology and the African-American experience / edited by Carroll Pursell.
p. cm.
"Published in cooperation with the Lemelson Center for the Study of Invention and Innovation at the Smithsonian Institution."
Includes bibliographical references and index.
ISBN-13: 978-0-262-16225-8 (hc. : alk. paper) — 978-0-262-66199-7 (pb. : alk. paper)
1. Technology—United States—History. 2. Technology—Social aspects—United States—History. 3. African Americans—History. I. Pursell, Carroll W. II. Lemelson Center.
T21.H38 2005
608.9′96′073—dc22 2004057866

# Contents

## Introduction

John Henry is America's best-known and most beloved African-American folk hero. I have chosen a phrase from the ballad of his life for the title of this book to emphasize the long experience that black people have had with technology in America, but also the ambiguity of that experience. In one sense John Henry exemplifies the familiar trope of the sheer physicality of the black male—all brawn and no brain—that Colson Whitehead calls in his wonderful novel *John Henry Days* "an ideal of black masculinity in a castrating country."[1] At the same time, however, steel driving was a precise and delicate operation, demanding a high level of skill from both the driver and the shaker who held the bit. Whitehead vividly describes what could happen if either worker made a mistake: a new, young, and inexperienced shaker assigned to Henry didn't perfectly make the "two quick shakes and a twist" that were required, and "the first blow shattered half the bones in the boy's hand and the second shattered the other half. There was no way he could stop his hammer from coming down a second time. He was swinging his next blow before the first struck the bit."

There are at least two other senses in which the John Henry story provides a mixed message about blacks and technology. Like some other American folk heroes (Mike Fink and Old Stormalong come to mind), he was celebrated for matching his craft skills against the industrial mechanization (in this case a steam drill) that was destroying those same skills. At the same time, however, he was at work building a railroad, the quintessential symbol of that same technological modernization. The second irony is that while John Henry won that race against the steam drill, he died with his hammer in his hand. Like John Henry, black Americans have done much of the hard work of building this country, using what tools they were allowed with a level of skill, often denied, that those tools demanded. In searching for the meaning of the John Henry ballad, Joel Dinerstein posits an "accommodation/resistance dialectic in a distinctive African American inquiry into technoprogress." He quotes Ralph Ellison to the point that "what we [blacks] have counterpoised against the necessary rage for progress in American life (and which we share with other Americans) will have proved to be at least as valuable as all our triumphs of technology."[2]

This book is situated at the intersection of African-American history and the history of technology. Neither field was one with which historians dealt a half century ago: a few words about slavery and the "failed experiment" of Reconstruction, and a mention of the cotton gin and McCormick's reaper would have sufficed. But over these years, along with a host of other specialties (most notably women's history), African-American history and the history of technology have outgrown their prehistories and produced large and significant literatures of their own.

Those prehistories followed a common enough pattern: in the words of one women's historian, they tended to be recuperative, liberal, and individualist. They were recuperative in the sense that they attempted to discover and bring to the front people and episodes that had been long hidden behind the American metanarrative of progress through the political efforts of white men. They were liberal in that they tended to buy into that predominant notion of inevitable progress, and wanted only for it to apply also to their subjects. And individualist because they celebrated the personal achievements of those persons who had met and overcome challenges and roadblocks to success.

Gradually, however, both fields have become more complex, more theoretical, more radical in the sense of looking at the social and cultural patterns that shape experiences and the meanings we take from them. Both technology and race are now understood to be socially constructed, not fixed categories but ones that are always contested and historically contingent. These more recent understandings are, of course, shared across a broad spectrum of historical subfields, making it both possible and fruitful for specialists to talk to each other across turf boundaries. The "linguistic turn," for example, and theories of social construction and of gender, are providing us with a common language.

I have spent virtually my entire professional career studying the history of technology, and in the last few years have seen the ways in which it can converge with other subfields within history. Ruth Schwartz Cowan's classic study, *More Work for Mother: The Ironies of Household Technology from the Open Hearth to the Microwave* (1983), melds women's history and the history of technology;[3] Theodore Steinberg's *Nature Incorporated: Industrialization and the Waters of New England* (1991) does the same for environmental history[4] and Arwen Mohun's *Steam Laundries: Gender, Technology, and Work in the United States and Great Britain, 1880–1940* (1999) for labor history.[5] I have no special expertise in African-American history, but even a cursory glance at the literature of the history of technology reveals the almost total lack of attention to matters of race, just as gender was once ignored.

In part, the absence of African-Americans in our histories of American technology is a result of the fact that the field, as it is practiced, privileges design: traditionally, it is the accomplishments of engineers and inventors which have been the focus of histories. This is at least in part because the field developed within schools of technology, and many of its earliest advocates and practitioners were trained as engineers. They studied what they knew best and celebrated their own kind. In its early years, the Society for the History of Technology alternated its annual meeting between the American Historical Association (AHA) and the American Association for the Advancement of Science (AAAS), and historians and engineers took turns being president of the organization.

It is hardly surprising that women and people of color took little part in this enterprise. Besides the obvious fact that both groups were under-represented among historians, even smaller proportions appeared in the ranks of engineers and inventors. For most of American history both groups had been effectively—and in the case of African-Americans, explicitly—barred from holding patents, and long after legal restrictions were lifted, habit, culture, and practical considerations kept their numbers very low indeed. Again not surprisingly, white men defended their exclusive access to the well-paid and prestigious professions. Writing in 1903, one engineer warned of the "danger" that comes when "we have the man who fires the boiler and pulls the throttle dubbed a locomotive or stationary engineer; we have the woman who fires the stove and cooks the dinner dubbed the domestic engineer, and it will not be long before the barefoot African, who pounds the mud into the brick molds, will be calling himself a ceramic engineer."[6] The comment was made, significantly, in response to a call to teach the history of engineering to engineering students. The "true" history of the field, apparently, would inoculate it against the unwarranted incursions of those unfit by virtue of their class, sex, and race. Engineering was (as it continues to be) largely a white, masculinist preserve. The role of race in the creation of the new engineer at the turn of the century is one that is entirely ignored, though the "management of men" (and in the arena of colonial development, this meant men of color) was perceived to be a critical component of the manly and successful professional engineer.

Since it could be comfortably assumed that, almost by definition, no people of color or women had an important role in designing the built environment, they could justifiably be left out of the history of technology. Early recuperative efforts by women's and African-American historians did, of course, discover that some from these groups had made contributions as inventors and engineers, but it must be admitted that the numbers were small and the contributions themselves (as was true also of those of white men) were usually routine. To try to locate women and people of color in those precise areas from which they had been effectively barred was to undertake a task largely doomed from the beginning.

But design is hardly the whole of technology, nor is it done only by those whom society has formally designated as designers. We can ask many questions of technologies: where did they come from, how do they work, what do they do, who owns, operates, maintains, repairs them, and what do they mean? Of these questions, engineers and inventors are integral to only the first. It should be obvious that any history of technology that leaves out more than half the American population (women, children, people of color) is a very partial history, in both senses of that word.

The historian Michael Adas has suggested that the habit of judging people by the level of their technologies goes back to the earliest contact between European explorers and the people of Africa and Asia. Even more than skin color, he asserts, Europeans looked at technology as "the measure of men."[7] Those peoples who had technologies most like the white Europeans' were thought to be higher on some sort of scale of civilization than were those who managed with technologies not recognized by the invaders. In our own century anthropologists have done something of the same thing, ranking "primitive" peoples in rank order based on the amount of energy they use. Since technology is what allows us to use energies other than our own, it came to seem that more and more powerful technologies meant higher levels of civilization.

We all live in a world that is saturated with technology—and this has always been true. The stock and trade of historians, however, is change through time, and perhaps inevitably those things that seem to change the least attract the least interest. Therefore, for example, traditional tools and skills attract less attention than new ones. Additionally, history still is popularly believed to belong only to the "public" sphere of human activities: politics, diplomacy, wars, and the world of business. "Private" affairs such as housework, the rearing of children, the relationship between the sexes, the entire area of "popular" as opposed to "high" culture, often has been thought somehow not as important. This prejudice is deeply imbricated with gender and racial assumptions of hierarchical worth. To the extent that the "public" sphere is reserved for white men, all else, by default, is the territory of those who are not white enough and not sufficiently manly.

One common thread that runs through the histories of technology and African-Americans both is that of social construction, the idea that categories are not inevitable and fixed but contested and contingent. We are coming to see that gender roles, racial categories, and technological designs are all the result of specific decisions, that they could be other than they are, and that they are subject to constant change. The Irish can change from being nonwhite to being white; computers can change from corporate mainframes to personal laptops; secretarial work can shift from a masculine to a feminine occupation.

The social construction of race is a critical element in understanding the historically evolving relationship between technology and African-Americans because the barring of black people from full participation in the nation's technological culture was only partly a matter of law and custom; it was also a matter of racial beliefs and meanings.[8] Definitions of masculinity, for example, have changed significantly throughout our history, but one consistent element has been a competence with tools. From the colonial artisan to the contemporary "handyman," the idea of the truly masculine man has included being good with tools and, ideally, owning them. To a remarkable degree, this masculine ideal even has been generalized to an example of American exceptionalism: "Yankee ingenuity." A necessary part of this belief is that those who are not American, or at least not fully citizens (women and people of color, for example) are seen as not being handy with tools, not having technical skills (or temperament, or instincts), not being good at or comfortable with technology (women are bad drivers, "slaves" broke their hoes).

In order to make this belief halfway plausible, it has been necessary to define technology in such a way that only the sorts of activities reserved for white men count as using "technology." Ruth Cowan has pointed out that the kitchen is, among other things, a complex technological network: the stove, the fuel for the stove, the pots, pans, and utensils needed for cooking (all of which must be maintained and repaired as well as simply used), and all the tools required for the serving and eating of food. The person (a woman usually; in some times and places, especially a woman of color) who works in the kitchen is not commonly though of as a skilled technician designing meals. As the materials collected in this book make patently clear, enslaved Africans and African-Americans have been skilled artisans (cooks and seamstresses as well as coopers and shoemakers) and engineers, inventors and "steel-driving men," manufacturers and Internet Web-site designers. To ignore all of this rich and varied historical experience is not only to distort the history of American technology, but to reinforce racial and gender stereotypes that continue to disadvantage so many Americans.

The construction of these stereotypes did not happen by accident, nor were they inevitable. The myth of "Negro disability" has been a critical tool in the continuing effort to keep African-Americans in a position of inferiority. One scholar has recently asked "how do men inhabit simultaneously their color and their gender?"[9] Another, studying one film genre, finds white men portrayed as "hypermasculine *and* superhuman, white *and* strikingly tanned, a conqueror of both primitive, nonwhite men *and* futuristic machine technology."[10] Gender and race powerfully interact to undergird hierarchical power and advantage, and technology plays an important role in that process. White males, for example, construct as the Other not only women, but boys as well; neither are manly and therefore are not to be trusted with "real" technology. It is not surprising that to characterize enslaved Africans, for example, as both effeminate and childlike is to set them apart and below real men, and this hierarchy can be made explicit and concrete through the withholding of access to technology. Women and children are not competent to operate complex and powerful machines, and because black men have been defined as "unmanly," neither are they. Little wonder then that Frederick Douglass, for example, believed that he had first to become a "man" before he could become his "own" man, and that the former was intimately connected with learning the trade of ship-caulking.

The ways in which technological skills can be used as weapons to denigrate those who presumably lack those skills can be seen in the many jokes told by electricians at the turn of the twentieth century. The point of these jokes was that both those who told and those who heard them shared an expertise in electricity and those who were the butt of the jokes did not. The unfortunate targets of these jokes were the Others: women, farmers, immigrants, and especially African-Americans. This technique of affirming one's own technical skills by identifying and then deriding those who lacked them was so powerful and universal that even college-educated black electricians used it against other African-Americans who were not skilled. The denial of access for African-Americans to critical aspects of technology was a material handicap to their full equality; the ritual insistence upon their presumed technical incapacity was an equally powerful and damaging cultural handicap.

The selection of documents in this collection reflect several goals. First, they are intended to demonstrate that enslaved Africans, and African-Americans after them, always have been deeply involved in the creation and use of a full range of technologies, and that they therefore must now be included in histories of American technology. Second, the selections are intended to be examples of the myriad sources available for such future projects. Advertisements for fugitive slaves, laws and ordinances, reminiscences, the black press, government reports, African-American jokes and folklore, protest pamphlets, and popular songs—the list could go on and on. Third and finally, these selections are intended to be a starting point for scholars who might wish to dig deeply into the subject. This collection is meant only to be suggestive, and the sooner it is surpassed, the better.

Part I covers the colonial period of American history. The documents illustrate ways in which enslaved Africans contributed technical skills that they brought with them to the New World, and once here they were taught new skills. From field hands to almanac makers, they exhibited the same range of competence held by white Americans.

Part II carries the story up to the Civil War. The major themes in this section are the persistence of artisanal skills, particularly in the South, and the simultaneous coming of the Industrial Revolution, particularly to the northern states. Black artisans, inventors, and entrepreneurs played important roles in the emerging national economy despite the persistence of slavery in the South and racial prejudice throughout the country.

Part III covers the years from the beginning of the Civil War to the turn of the twentieth century. The abandoned effort of Reconstruction and the construction of the era of segregation provided new forms for the old restrictions imposed on African-Americans but, as in the past, many continued to effectively use those technologies to which they gained access and to push for full participation in the new industrial order.

Part IV looks at the Progressive era, from the turn of the twentieth century through World War I. On the rhetorical level, these years were dominated by a debate among African-Americans over the proper nature and role of "industrial education." The long-standing concern that black ingenuity be documented through patent records and lists of inventions also continued to be evident, and a burgeoning black press took pains to report cases of successful black manufacturers and business people.

Part V covers the two interwar decades. Despite the Great Migration northward during the war, African-Americans remained a major presence in the South, largely engaged in agriculture but already beginning to feel the effects of the industrialization of that occupation which would drive so many thousands off farms after World War II. Those who had gone north around the war years found sufficient barriers to full and equal employment; women of color, as always, suffered the double handicap of their sex and their race. Of the many new consumer technologies that helped fuel the growing economy of the 1920s, none was more important than the automobile. While African-Americans used the automobile to escape the yoke of Jim Crow practice, it was found to be enmeshed in new forms of that same unequal treatment.

Part VI looks at the period from the beginning of World War II to the early 1970s, a time of postwar prosperity and growing demands for civil rights. As in World War I, the sudden need for war workers provided industrial opportunities for African-American workers from aviators and engineers to former domestics. Agricultural mechanization quickened its toll on farm workers' livelihoods, but new work sites, such as the burgeoning aerospace industry, put black workers in closer contact with the emerging high-technologies of the postwar era.

Part VII carries the story to the end of the twentieth century. The assertion of "black power" in the early 1970s raised fundamental issues of how African-Americans should position themselves with regard to the rapidly evolving technologies of power in America. And the deteriorating condition of emerging nations in Africa raised the issue of what, if any, obligations the African diaspora placed upon African-Americans. Access to emerging technologies was partly a matter of engineering education, but perhaps even more one of access to the technologies themselves. Of particular concern was the computer and the exploding importance of the Internet. And finally, since new technologies always embody a trade-off between winners and losers, how could communities of color protect themselves from being used as dumping grounds for the inevitable waste produced by "progress"? At the beginning of the new millennium, these questions are—as they have always been—answered in many ways, and never completely or finally.

## Notes

1. Colson Whitehead, *John Henry Days* (New York: Doubleday, 2001), 83, 189.

2. Joel Dinerstein, *Swinging the Machine: Modernity, Technology, and African American Culture between the Wars* (Amherst: University of Massachusetts Press, 2003), 123.

3. See also Nina E. Lerman, Arwen Palmer Mohun, and Ruth Oldenziel, "The Shoulders We Stand On and the View from Here: Historiography and Directions for Research," *Technology and Culture* 38 (January 1997), 9–30.

4. See also Joel Tarr and Jeffrey K. Stine, "At the Intersection of Histories: Technology and the Environment," *Technology and Culture* 39 (October 1998), 601–640.

5. See also Philip Scranton, "None-Too-Porous Boundaries: Labor History and the History of Technology," *Technology and Culture* 29 (October 1988), 722–743.

6. Discussion by H. W. Tyler of J. A. L. Waddell, "The Advisability of Instructing Engineering Students in the History of the Engineering Profession," *Proceedings of the Society for the Promotion of Engineering Education* 11 (1903), 201–202.

7. Michael Adas, *Machines as the Measure of Men: Science, Technology, and Ideologies of Western Dominance* (Ithaca: Cornell University Press, 1989).

8. The social construction of race is a vast subject, but a good starting point is the essays in David Theo Goldberg, ed., *Anatomy of Racism* (Minneapolis: University of Minnesota Press, 1990). For a discussion of white as a race, see David Roediger, *Towards the Abolition of Whiteness: Essays on Race, Politics, and Working Class History* (London: Verso, 1994), especially his introductory essay, "From the Social Construction of Race to the Abolition of Whiteness." See also his *The Wages of Whitness: Race and the Making of the American Working Class* (London: Verso, 1991).

9. Michael Uebel, "Men in Color: Introducing Race and the Subject of Masculinities," in *Race and the Subject of Masculinities*, ed. Harry Stecopoulos and Michael Uebel, 2 (Durham,

NC: Duke University Press, 1997). See also Gail Bederman, *Manliness & Civilization: A Cultural History of Gender and Race in the United States, 1880–1917* (Chicago: University of Chicago Press, 1995). For women, see Evelyn Brooks Higginbotham, "African-American Women's History and the Metalanguage of Race," *Signs* 17 (Winter 1992), 251–274.

10. Uebel, "Men in Color," 7.

# I COLONIAL ERA

Although the Americas to which the European settlers came in the sixteenth and seventeenth centuries was by no means unpopulated, the deadly combination of new diseases and more powerful technologies gave decided advantage to the new arrivals. Some Native American technologies were adopted, like the bark canoe and the cultivation of tobacco, but more often the colonists relied upon the medieval technologies they brought with them in their ships' cargoes or in their memories: plows and water mills, carpenters' tools and kitchenware. For several generations it was a struggle merely to reproduce these tools and to keep up artisanal skills in a land where farming was by far the most common occupation. Most work was done by hand, using only hand tools, and most travel was by foot or by water.

Particularly in the southern colonies, which stretched down from Pennsylvania and at that time included the British islands of the Caribbean, European settlers attempting to grow cash crops of rice, tobacco, indigo and later cotton, imported "servants" to do the hard work on plantations. Increasingly these were enslaved Africans whose "servitude" was not only for the length of their own lives, but passed on to their children. These unwilling immigrants too brought over their technologies, most often only in their memories. To an extent we have not yet measured, some of these African technologies joined the mix of tools and techniques put to work in the colonies. At the same time, since large southern plantations functioned somewhat as relatively autonomous communities, enslaved Africans were taught artisanal skills so that the buildings and boats, clothing and shoes, barrels for shipping, nails for construction, food for the workers, and all the other needs of the community could be met on-site. The South, with only a handful of cities, was considered a poor place for free artisans, black or white, to ply their skills. Enslaved Africans were, to a remarkable degree, the technologists of the southern colonies.

# 1 African Medicine in the New World

Cotton Mather on Smallpox Inoculation (1716)

An Account of the Method and Success of Inoculating the Small-Pox in Boston (1722)

## Cotton Mather on Smallpox Inoculation (1716)

Smallpox was one of the most common, virulent, and feared diseases of the eighteenth century in America as in Europe, striking down large numbers of Native Americans as well as colonists. During the New England epidemic of 1721, the Puritan preacher and colonial leader Cotton Mather attempted to convince the colony that people should be deliberately inoculated against the disease. It was a technique that he had learned in 1705 from Onesimus, whom he called "my Negro-man." In this letter to a friend in England, Mather described how he had found out about the procedure.

---

All that I shall now add, will be my Thanks to you, for comunicating to the Public in Dr *Halley*'s Transactions, ye Account which you had from Dr *Timonius*, at *Constantinople*, ye Method of obtaining and procuring ye *Small-Pox*, by *Insition*; which I perceive also by some in my Neighbourhood lately come from thence, has been for some time successfully practised there. I am willing to confirm you, in a favourable Opinion, of Dr *Timonius*'s Comunication; And therefore, I do assure you, that many months before I met with any Intimations of treating ye *Small-Pox*, with ye Method of Inoculation, any where in *Europe*; I had from a Servant of my own, an Account of its being practised in *Africa*. Enquiring of my Negro-man *Onesimus*, who is a pretty Intelligent Fellow, Whether he ever had ye *Small-Pox*; he answered, both, *Yes*, and, *No*; and then told me, that he had undergone an Operation, which had given him something of ye *Small-Pox*, & would forever præserve him from it; adding, That it was often used among ye *Guramantese*, & whoever had ye Courage to use it, was forever free from ye fear of the Contagion. He described ye Operation to me, and shew'd me in his Arm ye Scar, which it had left upon him; and his Description of it, made it the same, that afterwards I found related unto you by your *Timonius*.

This cannot but expire, in a Wonder, and in a request, unto my Dr *Woodward*. How does it come to pass, that no more is done to bring this operation, into experiment & into Fashion—in *England*? When there are so many Thousands of People, that would give many Thousands of Pounds, to have ye Danger and Horror of this frightful Disease well over with you. I beseech you, syr, to move in, and save more Lives than Dr *Sydenham*. For my own part, if I should live to see ye *Small-Pox* again enter into our City, I would immediately procure a Consult of our Physicians, to Introduce a Practice, which may be of so very happy a Tendency. But could we hear, that you have done it before us, how much would That embolden us!

---

Reprinted from: Letter to John Woodward, from George Lyman Kittredge's introduction to Increase Mather, "Several Reasons Proving That Inoculating or Transplanting the Small Pox, Is a Lawful Practice . . ." (1721).

## An Account of the Method and Success of Inoculating the Small-Pox in Boston (1722)

The next year a letter from an American to his friend in England added information on the background of the inoculation procedure.

---

Sir,

Gentleman well known in the City of *Boston*, had a *Garamantee* Servant, who first gave him an Account, of a Method frequently used in *Africa*, and which had been practis'd on himself, to procure an *easy Small-Pox*, and a perpetual security of neither *dying* by it, nor being again infected with it.

Afterwards he successively met with a Number of *Africans*; who all, in *their* plain Way, without any Combination, or Correspondence, agreed in *one Story, viz.* that in their Country (where they use to die like *Rotten Sheep*, when the *Small-Pox* gets among them) it is now become a *common Thing* to cut a Place or two in their Skin, sometimes one Place, and sometimes another, and put in a little of the Matter of the *Small-Pox*; after which, they, in a few Days, grow a *little Sick*, and a few *Small-Pox* break out, and by-and by they dry away; and that no Body ever dy'd of doing this, nor ever had the *Small-Pox* after it: Which last Point is confirm'd by their constant Attendance on the Sick in our Families, without receiving the Infection; and, so considerable is the Number of these in our Neighbourhood, that he had as evident Proof of the *Practice, Safety*, and *Success* of this Operation, as we have that there are *Lions* in *Africa*.

After this, he heard it affirmed, That it is no unusual Thing for our Ships on the Coast of *Guinea*, when they ship their Slaves, to find out by Enquiry which of the Slaves have not yet had the *Small-Pox*; and so carry them a-shore, in this Way to give it to them, that the poor Creatures may sell for a better Price; where they are often (inhumanly enough) to be dispos'd of.

Some Years after he had receiv'd his first *African* Informations, he found publish'd in our *Philosophical Transactions*, divers Communications from the *Levant*, which, to our Surprize, agreed with what he had from *Africa*.

First, That very valuable Person, Dr. *Emancel Timonius*, writes from *Constantinople*, in *December* 1713, That the Practice of procuring the *Small-Pox*, by a Sort of *Inoculation*, had been introduc'd among the *Constantinopolitans*, from the more Eastern and Northern *Asiaticks*, for about forty Years. At the first (he says) People were cautious and afraid; but the *happy Success* on Thousands of Persons, for (then) eight Years past, had put it *out of all Suspicion*. His Account is, That they who have this *Inoculation* practis'd upon them, are subject to very *slight Smptoms*, and sensible of but very little Sickness; nor do what *Small Pox* they have ever leave any *Scars* or *Pits* behind them.

---

Reprinted from: *An Account of the Method and Success of Inoculating the Small-Pox in Boston, in New-England, in a Letter from a Gentleman There, to His Friend in London* (1722).

# 2 New World Skills

## Runaway Slave Advertisements

Uncounted numbers of enslaved black workers fled their bondage during the colonial period, and their masters routinely advertised in newspapers for their return. These advertisements almost always described the fugitives in terms of their physical appearance, manner of dress, behavior, and craft skills, if any. The institution of large-scale agriculture in the southern American colonies led to the establishment of large plantations that operated, in many cases, as small communities that were relatively self-sufficient in terms of material culture. Large numbers of enslaved black workers received training for artisanal skills such as the making of barrels (coopers), shoes, buildings, clothing, and, of course, food preparation. As with free laborers, these skills were considered a part of the identity of the slave workers.

### *Virginia Gazette* (Parks), May 9 to May 16, 1745

North-Carolina, April 24, 1745

Ran away, on the 18th Instant, from the Plantation of the late Col. William Wilson, deceas'd, Two Slaves belonging to the Subscriber, the one a tall yellow Fellow, named Emanuel, about 6 Feet high, six or seven and Twenty Years of Age; hath a Scar on the outside of his left Thigh, which was cut with an Ax; he had on when he went away, a blue Jacket, an Ozenbrig Shirt and Trousers, and a Worsted Cap; he speaks pretty good English, and calls himself a Portugueze; is by Trade a Cooper, and took with him some Cooper's Tools. The other is a short, thick, well-set Fellow, stoops forward pretty much as he walks; does not speak so plain as the other; had on when he went away an Ozenbrig Pair of Trousers and Shirt, a white Negro Cotton Jacket, and took with him an Axe: They went away in a small Cannoe, and were seen at Capt. Pearson's, on Nuse River, the 18th Inst. and 'tis believ'd are gone towards Virginia. Whoever takes up the said Negros, and brings them to my House on Trent River, North-Carolina, or secures them so that I may have them again, shall have Four Pistoles Reward for each, paid by

Mary Wilson

Excerpted from: Lathan A. Windley, comp., *Runaway Slave Advertisements: A Documentary History from 1730s to 1790. Volume 1, Virginia and North Carolina* (Westport, Conn.: Greenwood Press, 1983), passim.

**_Virginia Gazette_ (Hunter), November 7, 1754**

Ran away from the Subscriber, living in King-William County, on the 16 Day of June, 1753, a Mulatto Wench, named Milly, about 26 Years of Age, of a middle Stature, long Visage, and freckled, has a drawling Speech, a down Look, and has been chiefly brought up to Carding and Spinning. She was first supposed to be harboured in New-Kent County by some of her Relations, but not hearing of her since the first Week after her Elopement, it is supposed she has either got to some of the neighbouring Provinces, or gone beyond Sea. Whoever will apprehend her, so that she may be had again, shall have Ten Pistoles Reward; or if she is beyond Sea, whoever declares the Name of the Skipper and Vessel that carried her, so that the Offender may be brought to Justice, shall have Twenty Pistoles Reward.
Ferdinando Leigh

**_Virginia Gazette_ (Purdie), April 11, 1766**
Warwick county, April 8, 1766

Run away from the subscriber, on or about the 10th of February last, a Virginia born Negro man named George America, about 5 feet 8 or 9 inches high, about 30 years old, of a yellow complexion, is a tolerable good shoemaker, and can do something of the house carpenters work, walks quick and upright, and has a scar on the back of his left hand; had on a cotton waistcoat and breeches osnabrugs shirt, and yarn stockings. As the said slave is outlawed, I do hereby offer a reward of 5 l. to any person that will kill and destroy him, and 40 s. if taken alive.
Thomas Watkins

**_Virginia Gazette_ (Purdie), May 2, 1766**

Run away from the subscriber, the 16th of February last, two Virginia born Negro men slaves, of a yellow complexion, about 5 feet 8 or 9 inches high; had on when they went away Negro cotton waistcoat and breeches, shoes and stockings, and osnabrugs shirt, and took with them several other clothes, and five Dutch Blankets. One named Charles, is a sawyer and shoemaker by trade, carried with him a set of shoemaker tools, is about 28 years of age, speaks slow, can read, and may probably procure a pass and get on board some vessel. The other named George, about the same age, is round shouldered, which causes him to stoop when he walks; they are both outlawed. Whoever brings, or safely conveys, the said slaves to me, in the upper end of Charles City county, shall have 5 l. reward for each, if taken in this colony, if out thereof 10 l.
Charles Floyd

### *Virginia Gazette* (Purdie & Dixon), April 16, 1767

Run away from the subscriber, near Williamsburg, last Saturday night, a Negro fellow named Bob, about 5 feet 7 inches high, about 26 years of age, was burnt when young, by which he has a scar on the wrist of his right hand, the thumb of his left hand burnt off, and the hand turns in; had on a double breasted dark coloured frieze jacet and yellow cotton breeches. He was lately brought home from Hartford county in North Carolina, where he has been harboured for three years past by one Van Pelt, who lives on Chinkopin creek; he passed for a freeman, by the name of Edward or Edmund Tamar, and has got a wife there. He is an extraordinary sawer, a tolerable good carpenter and currier, pretends to make shoes, and is a very good sailor. He has been gone for eight years, a part of which time he lived in Charleston, South Carolina. He can read and write; and, as he is a very artful fellow, will probably forge a pass. All masters of vessels are hereby cautioned from carrying him out of the colony, and any person from employing him. Whoever apprehends the said fellow, and conveys him to me, shall have 3 l. reward, if taken in this colony; if in North Carolina, 5 l. and if in any other province, 10 l.
William Trebell

### *Virginia Gazette* (Purdie & Dixon), August 10, 1769

Prince George, July 31, 1769

Run away from the subscriber, near the courthouse, the 1st of June last, a Virginia born Negro man named Ned, about 6 feet high, about 55 years of age, is a little round shouldered, stoops a little as he walks, his head gray, beard thin, a thin visage, and is a house carpenter, a wheelwright, and a very good sawer. He is outlawed. Whoever takes up the said Negro, and brings him to me, or confines him in jail well ironed, shall have 50 s. reward.
Burwell Green

## *A Profile of Runaway Slaves in Virginia and South Carolina from 1730 through 1787*

Lathan Algerna Windley

The African-American scholar Lathan Algerna Windley has made a careful count of the various characteristics reported for runaway enslaved blacks for a part of the South during the middle years of the eighteenth century. His list shows a wide diversity of craft skills the runaways practiced.

---

Excerpted from: Lathan Algerna Windley, *A Profile of Runaway Slaves in Virginia and South Carolina from 1730 through 1787* (New York: Garland Publishing, 1995), 171–173.

**Table 2.1**
Occupations of male runaway slaves

| Occupations | Virginia | South Carolina |
|---|---|---|
| House servant (or waitingman) | 68 | 50 |
| Ironwork | 3 | 0 |
| Plantation work (field slave, planter, etc.) | 14 | 12 |
| Public work (attended market, vendor, etc.) | 1 | 16 |
| Special (preacher, doctor, etc.) | 8 | 5 |
| Tradesman (shoemaker, blacksmith, etc.) | 58 | 61 |
| Waterman (ferryman, sailor, etc.) | 46 | 44 |
| Woodwork (carpenters, sawyers, coopers, etc.) | 70 | 103 |
| House servant and waterman | 5 | 3 |
| House servant and woodwork | 2 | 3 |
| House servant and plantation work | 4 | 2 |
| House servant, tradesman, and woodwork | 3 | 0 |
| House servant and tradesman | 2 | 2 |
| House servant, tradesman, and plantation | 1 | 0 |
| House servant, tradesman, and ironwork | 1 | 0 |
| House servant, tradesman, waterman, and woodwork | 1 | 0 |
| House servant and special | 1 | 0 |
| House servant, plantation work, and woodwork | 1 | 0 |
| House servant and public work | 0 | 1 |
| House servant, waterman, and tradesman | 0 | 1 |
| Plantation work and woodwork | 3 | 2 |
| Plantation work and tradesman | 4 | 0 |
| Plantation work and waterman | 3 | 2 |
| Plantation work, woodwork, and waterman | 1 | 0 |
| Plantation work, tradesman, and woodwork | 2 | 0 |
| Plantation work, tradesman, waterman, and woodwork | 1 | 0 |
| Public work, plantation work, and house servant | 0 | 1 |
| Public work and waterman | 0 | 1 |
| Waterman and ironwork | 1 | 0 |
| Waterman and tradesman | 6 | 2 |
| Waterman and woodwork | 4 | 3 |
| Waterman, woodwork, and tradesman | 1 | 0 |
| Waterman, woodwork, house servant, and plantation work | 0 | 2 |
| Woodwork and tradesman | 17 | 9 |
| Woodwork, tradesman, and special | 1 | 0 |
| Total | 333 | 325 |

## Advertisement for a Fugitive Slave (1769)

Thomas Jefferson

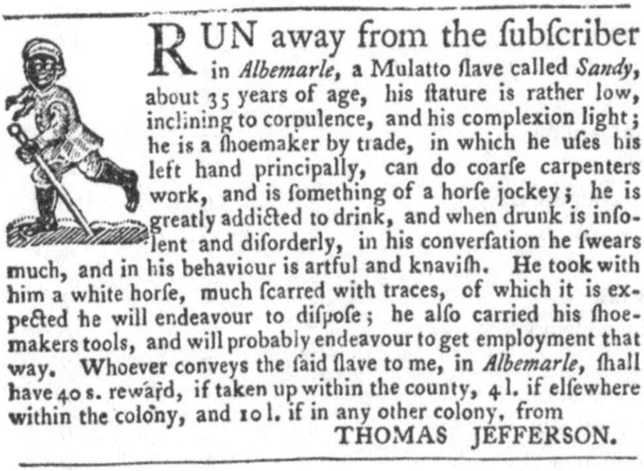

RUN away from the ſubſcriber in *Albemarle*, a Mulatto ſlave called *Sandy*, about 35 years of age, his ſtature is rather low, inclining to corpulence, and his complexion light; he is a ſhoemaker by trade, in which he uſes his left hand principally, can do coarſe carpenters work, and is ſomething of a horſe jockey; he is greatly addicted to drink, and when drunk is inſolent and diſorderly, in his converſation he ſwears much, and in his behaviour is artful and knaviſh. He took with him a white horſe, much ſcarred with traces, of which it is expected he will endeavour to diſpoſe; he alſo carried his ſhoemakers tools, and will probably endeavour to get employment that way. Whoever conveys the ſaid ſlave to me, in *Albemarle*, ſhall have 40 s. reward, if taken up within the county, 4 l. if elſewhere within the colony, and 10 l. if in any other colony, from

THOMAS JEFFERSON.

**Figure 2.1**
Perhaps the most famous, and possibly the most conflicted, early American slave owner was future president Thomas Jefferson. He was among those who placed advertisements for fugitive slaves who had marked craft skills. Reprinted from the *Virginia Gazette*, September 14, 1769.

## Letter from Benjamin Banneker to the Secretary of State, with His Answer (1792)

Benjamin Banneker, Thomas Jefferson

The best-known African-American scientist and mechanic of the colonial period was undoubtedly Benjamin Banneker (1731–1806), who made clocks, taught himself astronomy, and published an acclaimed almanac. When Thomas Jefferson, then Secretary of State, published his famous *Notes on the State of Virginia* (written in 1781), it contained arguments for the inferiority of Africans and compared them unfavorably to "white" people. Knowing Jefferson to be a fellow scientist, Banneker sent him a

**Figure 2.2**
Portrait of Benjamin Banneker, reproduced on the title page of his almanac for 1795, printed for John Fisher. (Variant spellings of names were common in the eighteenth century.)

Reprinted from: Benjamin Banneker, *Copy of a Letter from Benjamin Banneker to the Secretary of State with His Answer* (Philadelphia: Daniel Lawrence, 1792), 3–12. Figure reprinted from Silvio A. Bedini, *The Life of Benjamin Banneker* (New York: Charles Scribner's Sons, 1972), facing p. 102.

copy of his almanac while it was still in manuscript form. His own technical virtuosity was a powerful rebuttal to Jefferson's racist judgments.

---

**Copy of a Letter from Benjamin Banneker, &C.**
Maryland, Baltimore County, August 19, 1791

Sir,
I am fully sensible of the greatness of that freedom, which I take with you on the present occasion; a liberty which seemed to me scarcely allowable, when I reflected on that distinguished and dignified station in which you stand, and the almost general prejudice and prepossession, which is so prevalent in the world against those of my complexion.

I suppose it is a truth too well attested to you, to need a proof here, that we are a race of beings, who have long labored under the abuse and censure of the world; that we have long been looked upon with an eye of contempt; and that we have long been considered rather as brutish than human, and scarcely capable of mental endowments.

Sir, I hope I may safely admit, in consequence of that report which hath reached me, that you are a man far less inflexible in sentiments of this nature, than many others; that you are measurably friendly, and well disposed towards us; and that you are willing and ready to lend your aid and assistance to our relief, from those many distresses, and numerous calamities, to which we are reduced.

Now Sir, if this is founded in truth, I apprehend you will embrace every opportunity, to eradicate that train of absurd and false ideas and opinions, which so generally prevails with respect to us; and that your sentiments are concurrent with mine, which are, that one universal Father hath given being to us all; and that he hath not only made us all of one flesh, but that he hath also, without partiality, afforded us all the same sensations and endowed us all with the same faculties; and that however variable we may be in society or religion, however diversified in situation or color, we are all of the same family, and stand in the same relation to him.

Sir, if these are sentiments of which you are fully persuaded, I hope that you cannot but acknowledge, that it is the indispensable duty of those, who maintain for themselves the rights of human nature, and who possess the obligations of Christianity, to extend their power and influence to the relief of every part of the human race, from whatever burden or oppression they may unjustly labor under; and this, I apprehend, a full conviction of the truth and obligation of these principles should lead all to.

Sir, I have long been convinced, that if your love for yourselves, and for those inestimable laws, which preserved to you the rights of human nature, was founded on sincerity, you could not but be solicitous, that every individual, of whatever rank or distinction, might with you equally enjoy the blessings thereof; neither could you rest satisfied short of the most active effusion of your exertions, in order to their promotion from any state of degradation, to which the unjustifiable cruelty and barbarism of men may have reduced them.

Sir, I freely and cheerfully acknowledge, that I am of the African race, and in that color which is natural to them of the deepest dye; and it is under a sense of the most profound gratitude to the Supreme Ruler of the Universe, that I now confess to you, that I am not under that state of tyrannical thraldom, and inhuman captivity, to which too many of my brethren are doomed, but that I have abundantly tasted of the fruition of those blessings, which proceed from that free and unequalled liberty with which you are favored; and which, I hope, you will willingly allow you have mercifully received, from the immediate hand of that Being, from whom proceedeth every good and perfect Gift.

Sir, suffer me to recall to your mind that time, in which the arms and tyranny of the British crown were exerted, with every powerful effort, in order to reduce you to a state of Servitude: look back, I entreat you, on the variety of dangers to which you were exposed; reflect on that time, in which every human aid appeared unavailable, and in which even hope and fortitude wore the aspect of inability to the conflict, and you cannot but be led to a serious and grateful sense of your miraculous and providential preservation; you cannot but acknowledge, that the present freedom and tranquility which you enjoy you have mercifully received, and that it is the peculiar blessing of Heaven.

This, Sir, was a time when you clearly saw into the injustice of a state of slavery, and in which you had just apprehensions of the horrors of its condition. It was now that your abhorrence thereof was so excited, that you publicly held forth this true and invaluable doctrine, which is worthy to be recorded and remembered in all succeeding ages: "We hold these truths to be self-evident, that all men are created equal; that they are endowed by their Creator with certain unalienable rights, and that among these are, life, liberty, and the pursuit of happiness."

Here was a time, in which your tender feelings for yourselves had engaged you thus to declare, you were then impressed with proper ideas of the great violation of liberty, and the free possession of those blessings, to which you were entitled by nature; but, Sir, how pitiable is it to reflect, that although you were so fully convinced of the benevolence of the Father of Mankind, and of his equal and impartial distribution of these rights and privileges, which he hath conferred upon them, that you should at the same time counteract his mercies, in detaining by fraud and violence so numerous a part of my brethren, under groaning captivity and cruel oppression, that you should at the same time be found guilty of that most criminal act, which you professedly detected in others, with respect to yourselves.

I suppose that your knowledge of the situation of my brethren, is too extensive to need a recital here; neither shall I presume to prescribe methods by which they may be relieved, otherwise than by recommending to you and all others, to wean yourselves from those narrow prejudices which you have imbibed with respect to them, and as Job proposed to his friends, "put your soul in their souls' stead;" thus shall your hearts be enlarged with kindness and benevolence towards them; and thus shall you need neither the direction of myself or others, in what manner to proceed herein.

And now, Sir, although my sympathy and affection for my brethren hath caused my enlargement thus far, I ardently hope, that your candor and generosity will plead with you in my behalf, when I make known to you, that it was not originally my design; but having taken up my pen in order to direct to you, as a present, a copy of an Almanac, which I have calculated for the succeeding year, I was unexpectedly and unavoidably led thereto.

This calculation is the production of my arduous study, in this my advanced state of life; for having long have unbounded desires to become acquainted with the secrets of nature, I have had to gratify my curiosity herein, through my own assiduous application to Astronomical Study, in which I need not recount to you the many difficulties and disadvantages, which I have had to encounter.

And although I had almost declined to make my calculation for the ensuing year, in consequence of that time which I had allotted therefor, being taken up at the Federal Territory, by the request of Mr. Andrew Ellicott, yet finding myself under several engagements to Printers of this state, to whom I had communicated my design, on my return to my place of residence, I industriously applied myself thereto, which I hope I have accomplished with correctness and accuracy; a copy of which I have taken the liberty to direct to you, and which I humbly request you will favorably receive; and although you may have the opportunity of perusing it after its publication, yet I choose to send it to you in manuscript previous thereto, that thereby you might not only have an earlier inspection, but that you might also view it in my own hand writing.

And now, Sir, I shall conclude, and subscribe myself, with the most profound respect,

Your most obedient humble servant,
Benjamin Banneker

---

**To Mr. Benjamin Banneker.**
Philadelphia, August 30, 1791

Sir,

I Thank you, sincerely, for your letter of the 19th instant, and for the Almanac it contained. No body wishes more than I do, to see such proofs as you exhibit, that nature has given to our black brethren talents equal to those of the other colors of men; and that the appearance of the want of them, is owing merely to the degraded condition of their existence, both in Africa and America. I can add with truth, that no body wishes more ardently to see a good system commenced, for raising the condition, both of their body and mind, to what it ought to be, as far as the imbecility of their present existence, and other circumstances, which cannot be neglected, will admit.

I have taken the liberty of sending your Almanac to Monsieur de Condozett, Secretary of the Academy of Sciences at Paris, and Member of the Philanthropic Society, because I considered it as a document, to which your whole color had a right for their justification, against the doubts which have been entertained of them.
I am with great esteem, Sir,
Your most obedient
Humble Servant,
Thomas Jefferson

# II ANTEBELLUM YEARS

Roughly between the years 1763 and 1787, while the mainland colonies of America were struggling to redefine their place within the British Empire, and then creating a place without it, the British were themselves creating a revolution of a different sort. Within that brief quarter century new industrial devices from the steam engine to the iron-rolling mill, the transportation canal to the spinning jenny, were all introduced into the British industrial scene. Along with social inventions, like the profession of invention and the textile factory, these physical inventions ushered in the Industrial Revolution, quickly transforming how the British did and made many things; the net result was Great Britain becoming the "workshop of the world."

Word of these changes came slowly to America. The process of breaking away from imperial Britain, and perforce creating a new economic order to match the new political responsibilities, proved sufficiently distracting that when finally the new United States was born, its technology looked very much like it had throughout the colonial period. The need of establishing an economic base sufficient to secure and guarantee freedom from English rule and political liberties at home, however, made transportation and manufacturing innovations in the old country look attractive to the new one. As a matter of economic advantage and government policy, the United States and its constituent parts established a patent system, dug canals and built first turnpikes then railroads, begged, borrowed, or stole new inventions from abroad and added new ones (the result of "Yankee ingenuity"), built factories and furnaces, and finally began to rival the British themselves in the field of technological development.

The first center of this activity stretched along the Atlantic coast from Rhode Island to Delaware, where water power was abundant and cities were large enough to attract artisans, manufacturers, and inventors from Europe, particularly Great Britain. After the War of 1812 this activity spread to the old Northwest across the mountains, as far as the Mississippi. Only the South resisted this transformation. There the deep social and economic commitment to agriculture and the pervasive influence of a slave-labor system made investors reluctant to commit themselves to industrialization. There were exceptions—ironworks in Virginia, the widespread use of steam engines in processing Louisiana sugar cane, a railroad in South Carolina—but by and large as the North and Midwest adopted an economy of industrial capitalism, the South became more and more isolated; it was increasingly a technological colony of the North.

Enslaved Africans continued to receive training in the artisanal skills of a pre-industrial age, and for some those skills provided an economic base for an escape north to freedom. Others, either "free persons of color" or enslaved labor, found work in what industrial activities did take root in the South. Blacks who invented new technologies and tools were denied recognition of their ingenuity (unless they were considered free persons) or had to wait to gain patent rights until after the Civil War. The lives of enslaved Africans were replete with technologies but these, like those of their masters, while still effective were increasingly archaic.

# 3 The Persistence of Craft

*Advertisements for Runaway Slaves in Virginia, 1801–1820*
Daniel Meaders

*Life and Times of Frederick Douglass* (1882)
Frederick Douglass

*The Fugitive Blacksmith; or, Events in the History of James W. C. Pennington*

*Uncle Tom's Cabin* (1852)
Harriet Beecher Stowe

*A Journey in the Seaboard Slave States* (1856) and *A Journey in the Back Country* (1861)
Frederick Law Olmsted

*His Promised Land: The Autobiography of John P. Parker, Former Slave and Conductor on the Underground Railroad*
John P. Parker

**Layout of Parker's Phoenix Foundry (1884)**

**U.S. Patent to John Percial Parker for a Soil-Pulverizer (1890)**

**Tending a Cotton Gin (1853)**

## *Advertisements for Runaway Slaves in Virginia, 1801–1820*

Daniel Meaders

While the Industrial Revolution had been well introduced into the new United States by the end of the eighteenth century, craft skills, many of them several hundred years old, still produced most items and were necessary to the livelihoods of the people who made them in all parts of the country. Especially in the South, on the separate plantations and the small towns of the region, industrial products were sometimes brought in from Europe or the northern states but seldom made locally except on plantations. It is not surprising, then, that so long as slavery persisted in the South, escaped African slaves continued to be identified in important part by their technical skills.

**5/7/02**

Twenty Dollars Reward. Ran Away from the subscriber, on the 1st May, Patty, a likely Negro wench, about twenty years of age: she has been brought up in the house, is a good seamstress, & very capable: among a variety of clothes, she has a purple stuff petticoat, a fine blue cloth jacket, a corded dimity jacket and coat, a plain muslin gown, a handsome worked muslin do, a plain do, both well made, 2 white camel hair shawls with netting, fine white cotton stockings, several pair, and a pair of white silk do, a handsome black hat, neat shoes and other cloathing. She went off with a Negro fellow the property of Mr. Philip Fitzhugh: he is about 24 years of age, $5\frac{1}{2}$ feet high, well set, inclined to be corpulent, tawney complexion, lively countenance, and speaks distinctly, though quick; he is an excellent joiner. He took with him a pair of new brown cashmere pantaloons, a round upper jacket of the same cloth, a green broad cloth coat, with a blue velvet collar, a handsome swansdown waistcoat, with mettle buttons a new black hat, new shoes, fine white cotton stockings, green pantaloons, and other cloathing. The above reward will be paid to any person who shall lodge them in any jail, either in Maryland or Virginia, and ample compensation made for any other trouble or expence. Sarah Thornton, near Alexandria, May 7.

**3/22/03**

Fifteen Dollars Reward. Ran Away on Saturday, the 5th of this month, a bright Mulatto Man, named Stephen; about 28 or 30 years of age: He is about 5 feet 7 or 8 inches high, pleasant countenance, speaks rather slow, but very active and handy at any work; he

Excerpted from: Daniel Meaders, *Advertisements for Runaway Slaves in Virginia, 1801–1820* (New York: Garland Publishing, 1997), passim.

served an apprenticeship to a weaver, is a good workman at that business, and a remarkable good hand in a brickyard; in short, he is smart at anything he is set about; he is very fond of spirits of any kind. I cannot describe his dress. He never ran away before, so I expect he has been enticed off by some artful villain; if it be the case, and the man can be apprehended, I will give 30 Dollars on his conviction of the theft. Robert Boggess.

N.B. I forewarn all persons from harboring said fellow at their peril, as in that case they may expect to be prosecuted. Fairfax County, March 22.

---

**12/8/03**

Twenty Dollars Reward. Absconded from the service of Mr. Francis Keene, at Belvoir, on the 25th of November last, Negro David, (commonly called David Shanklin)—He is a dark mulatto man, about 30 years of age, 5 feet 10 or 11 inches high, a smart, active fellow, can saw well at the cross cut or whip saws, understands quarrying and blowing rock well, a good house servant, and, in short, handy at any business; he took with him a variety of clothes, none of which I can particularly describe (except a blue round waistcoat, trimmed with red, which he is very fond of wearing) being out of my service for 4 or 5 months, cutting wood at Velvoir. Said fellow ran away from me in 1796, got over to the City of Washington, & there obtained a pass from a man of the name of Mattingly (as he says) under which he passed as a free man, and hired himself to a Holt and Prout of that place—was apprehended there and carried to jail at the Eastern Branch, from whence I got him.

I do expect that he has formed connections in the town of Alexandria, and the city of Washington, and that he is now lurking about one of these places, as he has been seen but a few days ago in Alexandria. I will give ten dollars reward if taken in Alexandria or on this side the Potomac, or if on the other side the Potomac river the above reward, if delivered to me in Centerville, or to said Francis Keene, at Belvoir.

All masters of vessels or others, are hereby warned from harboring, secreting or carrying off said fellow at their perils, as I am determined to prosecute them with utmost rigor of the law. Francis Adams. Centerville, Fairfax County, Dec. 7th.

---

**4/8/11**

Twenty Dollars Reward, will be paid to any person who will deliver to me in Madison county, Negro Billy, who eloped on the 24th of last month. He is a likely fellow, about 30 years of age, 5 feet eight or nine inches high, remarkably straight, and has a scar on one of his cheeks, I believe the left, occasioned by a cat; he is a cooper by trade, and served his apprenticeship at the Occoquan Mills, and was purchased some years ago by Mr. Thos. Richards of a Mrs. Waggoner, who lived in the neighborhood of that place. He was seen eight days ago on his way to Alexandria, where he said he was going to get employment. I think it highly probable he may be found in the neighborhood of the above Mill. John B. Nooe.

**3/10/12**

Thirty Dollars Reward—Ran away from the subscriber in Sussex county, on the 15th of February last, a negro man named Simon, about six feet high, of a dark complexion, strong, stout, and of robust frame, about 36 years old, but from dissipation and intemperance looks rather older, had on Jacket and Trowsers of Virginia Cloth; his eyes are generally inflamed with intoxication, has a scar over one of his eyes, supposed to be the right eye; when spoken to makes quick answers, and is fluent and plausible in conversation; he will be found extremely cunning in expedients to evade detection; he is a first-rate Cooper, and understands something of the Brick laying Business, has a wife under the management of a Mr. Beverly Crump of New-Kent; it is probable therefore that he is now lurking about that county, it is also probable that he may endeavor to get employment at some of the mills or tobacco warehouses.

It is believed that his intention is to persuade his wife to elope with him to the northward. All masters of vessels and others are hereby forewarned from harbouring or carrying off said slave, under the penalty of the law.

The above reward, and all reasonable expences, will be paid to any person who will secure him in any jail in this taste, so that I get him again. George Blow. March 10.

**7/12/15**

$20 Reward. Runaway on the 18th ult. From the subscriber, living in Halifax County, Va. Daniel, a Negro Man of a black complection, about 40 years old, five feet ten or eleven inches high, remarkably well made, spry and active, when spoken to replies quick and distinct, and frequently with a down look. In the use of implements of husbandry, he far exceeds ordinary fellows—the coat he wore away was of blue broadcloth. There is very little doubt but he has obtained a free pass, with which he will attempt to reach the State of Ohio or Pennsylvania. Whoever will apprehend said negro and confine him in jail so that I get him again, or will bring him to me, shall receive the above reward, together with all reasonable expences and charges. John Wimbish.

**2/2/19**

200 Dollars Reward. Absconded on the 9th day of Nov. last, from Mr. James Atkinson (blacksmith), of Alexandria, (D.C.) a bright mulatto slave, named Frederick, commonly known by his associates as Frederick Bankhead; he is about 25 years of age, 5 feet 8 or 10 inches high, active and intelligent, stammers a little when spoken to, with a large scar on his upper lip, gray eyes and curly hair, which he generally keeps cut and combed in fashion.

This fellow has been working at the blacksmith's trade, with the best smiths, for the last six years; but in his youth he was accustomed to waiting in the house and attending on horses, which last employment he may have preferred to his trade. His dress was a sailor's jacket and trowsers, a fur hat, and a handkerchief tied loosely about his neck, with a breast pin, and he commonly walked with a stick. It is believed he is still lurking about the District of Columbia.

The subscriber will give the above reward if secured in any jail in the United States, so that he can get him again; or, of he should be delivered to the subscriber, (living in Virginia, Fairfax county, Sudley Mill) the above reward will be given, and all reasonable expenses paid. William Robinson. February 2.

## *Life and Times of Frederick Douglass* (1882)

Frederick Douglass

Escaped slave and abolitionist leader Frederick Douglass was one of the many southern blacks trained in a craft skill, ship caulking in his case. Early in his captivity he became aware of the varied technologies of the typical plantation, and his acquisition of a trade was critical in preparing him, both psychologically and economically, for freedom.

---

Col. Lloyd's plantation . . . was a little nation by itself, having its own language, its own rules, regulations, and customs. The troubles and controversies arising here were not settled by the civil power of the State. The overseer was the important dignitary. He was generally accuser, judge, jury, advocate, and executioner. The criminal was always dumb, and no slave was allowed to testify other than against his brother slave. . . .

I found myself here, there was no getting away, and naught remained for me but to make the best of it. Here were plenty of children to play with and plenty of pleasant resorts for boys of my age and older. The little tendrils of affection, so rudely broken from the darling objects in and around my grandmother's home, gradually began to extend and twine themselves around the now surroundings. Here, for the first time, I saw a large windmill, with its wide-sweeping white wings, a commanding object to a child's eye. This was situated on what was called Long Point—a tract of land dividing Miles river from the Wye. I spent many hours here watching the wings of this wondrous mill. In the river, or what was called the "Swash," at a short distance from the shore, quietly lying at anchor, with her small row boat dancing at her stern, was a large sloop, the Sally Lloyd, called by that name in honor of the favorite daughter of the Colonel. These two objects, the sloop and mill, as I remember, awakened thoughts, ideas, and wondering. Then here were a great many houses, human habitations full of the mysteries of life at every stage of it. There was the little red house up the road, occupied by Mr. Seveir, the overseer. A little nearer to my old master's stood a long, low, rough building literally alive with slaves of all ages, sexes, conditions, sizes, and colors. This was called the long quarter. Perched upon a hill east of our house, was a tall, dilapidated old brick building, the architectural dimensions of which proclaimed its creation for a different purpose, now occupied by slaves, in a similar manner to the long quarters. Besides these, there were numerous other slave houses and huts scattered around in the neighborhood, every nook and corner of which were completely occupied.

---

Excerpted from: Frederick Douglass, *Life and Times of Frederick Douglass* (Hartford: Park Publishing Co., 1882), 43–45, 114–115, 222–231, 256–257.

Old master's house, a long brick building, plain but substantial, was centrally located, and was an independent establishment. Besides these houses there were barns, stables, store-houses, tobacco-houses, blacksmith shops, wheelwright shops, cooper shops; but above all there stood the grandest building my young eyes had ever beheld, called by every one on the plantation the *great* house. This was occupied by Col. Lloyd and his family. It was surrounded by numerous and variously-shaped out-buildings. There were kitchens, wash-houses, dairies, summer-houses, green-houses, hen-houses, turkey-houses, pigeon-houses, and arbors of many sizes and devices, all neatly painted or whitewashed, interspersed with grand old trees, ornamental and primitive, which afforded delightful shade in summer and imparted to the scene a high degree of stately beauty. The *great* house itself was a large white wooden building with wings on three sides of it. In front a broad portico extended the entire length of the building, supported by a long range of columns, which gave to the Colonel's home an air of great dignity and grandeur. It was a treat to my young and gradually opening mind to behold this elaborate exhibition of wealth, power, and beauty.

The carriage entrance to the house was by a large gate, more than a quarter of a mile distant. The intermediate space was a beautiful lawn, very neatly kept and cared for. It was dotted thickly over with trees and flowers. The road or lane from the gate to the great house was richly paved with white pebbles from the beach, and in its course formed a complete circle around the lawn. . . .

Here was transacted the business of twenty or thirty different farms, which, with the slaves upon them, numbering, in all, not less than a thousand, all belonged to Col. Lloyd. Each farm was under the management of an overseer, whose word was law. . . .

I was too young to think of running away immediately; besides, I wished to learn to write before going, as I might have occasion to write my own pass. I now not only had the hope of freedom, but a foreshadowing of the means by which I might some day gain that inestimable boon. Meanwhile I resolved to add to my educational attainments the art of writing.

After this manner I began to learn to write. I was much in the ship-yard—Master Hugh's, and that of Durgan & Bailey, and I observed that the carpenters, after hewing and getting ready a piece of timber to use, wrote on it the initials of the name of that part of the ship for which it was intended. When, for instance, a piece of timber was ready for the starboard side, it was marked with a capital "S." A piece for the larboard side was marked "L."; larboard forward was marked "L. F.;" larboard aft was marked "L. A."; starboard aft, "S. A."; and starboard forward, "S. F." I soon learned these letters, and for what they were placed on the timbers.

My work now was to keep fire under the steam-box, and to watch the ship-yard while the carpenters had gone to dinner. This interval gave me a fine opportunity for copying the letters named. I soon astonished myself with the case with which I made the letters, and the thought was soon present, "If I can make four letters I can make more." Having made these readily and easily, when I met boys about the Bethel church or on any of our play-grounds, I entered the lists with them in the art of writing, and would make the letters which I had been so fortunate as to learn, and ask them to "beat that if they could." With play-mates for my teachers, fences and pavements for my copy-books, and chalk for my pen and ink, I learned to write. I however adopted, afterward, various methods for improving my hand. The most successful was copying the *italics* in Webster's spelling-book until I could make them all without looking on the book. By this time my little "Master Tommy" had grown to be a big boy, and had written over a number of copy-books and brought them home. They had been shown to the neighbors, had elicited due praise, and had been laid carefully away. Spending parts of my time both at the ship-yard and the house, I was often the lone keeper of the latter as of the former. When my mistress left me in charge of the house I had a grand time. I got Master Tommy's copy-books and a pen and ink, and in the ample spaces between the lines I wrote other lines as nearly like his as possible. The process was a tedious one, and I ran the risk of getting a flogging for marking the highly-prized copy-books of the oldest son. In addition to these opportunities, sleeping as I did in the kitchen loft, a room seldom visited by any of the family, I contrived to get a flour-barrel up there and a chair, and upon the head of that barrel I have written, or endeavored to write, copying from the Bible and the Methodist hymn-book, and other books which I had accumulated, till late at night, and when all the family were in bed and asleep. I was supported in my endeavors by renewed advice and by holy promises from the good father Lawson, with whom I continued to meet and pray and read the Scriptures. Although Master Hugh was aware of these meetings, I must say for his credit that he never executed his threats to whip me for having thus innocently employed my leisure time....

Very soon after I went to Baltimore to live, Master Hugh succeeded in getting me hired to Mr. William Gardiner, an extensive ship-builder on Fell's Point. I was placed there to learn to calk, a trade of which I already had some knowledge, gained while in Mr. Hugh Auld's ship-yard. Gardiner's, however, proved a very unfavorable place for the accomplishment of the desired object. Mr. Gardiner was that season engaged in building two large man-of-war vessels, professedly for the Mexican government. These vessels were to be launched in the month of July of that year, and in failure thereof Mr. Gardiner would forfeit a very considerable sum of money. So, when I entered the ship-yard, all was hurry and driving. There were in the yard about one hundred men; of these, seventy or eighty were regular carpenters—privileged men. There was no time for a raw hand to learn anything. Every man had to do that which he knew how to do, and in entering the yard Mr. Gardiner had directed me to do whatever the carpenters told me to do. This was placing me at the beck and call of about seventy-five men. I was to regard all these as my masters. Their word was to be my law. My situation was a trying one. I was called a dozen ways in the space of a single minute. I needed a dozen pairs of hands. Three or four voices would strike my ear at the same moment. It was "Fred, come help me to cant this timber here,"—"Fred, come carry this timber yonder,"—"Fred, bring that roller here,"—"Fred, go get a fresh can of water," "Fred, come help saw off the end of this timber,"—"Fred, go quick and get the crow-bar,"—"Fred, hold on the end of this fall,"—"Fred, go to the blacksmith's shop and get a new punch,"—"Halloo, Fred! run and bring me a cold-chisel,"—"I say, Fred, bear a hand, and get up a fire under the steam-box as quick as lightning,"—"Hullo, nigger! come turn this grindstone"—"Come, come; move, move! and *bowse* this timber forward,"—"I say, darkey, blast your eyes! why don't you heat up some pitch?"—"Halloo! halloo! halloo! (three voices at the same time)"—"Come here; go there; hold on where you are. D—n you, if you move I'll knock your brains out!" Such, my dear reader, is a glance at the school which was mine during the first eight months of my stay at Gardiner's ship-yard. At the end of eight months Master Hugh refused longer to allow me to remain with Gardiner. The circumstance which led to this refusal was the committing of an outrage upon me, by the white apprentices of the ship-yard. The fight was a desperate one, and I came out of it shockingly mangled. I was out and bruised in sundry places, and my left eye was nearly knocked out of its socket. The facts which led to this brutal outrage upon me illustrate a phase of slavery which was destined to become an important element in the overthrow of the slave system, and I may therefore state them with some minuteness. That phase was this—the conflict of slavery with the interests of white mechanics and laborers. In the country this conflict was not so apparent; but in cities, such as Baltimore, Richmond, New Orleans, Mobile, etc., it was seen pretty clearly: The slaveholders, with a craftiness peculiar to themselves, by encouraging the enmity of the poor laboring white man against the blacks, succeeded in making the said white man almost as much a slave as the black slave himself. The difference between the white slave and the black slave was this: the latter belonged to one slave-holder, and the former belonged to the slave-holders collectively. The white slave had taken from him by indirection what the black slave had taken from him directly and without ceremony. Both were plundered, and by the same plunderers. The slave was robbed by his master of all his earnings, above what was required for his bare physical necessities, and the white laboring man was robbed by the slave system of the just results of his labor, because he was flung into competition with a class of laborers who worked without wages. The slaveholders blinded them to this competition by keeping alive their prejudice against the slaves as *men*—not against them as *slaves*. They appealed to their pride, often denouncing emancipation as tending to place the white working man on an equality with negroes, and by this means they succeeded in drawing off the minds of the poor whites from the real fact, that by the rich slave master they were already regarded as but a single remove from equality with the slave. The impression was cunningly made that slavery was the only power that could prevent the laboring white man from falling to the level of the slave's poverty and degradation. To make this enmity deep and broad between the slave and the poor white man, the latter was allowed to abuse and whip the former without hindrance. But, as I have said, this state of affairs prevailed *mostly* in the country. In the city of Baltimore there were not unfrequent murmurs that educating slaves to be mechanics might, in the end, give slave-masters power to dispense altogether with the services of the poor white man. But with characteristic dread of offending the slaveholders, these poor white mechanics in Mr. Gardiner's ship-yard, instead of applying the natural, honest remedy for the apprehended evil, and objecting at once to work there by the side of slaves, made a cowardly attack upon the free colored mechanics, saying they were eating the bread which should be eaten by American freemen, and swearing that they would not work with them. The feeling was *really* against having their labor brought into competition with that of the colored freeman, and aimed to prevent him from serving himself, in the evening of life, with the trade with which he had served his master, during the more vigorous portion of his days. Had they succeeded in driving the black freemen out of the ship-yard, they would have determined also upon the removal of the black slaves. The feeling was very bitter toward all colored people in Baltimore about this time (1836), and they—free and slave—suffered all manner of insult and wrong.

Until a very little while before I went there, white and black carpenters worked side by side in the ship-yards of Mr. Gardiner, Mr. Duncan, Mr. Walter Price, and Mr. Robb. Nobody seemed to see any impropriety in it. Some of the blacks were first-rate workmen and were given jobs requiring the highest skill. All at once, however, the white carpenters knocked off and swore that they would no longer work on the same stage with negroes. Taking advantage of the heavy contract resting upon Mr. Gardiner to have the vessels for Mexico ready to launch in July, and of the difficulty of getting other hands at that season of the year, they swore they would not strike another blow for him unless he would discharge his free colored workmen. Now, although this movement did not extend to me *in form*, it did reach me in *fact*. The spirit which it awakened was one of malice and bitterness toward colored people *generally*, and I suffered with the rest, and suffered severely. My fellow-apprentices very soon began to feel it to be degrading to work with me. They began to put on high looks and to talk contemptuously and maliciously of "the niggers," saying that they would take the "country," that they "ought to be killed." Encouraged by workmen who, knowing me to be a slave, made no issue with Mr. Gardiner about my being there, these young men did their utmost to make it impossible for me to stay. They seldom called me to do anything without coupling the call with a curse, and Edward North, the biggest in everything, rascality included, ventured to strike me, whereupon I picked him up and threw him into the dock. Whenever any of them struck me I struck back again, regardless of consequences. I could manage any of them *singly*, and so long as I could keep them from combining I got on very well. In the conflict which ended my stay at Mr. Gardiner's I was beset by four of them at once—Ned North, Ned Hayes, Bill Stewart, and Tom Humphreys. Two of them were as large as myself, and they came near killing me, in broad daylight. One came in front, armed with a brick; there was one at each side and one behind, and they closed up all around me. I was struck on all sides; and while I was attending to those in front I received a blow on my head from behind, dealt with a heavy hand-spike. I was completely stunned by the blow, and fell heavily on the ground among the timbers. Taking advantage of my fall they rushed upon me and began to pound me with their fists. I let them lay on for awhile after I came to myself, with a view of gaining strength. They did me little damage so far; but finally getting tired of that sport I gave a sudden surge, and despite their weight I rose to my hands and knees. Just as I did this one of their number planted a blow with his boot in my left eye, which for a time seemed to have burst my eye-ball. When they saw my eye completely closed, my face covered with blood, and I staggering under the stunning blows they had given me, they left me. As soon as I gathered strength I picked up the hand-spike and madly enough attempted to pursue them; but here the carpenters interfered and compelled me to give up my pursuit. It was impossible to stand against so many.

Dear reader, you can hardly believe the statement, but it is true, and therefore I write it down; no fewer than fifty white men stood by and saw this brutal and shameful outrage committed, and not a man of them all interposed a single word of merey. There were four against one, and that one's face was beaten and battered most horribly, and no one said, "that is enough;" but some cried out, "Kill him! kill him! kill the d—n nigger! knock his brains out! he struck a white person!" I mention this inhuman outcry to show the character of the men and the spirit of the times at Gardiner's ship-yard; and, indeed, in Baltimore generally, in 1836. As I look back to this period, I am almost amazed that I was not murdered outright, so murderous was the spirit which prevailed there. On two other occasions while there I came near losing my life, on one of which I was driving bolts in the hold through the keelson with Hays. In its course the bolt bent. Hays cursed me, and said that it was my blow which bent the bolt. I denied this and charged it upon him. In a fit of rage he seized an adze and darted toward me. I met him with a maul and parried his blow, or I should have lost my life.

After the united attack of North, Stewart, Hayes, and Humphreys, finding that the carpenters were as bitter toward me as the apprentices, and that the latter were probably set on by the former, I found my only chance for life was in flight. I succeeded in getting away without an additional blow. To strike a white man was death by lynch law, in Gardiner's ship-yard; nor was there much of any other law toward the colored people at that time in any other part of Maryland....

Master Hugh, on finding he could get no redress for the cruel wrong, withdrew me from the employment of Mr. Gardiner and took me into his own family, Mrs. Auld kindly taking care of me and dressing my wounds until they were healed and I was ready to go to work again.

While I was on the Eastern Shore, Master Hugh had met with reverses which overthrew his business; and he had given up ship-building in his own yard, on the City Block, and was now acting as foreman of Mr. Walter Price. The best he could do for me was to take me into Mr. Price's yard, and afford me the facilities there for completing the trade which I began to learn at Gardiner's. Here I rapidly became expert in the use of calker's tools, and in the course of a single year, I was able to command the highest wages paid to journeymen calkers in Baltimore.

The reader will observe that I was now of some pecuniary value to my master. During the busy season I was bringing six and seven dollars per week. I have sometimes brought him as much as nine dollars a week, for wages were a dollar and a half per day.

After learning to calk, I sought my own employment made my own contracts, and collected my own earnings—giving Master Hugh no trouble in any part of the transactions to which I was a party.

Here, then, were better days for the Eastern Shore *slave.* I was free from the vexatious assaults of the apprentices at Gardiner's, and free from the perils of plantation life, and once more in favorable condition to increase my little stock of education, which had been at a dead stand since my removal from Baltimore. I had on the Eastern Shore been only a teacher, when in company with other slaves, but now there were colored persons here who could instruct me. Many of the young calkers could read, write, and cipher. Some of them had high notions about mental improvement, and the free ones on Fell's Point organized what they called the "East Baltimore Mental Improvement Society." To this society, notwithstanding it was intended that only free persons should attach themselves, I was admitted, and was several times assigned a prominent part in its debates. I owe much to the society of these young men.

The reader already knows enough of the *ill* effects of good treatment on a slave to anticipate what was now the case in my improved condition. It was not long before I began to show signs of disquiet with slavery, and to look around for means to get out of it by the shortest route. I was living among *freemen,* and was in all respects equal to them by nature and attainments. *Why should I be a slave?* There was *no* reason why I should be the thrall of any man. Besides, I was now getting, as I have said, a dollar and fifty cents per day. I contracted for it, worked for it, collected it; it was paid to me, and it was *rightfully* my own; and yet upon every returning Saturday night, this money—my own hard earnings, every cent of it—was demanded of me and taken from me by Master Hugh. He did not earn it; he had no hand in earning it; why, then should he have it? I owed him nothing. He had given me no schooling, and I had received from him only my food and raiment; and for these my services were supposed to pay from the first. . . .

New Bedford, therefore, which at that time was really the richest city in the Union, in proportion to its population, took me greatly by surprise, in the evidences it gave of its solid wealth and grandeur. I found that even the laboring classes lived in better houses, that their houses were more elegantly furnished, and were more abundantly supplied with conveniences and comforts, than the houses of many who owned slaves on the Eastern Shore of Maryland. This was true not only of the white people of that city, but it was so of my friend, Mr. Johnson. He lived in a nicer house, dined at a more ample board, was the owner of more books, the reader of more newspapers, was more conversant with the moral, social and political condition of the country and the world, than nine-tenths of the slaveholders in all Talbot county. I was not long in finding the cause of the difference, in these respects, between the people of the north and south. It was the superiority of educated mind over mere brute force. I will not detain the reader by extended illustrations as to how my understanding was enlightened on this subject. On the wharves of New Bedford I received my first light. I saw there industry without bustle, labor without noise, toil—honest, earnest, and exhaustive—

without the whip. There was no loud singing or hallooing, as at the wharves of southern ports when ships were loading or unloading; no loud cursing or quarreling; everything went on as smoothly as well-oiled machinery. One of the first incidents which impressed me with the superior mental character of labor in the north over that of the south, was the manner of loading and unloading vessels. In a southern port twenty or thirty hands would be employed to do what five or six men, with the help of one ox, would do at the wharf in New Bedford. Main strength—human muscle—unassisted by intelligent skill, was slavery's method of labor. With a capital of about sixty dollars in the shape of a good-natured old ox, attached to the end of a stout rope, New Bedford did the work of ten or twelve thousand dollars, represented in the bones and muscles of slaves, and did it far better. In a word, I found everything managed with a much more scrupulous regard to economy, both of men and things, time and strength, than in the country from which I had come. Instead of going a hundred yards to the spring, the maid-servant had a well or pump at her elbow. The wood used for fuel was kept dry and snugly piled away for winter. Here were sinks, drains, self-shutting gates, pounding-barrels, washing-machines. . . .

## *The Fugitive Blacksmith; or, Events in the History of James W. C. Pennington*

James C. Pennington, who began life enslaved but escaped to the North and became a preacher, was another who had mastered a technical skill.

---

At this time my days were extremely dreary. When I was nine years of age, myself and my brother were hired out from home; my brother was placed with a pump-maker, and I was placed with a stone-mason. We were both in a town some six miles from home. As the men with whom we lived were not slaveholders, we enjoyed some relief from the peculiar evils of slavery. Each of us lived in a family where there was no other negro.

The slaveholders in that state often hire the children of their slaves out to non-slaveholders, not only because they save themselves the expense of taking care of them, but in this way they get among their slaves useful trades. They put a bright slave-boy with a tradesman, until he gets such a knowledge of the trade as to be able to do his own work, and then he takes him home. I remained with the stonemason until I was eleven years of age: at this time I was taken home. This was another serious period in my childhood; I was separated from my older brother, to whom I was much attached; he continued at his place, and not only learned the trade to great perfection, but finally became the property of the man with whom he lived, so that our separation was permanent, as we never lived nearer after, than six miles. My master owned an excellent blacksmith, who had obtained his trade in the way I have mentioned above. When I returned home at the age of eleven, I was set about assisting to do the mason-work of a new smith's shop. This being done, I was placed at the business, which I soon learned, so as to be called a "first-rate blacksmith." I continued to work at this business for nine years, or until I was twenty-one, with the exception of the last seven months.

In the spring of 1828, my master sold me to a Methodist man, named ———, for the sum of seven hundred dollars. It soon proved that he had not work enough to keep me employed as a smith, and he offered me for sale again. On hearing of this, my old master re-purchased me, and proposed to me to undertake the carpentering business. I had been working at this trade six months with a white workman, who was building a large barn when I left. I will now relate the abuses which occasioned me to fly.

---

Excerpted from: James W. C. Pennington, *The Fugitive Blacksmith; or, Events in the History of James W. C. Pennington, Pastor of a Presbyterian Church, New York, Formerly a Slave in the State of Maryland, United States*, 3rd ed. (London: Charles Gilpen, 5, Bishopgate Without, 1850), 4–9.

Three or four of our farm hands had their wives and families on other plantations. In such cases, it is the custom in Maryland to allow the men to go on Saturday evening to see their families, stay over the Sabbath, and return on Monday morning, not later than "half-an-hour by sun." To overstay their time is a grave fault, for which, especially at busy seasons, they are punished.

One Monday morning, two of these men had not been so fortunate as to get home at the required time: one of them was an uncle of mine. Besides these, two young men who had no families, and for whom no such provision of time was made, having gone somewhere to spend the Sabbath, were absent. My master was greatly irritated, and had resolved to have, as he said, "a general whipping-match among them."

Preparatory to this, he had a rope in his pocket, and a cowhide in his hand, walking about the premises, and speaking to every one he met in a very insolent manner, and finding fault with some without just cause. My father, among other numerous and responsible duties, discharged that of shepherd to a large and valuable flock of Merino sheep. This morning he was engaged in the tenderest of a shepherd's duties; —a little lamb, not able to go alone, lost its mother; he was feeding it by hand. He had been keeping it in the house for several days. As he stooped over it in the yard, with a vessel of new milk he had obtained, with which to feed it, my master came along, and without the least provocation, began by asking, "Bazil have you fed the flock?"

"Yes, sir."

"Were you away yesterday?"

"No, sir."

"Do you know why these boys have not got home this morning yet?"

"No, sir, I have not seen any of them since Saturday night."

"By the Eternal, I'll make them know their hour. The fact is, I have too many of you; my people are getting to be the most careless, lazy, and worthless in the country."

"Master," said my father, "I am always at my post; Monday morning never finds me off the plantation."

"Hush Bazil! I shall have to sell some of you; and then the rest will have enough to do; I have not work enough to keep you all tightly employed; I have too many of you."

All this was said in an angry, threatening, and exceedingly insulting tone. My father was a high-spirited man, and feeling deeply the insult, replied to the last expression,— "If I am one too many, sir, give me a chance to get a purchaser, and I am willing to be sold when it may suit you."

"Bazil, I told you to hush!" and suiting the action to the word, he drew forth the "cowhide" from under his arm, fell upon him with most savage cruelty, and inflicted fifteen or twenty severe stripes with all his strength, over his shoulders and the small of his back. As he raised himself upon his toes, and gave the last stripe, he said, "By the * * * I will make you know that I am master of your tongue as well as of your time!"

Being a tradesman, and just at that time getting my breakfast, I was near enough to hear the insolent words that were spoken to my father, and to hear, see, and even count the savage stripes inflicted upon him.

Let me ask any one of Anglo-Saxon blood and spirit, how would you expect a *son* to feel at such a sight?

This act created an open rupture with our family—each member felt the deep insult that had been inflicted upon our head; the spirit of the whole family was roused; we talked of it in our nightly gatherings, and showed it in our daily melancholy aspect. The oppressor saw this, and with the heartlessness that was in perfect keeping with the first insult, commenced a series of tauntings, threatenings, and insinuations, with a view to crush the spirit of the whole family.

Although it was some time after this event before I took the decisive step, yet in my mind and spirit, I never was a *Slave* after it.

Whenever I thought of the great contrast between my father's employment on that memorable Monday morning, (feeding the little lamb,) and the barbarous conduct of my master, I could not help cordially despising the proud abuser of my sire; and I believe he discovered it, for he seemed to have diligently sought an occasion against me. Many incidents occurred to convince me of this, too tedious to mention; but there is one I will mention, because it will serve to show the state of feeling that existed between us, and how it served to widen the already open breach.

I was one day shoeing a horse in the shop yard. I had been stooping for some time under the weight of the horse, which was large, and was very tired; meanwhile, my master had taken his position on a little hill just in front of me, and stood leaning back on his cane, with his hat drawn over his eyes. I put down the horse's foot, and straightened myself up to rest a moment, and without knowing that he was there, my eye caught his. This threw him into a panic of rage; he would have it that I was watching him. "What are you rolling your white eyes at me for, you lazy rascal?" He came down upon me with his cane, and laid on over my shoulders, arms, and legs, about a dozen severe blows, so that my limbs and flesh were sore for several weeks; and then after several other offensive epithets, left me.

This affair my mother saw from her cottage, which was near; I being one of the oldest sons of my parents, our family was now mortified to the lowest degree. I had always aimed to be trustworthy; and feeling a high degree of mechanical pride, I had aimed to do my work with dispatch and skill; my blacksmith's pride and taste was one thing that had reconciled me so long to remain a slave. I sought to distinguish myself in the finer branches of the business by invention and finish; I frequently tried my hand at making guns and pistols, putting blades in penknives, making fancy hammers, hatchets, sword-canes, &c., &c. Besides I used to assist my father at night in making straw-hats and willow-baskets, by which means we supplied our family with little articles of food, clothing and luxury, which slaves in the mildest form of the system never get from the master; but after this, I found that my mechanic's pleasure and pride were gone. I thought of nothing but the family disgrace under which we were smarting, and how to get out of it.

## *Uncle Tom's Cabin* (1852)

Harriet Beecher Stowe

Harriet Beecher Stowe's 1852 novel *Uncle Tom's Cabin* was arguably the most powerful antislavery tract of the antebellum period. The characters of wise and gentle Uncle Tom and brave fugitive mother Eliza, who escaped across the ice floes to freedom, are well known even to people who have not read the novel. Less celebrated is Eliza's husband George, a mechanic and inventor. In this selection, race trumps economics as George's owner refuses to allow him to continue working at a job that acknowledged and utilized his technological skill and ingenuity. George's owner clearly believed, as did others who sought to enforce the enslavement of African-Americans, that any admission of a slave's mechanical competence would powerfully undercut white claims of superiority and justifications of racist assumptions.

### Chapter II
The Mother

Eliza had been brought up by her mistress from girlhood as a petted and indulged favourite.

The traveller in the south must often have remarked that peculiar air of refinement, that softness of voice and manner, which seems in many cases to be a particular gift to the quadroon and mulatto women. These natural graces in the quadroon are often united with beauty of the most dazzling kind, and in almost every case with a personal appearance prepossessing and agreeable. Eliza, such as we have described her, is not a fancy sketch, but taken from remembrance, as we saw her years ago in Kentucky. Safe under the protecting care of her mistress, Eliza had reached maturity without those temptations which make beauty so fatal an inheritance to a slave. She had been married to a bright and talented young mulatto man, who was a slave on a neighbouring estate and bore the name of George Harris.

This young man had been hired out by his master to work in a bagging factory, where his adroitness and ingenuity caused him to be considered the first hand in the place. He had invented a machine for the cleaning of the hemp which, considering the education and circumstances of the inventor, displayed quite as much mechanical genius as Whitney's cotton-gin.[1]

He was possessed of a handsome person and pleasing manners, and was a general favourite in the factory. Nevertheless, as this young man was in the eye of the law

Excerpted from: Harriet Beecher Stowe, *Uncle Tom's Cabin: A Tale of Life among the Lowly* [1852] (New York: Eaton & Mains, 1897), 17–21; figure reprinted from p. 19.

not a man, but a thing, all these superior qualifications were subject to the control of a vulgar, narrow-minded, tyrannical master. This same gentleman, having heard of the fame of George's invention, took a ride over to the factory, to see what this intelligent chattel had been about. He was received with great enthusiasm by the employer, who congratulated him on possessing so valuable a slave.

He was waited upon over the factory, shown the machinery by George, who, in high spirits, talked so fluently, held himself so erect, looked so handsome and manly, that his master began to feel an uneasy consciousness of inferiority. What business had his slave to be marching round the country, inventing machines, and holding up his head among gentlemen? He'd soon put a stop to it. He'd take him back, and put him to hoeing and digging, and "see if he'd step about so smart." Accordingly, the manufacturer and all hands concerned were astounded when he suddenly demanded George's wages and announced his intention of taking him home.

"But, Mr. Harris" remonstrated the manufacturer, "isn't this rather sudden?"

"What if it is? —isn't the man *mine*?"

"We would be willing, sir, to increase the rate of compensation."

"No object at all, sir. I don't need to hire any of my hands out, unless I've a mind to."

"But, sir, he seems peculiarly adapted to this business."

"Dare say he may be; never was much adapted to anything that I set him about, I'll be bound."

"But only think of his inventing this machine," interposed one of the workmen, rather unluckily.

"Oh yes! —a machine for saving work, is it? He'd invent that, I'll be bound; let a nigger alone for that, any time. They are all labour-saving machines themselves, every one of 'em. No, he shall tramp!"

George had stood like one transfixed at hearing his doom thus suddenly pronounced by a power that he knew was irresistible. He folded his arms tightly, pressed in his lip, but a whole volcano of bitter feelings burned in his bosom, and sent streams of fire through his veins. He breathed short, and his large dark eyes flashed like live coals; and he might have broken out into some dangerous ebullition had not the kindly manufacturer touched him on the arm and said, in a low tone,

"Give way, George; go with him for the present. We'll try to help you yet."

The tyrant observed the whisper, and conjectured its import, though he could not hear what was said; and he inwardly strengthened himself in his determination to keep the power he possessed over his victim.

George was taken home and put to the meanest drudgery of the farm. He had been able to repress every disrespectful word; but the flashing eye, the gloomy and troubled brow, were part of a natural language that could not be repressed—indubitable signs, which showed too plainly that a man could not become a thing.

It was during the happy period of his employment in the factory that George had seen and married his wife. During that period—being much trusted and favoured by his employer—he had full liberty to come and go at discretion. The

**Figure 3.1**
Illustration titled "'Give way, George; Go.'"

marriage was highly approved of by Mrs. Shelby, who, with a little womanly complacency in match-making, felt pleased to unite her handsome favourite with one of her own class who seemed in every way suited to her; and so they were married in her mistress's great parlour, and her mistress herself adorned the bride's beautiful hair with orange-blossoms, and threw over it the bridal veil, which certainly could scarce have rested on a fairer head; and there was no lack of white gloves, and cake, and wine—of admiring guests to praise the bride's beauty and her mistress's indulgence and liberality. For a year or two Eliza saw her husband frequently, and there was nothing to interrupt their happiness, except the loss of two infant children, to whom she was passionately attached, and whom she mourned with a grief so intense as to call for gentle remonstrance from her mistress, who sought, with maternal anxiety, to direct her naturally passionate feelings within the bounds of reason and religion.

After the birth of little Harry, however, she had gradually become tranquillised and settled; and every bleeding tie and throbbing nerve, once more entwined with that little life, seemed to become sound and healthful, and Eliza was a happy woman up to the time that her husband was rudely torn from his kind employer and brought under the iron sway of his legal owner.

The manufacturer, true to his word, visited Mr. Harris a week or two after George had been taken away, when, as he hoped, the heat of the occasion had passed away, and tried every possible inducement to lead him to restore him to his former employment.

"You needn't trouble yourself to talk any longer," said he, doggedly; "I know my own business, sir."

"I did not presume to interfere with it, sir. I only thought that you might think it for your interest to let your man to us on the terms proposed."

"Oh, I understand the matter well enough. I saw your winking and whispering the day I took him out of the factory; but you don't come it over me that way. It's a free country, sir; the man's *mine*, and I do what I please with him—that's it!"

And so fell George's last hope; —nothing before him but a life of toil and drudgery, rendered more bitter by every little smarting vexation and indignity which tyrannical ingenuity could devise.

A very humane jurist once said: "The worst use you can put a man to is to hang him." No; there is another use that a man can be put to that is WORSE!

### Note

1. A machine of this description was really the invention of a young coloured man in Kentucky.

## *A Journey in the Seaboard Slave States (1856)* and *A Journey in the Back Country (1861)*

Frederick Law Olmsted

On the eve of the Civil War, Frederick Law Olmsted traveled widely through the South, from the Carolinas to Texas and Kentucky to Louisiana, observing and recording a society fundamentally shaped by slavery. In his travels he came across many examples of enslaved Africans who not only were proficient with the technologies of their time and place, but were acknowledged to be so by white masters and strangers alike. The tension between admission and denial of technological skill and ingenuity hints at the high stakes at risk in allowing the enslaved to find pride in their work.

### *A Journey in the Seaboard Slave States*

From the settlement, we drove to the "mill"—not a flouring mill, though I believe there is a run of stones in it—but a monster barn, with more extensive and better machinery for threshing and storing rice, driven by a steam-engine, than I have ever seen used for grain on any farm in Europe or America before. Adjoining the mill-house were shops and sheds, in which black-smiths, carpenters, and other mechanics—all slaves, belonging to Mr. X.—were at work. He called my attention to the excellence of their workmanship, and said that they exercised as much ingenuity and skill as the ordinary mechanics that he was used to employ in New England. He pointed out to me some carpenter's work, a part of which had been executed by a New England mechanic, and a part by one of his own hands, which indicated that the latter was much the better workman.

I was gratified by this, for I had been so often told, in Virginia, by gentlemen, anxious to convince me that the negro was incapable of being educated or improved to a condition in which it would be safe to trust him with himself—that no negro-mechanic could ever be taught, or induced to work carefully or nicely—that I had begun to believe it might be so.

We were attended through the mill-house by a respectable-looking, orderly, and gentlemanly-mannered mulatto, who was called, by his master, "the watchman." His duties, however, as they were described to me, were those of a steward, or intendant. He carried, by a strap at his waist, a very large number of keys, and had charge of all the stores of provisions, tools, and materials of the plantations, as well as of all their produce, before it was shipped to market. He weighed and measured out all the

Excerpted from: Frederick Law Olmsted, *A Journey in the Seaboard Slave States, with Remarks on Their Economy* (New York: Dix and Edwards, 1856), 425–428; and from *A Journey in the Back Country* (New York: Mason Bros., 1861), 78, 151, 180–181, 381–383.

rations of the slaves and the cattle; super-intended the mechanics, and himself made and repaired, as was necessary, all the machinery, including the steam-engine.

In all these departments, his authority was superior to that of the overseer. The overseer received his private allowance of family provisions from him, as did also the head-servant at the mansion, who was his brother. His responsibility was much greater than that of the overseer; and Mr. X. said, he would trust him with much more than he would any overseer he had ever known.

Anxious to learn how this trustworthiness and intelligence, so unusual in a slave, had been developed or ascertained, I inquired of his history, which was, briefly, as follows.

Being the son of a favorite house-servant, he had been, as a child, associated with the white family, and received by chance something of the early education of the white children. When old enough, he had been employed, for some years, as a waiter; but, at his own request, was eventually allowed to learn the blacksmith's trade, in the plantation-shop. Showing ingenuity and talent, he was afterwards employed to make and repair the plantation cotton-gins. Finally, his owner took him to a steam-engine builder, and paid $500 to have him instructed as a machinist. After he had become a skillful workman, he obtained employment, as an engineer; and for some years continued in this occupation, and was allowed to spend his wages for himself. Finding, however, that he was acquiring dissipated habits, and wasting all his earnings, Mr. X. eventually brought him, much against his inclinations, back to the plantations. Being allowed peculiar privileges, and given duties wholly flattering to his self-respect, he soon became contented; and, of course, was able to be extremely valuable to his owner.

I have seen another slave-engineer. The gentleman who employed him told me that he was a man of talent, and of great worth of character. He had desired to make him free, but his owner, who was a member of the Board of Brokers, and of Dr. ———'s Church, in New York, believed that Providence designed the negro race for slavery, and refused to sell him for that purpose. He thought it better that he (his owner) should continue to receive two hundred dollars a year for his services while he continued able to work, and then he should feel responsible that he did not starve, or come upon the public for a support, in his old age. The man himself, having light and agreeable duties, well provided for, furnished with plenty of spending money in gratuities by his employer, patronized and flattered by the white people, honored and looked up to by those of his own color, was rather indifferent in the matter; or even, perhaps, preferred to remain a slave, to being transported for life, to Africa. . . .

Near the first quarters we visited there was a large black-smith's and wheelwright's shop, in which a number of mechanics were at work. Most of them, as we rode up, were eating their breakfast, which they warmed at their fires. Within and around the shop there were some fifty plows which they were putting in order. The manager inspected the work, found some of it faulty, sharply reprimanded the workmen for not getting on faster, and threatened one of them with a whipping for not paying closer attention to the directions which had been given him. He told me that he once employed a white man from the North, who professed to be a first-class workman, but he soon found he could not do nearly as good work as the negro mechanics on the estate, and the latter despised him so much, and got such high opinions of themselves in consequence of his inferiority, that he had been obliged to discharge him in the midst of his engagement. . . .

"Two year ago," he continued, after taking his dram, as we sat by the fire in the north room, "when I had a carpenter here to finish off this house, I told one of my boys he must come in and help him. I reckoned he would larn quick, if he was a mind to. So he come in, and a week arterwards he fitted the plank and laid this floor, and now you just look at it; I don't believe any man could do it better. That was two year ago, and now he's as good a carpenter as you ever see. I bought him some tools after the carpenter left, and he can do any thing with 'em—make a table or a chest of drawers or any thing. I think niggers is somehow nat'rally ingenious; more so 'n white folks. They is wonderful apt to any kind of slight." . . .

"Are there many persons here who have as bad an opinion of slavery as you have?"

"I reckon you never saw a conscientious man who had been brought up among slaves who did not think of it pretty much as I do—did you?"

"Yes, I think I have, a good many."

"Ah, self-interest warps men's minds wonderfully, but I don't believe there are many who don't think so, sometimes—it's impossible, I know, that they don't."

Were there any others in this neighborhood, I asked, who avowedly hated slavery? He replied that there were a good many mechanics, all the mechanics he knew, who felt slavery to be a great curse to them, and who wanted to see it brought to an end in some way. The competition in which they were constantly made to feel themselves engaged with slave-labor was degrading to them, and they felt it to be so. He knew a poor, hard-working man who was lately offered the services of three negroes for six years each if he would let them learn his trade, but he refused the proposal with indignation, saying he would starve before he helped a slave to become a mechanic.[1] There was a good deal of talk now among them about getting laws passed to prevent the owners of slaves from having them taught trades, and to prohibit slave-mechanics from being hired out. He could go out to-morrow, he supposed, and in the course of a day get two hundred signatures to a paper alleging that slavery was a curse to the people of Mississippi, and praying the Legislature to take measures to relieve them of it as soon as practicable. (The county contains three times as many slaves as whites.) . . .

***A Journey in the Back Country***
Character and Manners

**Formula for Justifying Slavery**
SINCE the growth of the cotton demand has doubled the value of slave labor, and with it the pecuniary inducement to prevent negroes from taking care of themselves, hypotheses and easy methods for justifying their continued slavery have been multiplied. I have not often conversed with a planter about the condition of the slaves,

that he did not soon make it evident, that a number of these were on service in his own mind, naively falling back from one to another, if a few inquiries about matters of fact were addressed him without obvious argumentative purpose. The beneficence of slavery is commonly urged by an exposition not only of the diet, and the dwellings, and the jollity, and the devotional eloquence of the negroes, but also by demonstrations of the high mental attainments to which individuals are already found to be arriving. Thus there is always at hand some negro mathematician, who is not merely held to be far in advance of the native Africans, but who beats most white men in his quickness and accuracy in calculation, and who is at the same time considered to be so thoroughly trustworthy, that he is constantly employed by his master as an accountant and collecting agent; or some negro whose reputation for ingenuity and skill in the management and repair of engines, sugar-mills, cotton presses, or other machinery, is so well established that his services are more highly valued, throughout a considerable district, than any white man's; or some negro who really manages his owner's plantation, his agricultural judgment being deferred to, as superior to that of any overseer or planter in the county. Scarcely a plantation did I visit on which some such representative black man was not acknowledged and made a matter of boasting by the owner, who, calling attention perhaps to the expression of intelligence and mien of self-confidence which distinguished his premium specimen, would cheerfully give me a history of the known special circumstances, practically constituting a special mental feeding by which the phenomenon was to be explained. Yet it might happen that the same planter would presently ask, pointing to the brute-like countenance of a moping field hand, what good would freedom be to such a creature? And this would be one who had been provided from childhood with food, and shelter, and clothing with as little consideration of his own therefor as for the air he breathed; who had not been allowed to determine for himself with whom he should associate; with what tools and to what purpose he should labor; who had had no care on account of his children; who had no need to provide for old age; who had never had need to count five-and-twenty; the highest demand upon whose faculties had been to discriminate between cotton and crop-grass, and to strike one with a hoe without hitting the other; to whose intelligence, though living in a civilized land, the pen and the press, the mail and the telegraph, had contributed nothing; who had no schooling as a boy; no higher duty as a man than to pick a given quantity of cotton between dawn and dark; and of whom, under this training and these confinements, it might well be wondered that he was found able to understand and to speak the language of human intelligence any more than a horse.

## Note

1. At Wilmington, North Carolina, on the night of the 27th of July (1857), the frame-work of a new building was destroyed by a number of persons and a placard attached to the disjointed lumber, stating that a similar course would be pursued in all cases, against edifices that should be erected by negro contractors or carpenters, by one of which class of men

the house had been constructed. There was a public meeting called a few days afterwards, to take this outrage into consideration, which was numerously attended. Resolutions were adopted, denouncing the act, and the authorities were instructed to offer a suitable reward for the detection and conviction of the rioters. "The impression was conveyed at the meeting," says the Wilmington Herald, "that the act had been committed by members of an organized association, said to exist here, and to number some two hundred and fifty persons, and possibly more, who, as was alleged, to right what they considered a grievance in the matter of negro competition with white labor, had adopted the illegal course of which the act in question was an illustration." Proceedings of a similar significance have occurred at various points, especially in Virginia.

## *His Promised Land: The Autobiography of John P. Parker, Former Slave and Conductor on the Underground Railroad*

John P. Parker

John Parker learned the craft of iron founding and used it to buy his freedom and establish a flourishing business in the North, where he became not only a respected entrepreneur but also an important figure in the Underground Railroad.

But my kind old master knew of the time to come for me, so again he advised me to learn a trade.

This time he placed me with a friend of his, the owner of an iron foundry. I was to learn the trade of an iron molder. It was [a] natural bent, so I went at it with a will, so that I was soon a full-fledged molder.

Being of an inventive turn of mind, as you will see later, I soon rigged up my bench so I could do more and better work than any man in the shop. This fact naturally caused some ill feeling among the other workmen towards me.

My master gave me what I made, so I very foolishly spent my money on myself. I remember I paid $20 for a hat. My extravagance caused the foreman to complain to the superintendent, who warned me that I was playing a game that would eventually lead me into trouble. Instead of heeding his advice, I went right on squandering my earnings, heaping up trouble for myself.

By this time I must have become not only extravagant but quite impudent. One morning I arrived at the foundry a little late. I had on my good clothes, of which I was exceedingly proud. The superintendent, who was on the crane lifting a heavy casting, called to me to come help. I called to him as soon as I put on my overalls that I would come. This apparent insubordination threw him into a rage.

By this time I was angry myself, so I told him I would come when I changed my clothes and not before. This so angered him he lashed at me and struck me in the face. I struck back and the next thing I knew, I was in a regular knock-down-and-drag-out fistfight. That was the end of me. That night my friend and master told me he was going to send me to a friend of his in New Orleans, who was also the owner of an iron foundry, until my trouble would blow over.

Monday morning I was in New Orleans, at the iron foundry. I was put to work under a foreman who did not know his job. I knew then that my position was hopeless, because I knew I was a good workman, and would put him to shame, through the kind of work the foreman was turning out. Sure enough I lasted just one week. Saturday

Excerpted from: Stuart Seely Sprague, ed., *His Promised Land: The Autobiography of John P. Parker, Former Slave and Conductor on the Underground Railroad* (New York: W. W. Norton, 1996), 64–70. Reprinted by permission of W. W. Norton & Company, Inc.

night I was dismissed. But dismissal was nothing to what followed, since I was told that my master had sent word, if I did not get along, I was to be sold as a field hand.

When Mr. Jennings, the man's name, went out of his office, I followed close on his heels, determined to make one more appeal to my good master in Mobile. When I arrived in Mobile the next morning, my doctor friend absolutely refused to hear my side of the case. The only promise I could get out of him was that on Wednesday, I would be sent back to Mr. Jennings in New Orleans. I knew that meant the cotton fields of Alabama would see my finish.

Seeing my pleadings with the doctor were futile, I decided to take my case in my own hands. Among the people I knew in Mobile was a widow named Mrs. Ryder. She was a patient of the doctor's, so I had been to her home a number of times. To her I now went asking her to buy my freedom, letting me pay her back from my earnings.

She was frank enough to tell me that the proposition did not appeal to her, because I was always in trouble, and could not keep a job. From her I went to several other wealthy people, but received absolutely no encouragement. I was a dog with a bad name, which was in fact very bad. Tuesday night saw me in the depths of despair. I begged the doctor [to] give me a week, but he was immovable. It was Wednesday for New Orleans, and that was final.

Wednesday morning when I went to make a final plea to Mrs. Ryder, she refused to see me. I was desperate, so I held on, until she finally agreed to see me. She apparently was not interested in me, still holding to the fact that however good a workman I might be, I could not hold my job. I finally made her the proposition that while I was sure I could pay her back in two years, I would stay on another year if she would only buy my freedom.

It was my persistency that finally won her consent, rather than her cupidity, for she agreed to release me as soon as I had paid her in full. My contract all signed and agreed to called for the payment of $1,800, with interest, to be paid at the rate of $10 per week. That day I became the slave of Mrs. Ryder. My friend the doctor was pleased with the deal, as he knew I would have a good home. So we parted as good friends, which he always was to me.

Mrs. Ryder gave me a free hand to go where I wanted to and do as I pleased. She was just as good to me as the doctor, but she had no library, which was a direct loss to me. Fortunately, there was a vacancy in a Mobile foundry that was very busy, and short of molders. I was employed by piecework, so the more castings I turned out the greater my pay. That plan suited me fine.

Long before the other workmen were around in the morning, I was hard at work over my molds. The days were too short. On Sundays, I met the steamers with my wheelbarrow, ready to deliver packages or trunks. In addition to my work in the foundry, I ran a regular three-ball pawnshop, buying and selling anything and everything offered me.

Each week I not only paid my installment [but also] frequently doubled it, so at the end of six months I had made a very substantial payment on my contract. Mrs. Ryder was pleased and I was more delighted at the prospect of my early freedom. At this time I was 18, strong as an ox, and working like a steam engine, under high pressure. Another six months would see me in sight of the end of slavery.

I had been quietly working for some time on a new idea of a circular harrow or clod smasher, which was a very important farm implement of that period with so much new land to break up. Being handy with tools, on my own time I secretly made a model. It looked so good I showed it to the superintendent, who took it so much to heart, I never saw my model again. I went to the owner of the foundry, who in turn called in the superintendent. In my presence the superintendent claimed both the idea and model were his, and that I had nothing to do with the development of the machine. The words were hardly out of his mouth when I had him by the throat. If I had been normal, I never would have done such a senseless thing like that. But I had hopes that my invention would not only pay me out of slavery, but give a start when I was free. As it was, the treachery of the man was more than my overwrought nerves would stand.

I hurried home and told Mrs. Ryder what I had done and why. While she was sympathetic at the same time she was so practical, she knew what would happen to me. She advised me to quit the foundry trade, which I was compelled to do because I was not wanted in the two foundries in Mobile. Once on the street, my position was hopeless. Even my trading schemes failed, so I was running behind [in] my contract.

But Mrs. Ryder was a good woman. She encouraged me all she could, never saying a word to me about my lapse in payments. To add to my cup of bitterness, passing by the foundry, I saw at least seven of my clod crushers packed and ready for shipment. I stopped and counted them over and over again. The profits on that shipment would have practically wiped out my indebtedness, but I passed on a slave and a beggar.

Then I had a real break in my string of ill luck. A new foundry was starting. They needed molders and I applied for a job and got it at once. The next morning I was around as soon as the doors were opened, once more alert and hopeful. The first week I lived in that shop. Early and late I was at my bench. Every penny went to my benefactress. In exactly 18 months after I had entered into the contract with Mrs. Ryder, I made my last payment to her, starting forth a free man.

She wanted me to stay on, as I had been an exceedingly handy man about the house. But I had other plans. I wanted to get on in life, and I knew with all her best wishes Mobile was a poor place for me to stop. As soon as my free papers were signed, I asked for a passport to Jeffersonville, Indiana, where I had been told there were iron foundries.

Then I did a foolish thing. Being free, I went around to gloat over the man who stole my clod smasher. Upon entering the shop the first thing that struck my eye was a row of my machines, with a lot of castings for more on the floor. The superintendent was not pleased to see me, for the wretch's conscience troubled him for his defrauding me.

However, I told him I was free, and was leaving for the north. He demanded to see my papers. I told him they were sewn in my clothes, that they had been passed upon by the authorities and they were in order. He was determined to cause me trouble. So rather than give him that satisfaction, I ruined my new clothes, by ripping out the lining.

After looking at them a few minutes, he sneeringly put them in his jacket, saying they were not worth the paper they were written on. For once I controlled my temper, knowing Mrs. Ryder would make good any irregularities that might exist in the papers already issued.

Sure enough Mrs. Ryder did make good the fault, and again I sewed the papers in my clothes. Now I determined on my revenge on the man who had caused me all the trouble. I knew he was the last man to leave the foundry at night. Biding my time, I stepped through the door into the shop in time to catch him alone. I took off my coat and vest. He was no coward and knew what was coming. I told him I was leaving Mobile forever. Before I left I was going to give him a good beating or he was going to perform [one] upon me.

It was a fight to a finish. The molds were off the floor so we had ample space to fight. I had been through too many rough-and-tumble fights not to know the tricks of combat. With the notice I had given my man, it was impossible for me to get in the first blow, which is a very [important] point in this sort of contest. It was now a fight man to man. As I have said, my man was no coward, was strong, and willing to fight. We plunged at each other again and again, our arms working like flails. I only had one thing in mind and that was his treachery, and this was my last chance at him.

I am very sure my will to beat him kept me on my feet. In one of my rushes, my opponent's impetus carried him over me, throwing him heavily to the hard floor. I was on him like a flash. When he staggered up I hit him fairly on the jaw, knocking him down again. Then I knew he was done for, but he was game and came back for more. But the animal in me knew no pity. As he arose I swung hard on his jaw. He trembled all over. Then I hit him again with every ounce of vengeance I could muster. This time he went down for good. I gloated over his bruised face, discolored eyes. As a free man, I had met him fairly and asserted my superiority ... [over] a contemptible foe.

Hurrying home, I found Mrs. Ryder at home. I told her frankly of my fight and the necessity of my catching the New Orleans steamer that evening. She urged me to stay and meet the difficulty, but I made up my mind to go.

I bade her goodbye, with regrets because of her almost motherly kindness. I then went to see my old master the doctor, because I did owe so much to him. He knew I had been fighting again. Without knowing any of the details, he advised me to leave Mobile at once and not stop until I had reached one of the free states.

So far as my masters were concerned, most of my life as a slave was a pleasant one, so far as my bodily wants were concerned. But I hated the injustices and restraints against my own initiative more than it is possible for words to express. To me that was the great curse of slavery. If I had submitted, I presume I would have been a good house servant, but my independent nature would not permit [me] to do so humbly. I never saw either my benefactor or benefactress again.

That night I left quietly for New Orleans. The next day I was safely aboard an upriver Mississippi River packet on my way to Jeffersonville, Indiana. On the way up the river I tried to locate various points of my previous adventure, but I was unable to do so. At Jeffersonville I went to work as a molder. I found my work agreeable; but for my wandering desires, I supposed I would have stayed there all my life. But Cincinnati lured my away to other adventures.

## Layout of Parker's Phoenix Foundry (1884)

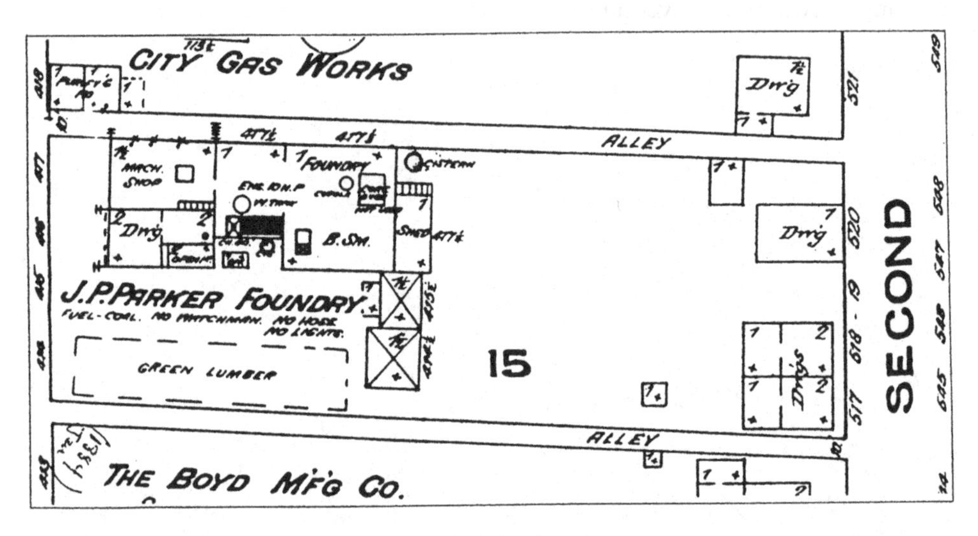


**Figure 3.2**
After the Civil War, insurance companies increasingly relied on maps produced by the Sanborn company to guide them in determining premiums. Maps such as this one of John P. Parker's Phoenix Foundry in Ripley, Ohio, in 1884 are often the sole remaining visual evidence of the properties they depict. Reprinted from Sanford Insurance Co. files.

## U.S. Patent to John Percial Parker for a Soil-Pulverizer (1890)

In 1890, John Parker finally received a patent for the soil pulverizer he had invented while still enslaved.

**United States Patent Office**

**John Percial Parker, of Ripley, Ohio**
Soil-Pulverizer

Specification Forming Part of Letters Patent No. 442,538, Dated December 9, 1890.
Application Filed August 7, 1890. Serial No. 361,246. (No model.)

To all whom it may concern:

Be it known that I, John Percial Parker, a citizen of the United States, and a resident of Ripley, in the county of Brown and State of Ohio, have invented certain new and useful Improvements in Soil-Pulverizers; and I do hereby declare that the following is a full, clear, and exact description of the invention, which will enable others skilled in the art to which it appertains to make and use the same, reference being had to the accompanying drawings, which form a part of this specification.

My invention relates to improvements in apparatus for cutting up and pulverizing soil for agricultural purposes of that class known as "revolving" or "rotary" pulverizers.

Heretofore a rotary soil-pulverizer has been constructed which comprised a series of disks mounted upon a shaft or axle revolving in a suitable frame, said disks being provided with a series of lateral projections near their peripheries, whereby the soil was crushed and pulverized. This construction of apparatus was very effective in use; but still there were disadvantages which my invention is intended to obviate.

The object of the invention is to provide a simple and economical apparatus of the above description which will effectually and rapidly crush clods, cut up corn-stalks and other like objects, and pulverize the soil so as to leave it in the best possible condition for agricultural purposes.

The invention consists in the novel construction and combination of parts, herein-after fully described, and specifically pointed out in the claim....

Reprinted from: U.S. Patent No. 442,538, dated December 9, 1890, to John Percial Parker, of Ripley, Ohio, for a Soil-Pulverizer.

The operation will be readily understood. As the apparatus is drawn across a field, the disks will rotate and the blades and teeth will thoroughly and effectually cut up, crush, and pulverize the soil so as to fit it for agricultural purposes.

It will be seen that the outer edge of the rim between the teeth forms a cutter as well as the outer edges of the projections, which are both cutters and crushers.

Having thus described my invention, what I claim is—

In an apparatus for crushing and pulverizing soil and cutting cornstalks and other obstacles, the combination, with a frame and a revolving shaft or axle, of a series of disks or wheels mounted upon the shaft, the outer edges of which form cutters, and the laterally-extending ogee-shaped projections having outwardly-extending blade and peripheral projections or teeth, substantially as described.

In testimony that I claim the foregoing as my own I have hereunto affixed my signature in presence of two witnesses.

John Percial Parker

Witnesses:

Marshall Creekbaum

J. M. Criswell

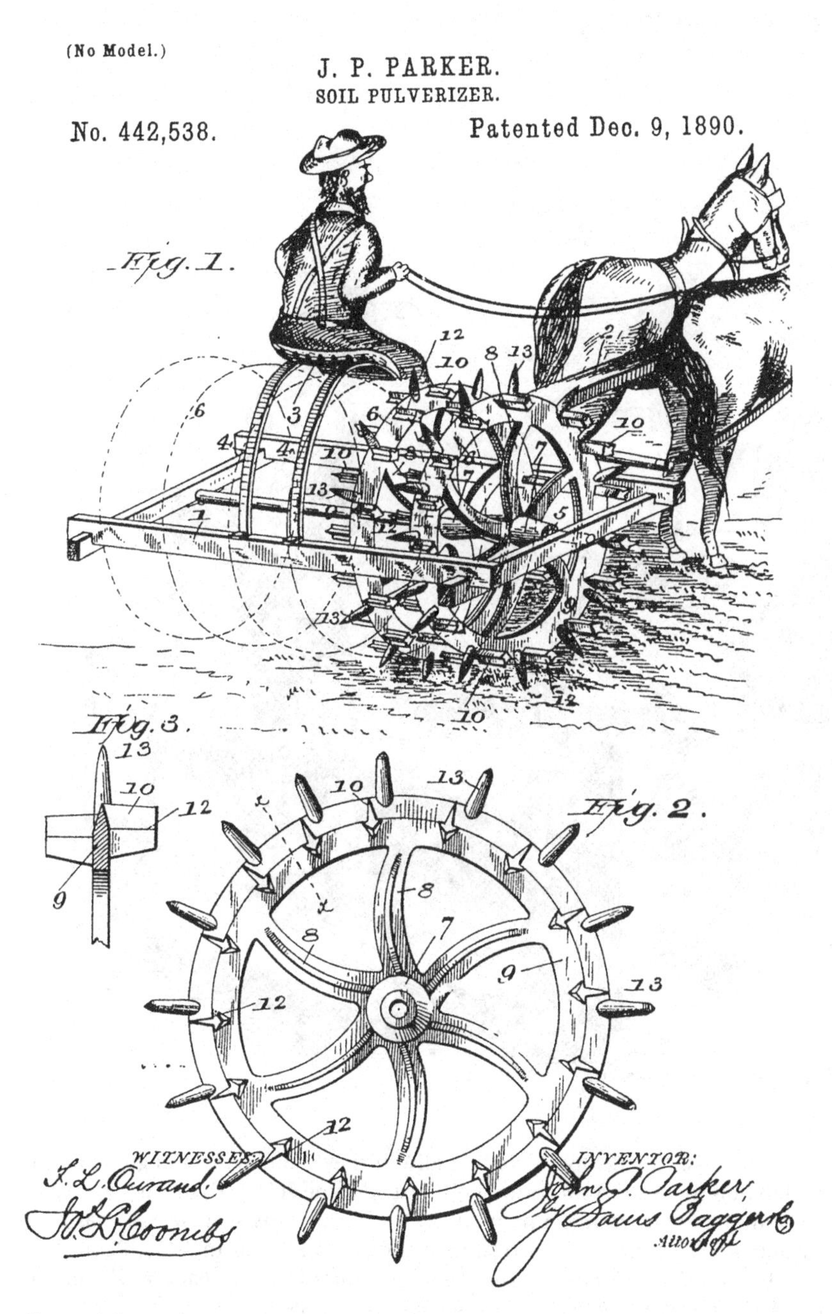


Figure 3.3
Patent drawing for John P. Parker's Soil-Pulverizer.

## Tending a Cotton Gin (1853)

**Figure 3.4**
The introduction of labor-saving devices including Eli Whitney's famous cotton gin reduced the time required to pull the seeds from cotton bolls turned some enslaved Africans into machine tenders, not unlike their white counterparts in the North. One such worker was pictured in an illustrated volume that recorded America's first world's fair in 1853. Reprinted from B. Silliman, Jr., and C. R. Goodrich, *The World of Science, Art, and Industry Illustrated from Examples in the New-York Exhibition, 1853–54* (New York, 1854), 8.

# 4 The New Industrial Age

*Notes on North America, Agricultural, Economical, and Social* (1851)
James F. W. Johnston

**Scenes from Oak Lawn, Louisiana Plantation (1864)**

**Slave Labor as Reported in *Nile's Weekly Register* (1849) and *DeBow's Southern and Western Review* (1851)**

*The History of the First Locomotives in America* (1874)
William H. Brown

**Advertisement in *The Liberator* Seeking Colored Inventors (1834)**

**U.S. Patent to Norbert Rillieux for an "Improvement in Sugar-Works" (1843)**

**U.S. Patent to Norbert Rillieux for an "Improvement in Sugar-Making" (1846)**

**The Confederate Patent Act (1861)**

## *Notes on North America, Agricultural, Economical, and Social* (1851)

James F. W. Johnston

The South did not embrace the advancing Industrial Revolution until after the Civil War, but for decades beforehand observers speculated about the possibility and possible effects of a regional shift from a sole dependence on agriculture to an economy in which manufacturing played a larger part. In 1851 a touring Scottish chemist recorded his observations and speculations.

In the great increase of slaves employed in the sugar culture in this one State of Louisiana—from 63,000 to 126,000 in five years—we see the direction taken by the slaves from the more eastern States, and we understand more clearly the meaning of Mr Meade, when he said, that "Virginia had a slave population of half a million, whose value was chiefly dependent upon southern demand." Were the slaves of this and other States sold bodily, so to speak, to the south, there would be a hope of clearing State after State of the severe infliction; but when only the increase is sold, Virginia is to Louisiana and Texas what Africa is to Cuba and Brazil; and the more the African traffic is put down by England, the more profitable will the internal slave-trade become to southern America![1]

There is another aspect of this question which awakens gloomy apprehensions as to the future of the American slave. The introduction of the cotton manufacture into the slave States—Virginia, Kentucky, North and South Carolina, Georgia, Tennessee, and Mississippi—in which there are some hundreds of factories, consuming already from 300,000 to 400,000 bales of cotton a-year—has brought a new use of his slaves within the reach of the southern planter. The same power which compels them to toil in gangs under a burning sun, will constrain them to waste life in the factories, if it can be done profitably to the master. The great difficulty of the manufacturers in the New England States, is the question of labour—the scarcity of work-people, the high wages they demand, and the delicacy required to manage them. In the south these difficulties vanish. Slave labour is easily obtained, and the slave obeys as mechanically as the machine he superintends. A great and rapid extension of the factory system is therefore looked for in the south, and many predict that the manufacturers of the eastern States will sink before them.

But whether the latter result follow or not, the prospect is anything but cheering to the friends of free labour. If to the cotton-culture—hitherto the great slave-multiplier—be added that of sugar, as a profitable employment, and to both the use

Excerpted from: James F. W. Johnston, *Notes on North America, Agricultural, Economical, and Social* (Edinburgh: William Blackwood and Sons, 1851), II, 364–367.

of slaves in cotton and other factories, it cannot be doubted that a new and great stimulus will be given to the breeding and traffic in slaves, and a stronger attachment created towards those domestic institutions by which slavery is established and made legal.

And if in free England the factory system has been productive of so many evils, physical, moral, and social, who shall say to what new forms of oppression and misery it may give rise in vast workshops peopled by human beings who have no civil rights, and who are superintended by others whose immediate profit may be the greatest when their sufferings are rendered the most unbearable?

In the preceding chapter, I have spoken of the direct influence—political, religious, and educational—which the institutions of the United States are destined to exercise upon our own, and of the gradual assimilation which, should peace and progress continue among them, may be expected to take place between their institutions and ours.

But this rapid extension of the cotton manufacture in the southern States, and the employment of slave labour in their factories, besides the influence it is likely to have upon the future condition of the slaves and of the slave question in the United States, can scarcely fail to affect in a marked manner the future comfort and condition of our home manufacturing population. If the labour of coloured slaves, so employed, really prove cheaper than that of free white men, then either our manufactures must decline and decrease, or the condition and emoluments of our workmen must be gradually reduced to the level of those of the slave operatives of the American factories. The possibility of such a result is melancholy and disheartening, at a time when so many are anxious rather to improve and elevate than further to depress our labouring people.

But we have, as an encouragement, the assertion made by many, that free labour, even in equal circumstances, is cheaper than slave labour. How much more ought it to be so, when the free labourers are white men of English blood, enlightened by some measure of education, and assisted by all the aids of a constantly advancing mechanical skill? Though our home property may not ameliorate the condition of the unhappy chattels who are destined to labour in the factories of the southern States, we may, nevertheless, still hope that *their* condition, whatever it be, will not materially depress that of any class of labourers in our own more favoured country.

In any event, whether they do or do not come into direct competition with our home labourers, it cannot be a matter of indifference to us, either on the score of interest or of humanity, that the actual condition of the slave population of the United States should be sustained and ameliorated, rather than still further or for a longer period depressed; and if the maintaining of the existing Union will promote that end, we ought to wish and work for its maintenance. It is true that, supposing the Union indefinitely perpetuated, the additional encouragements to slavery presented by the sugar-traffic and by the factory system would not be removed, nor could northern intelligence and energy be prevented from lending itself to the extension of these newer branches of industry, through the more abundant and obedient labour of the south. Still the public opinion of the northern States, and the annual discussions and legislation of Congress, would operate as powerful salutary restraints, and would check the evils of a bad system as much, probably, as any other we can now contemplate.

## Note

1. In the whole Union, during the last ten years, the slaves have increased by 808,000, or 80,000 a-year. In 1860 they will number about 4,500,000.

## Scenes from Oak Lawn, Louisiana Plantation (1864)

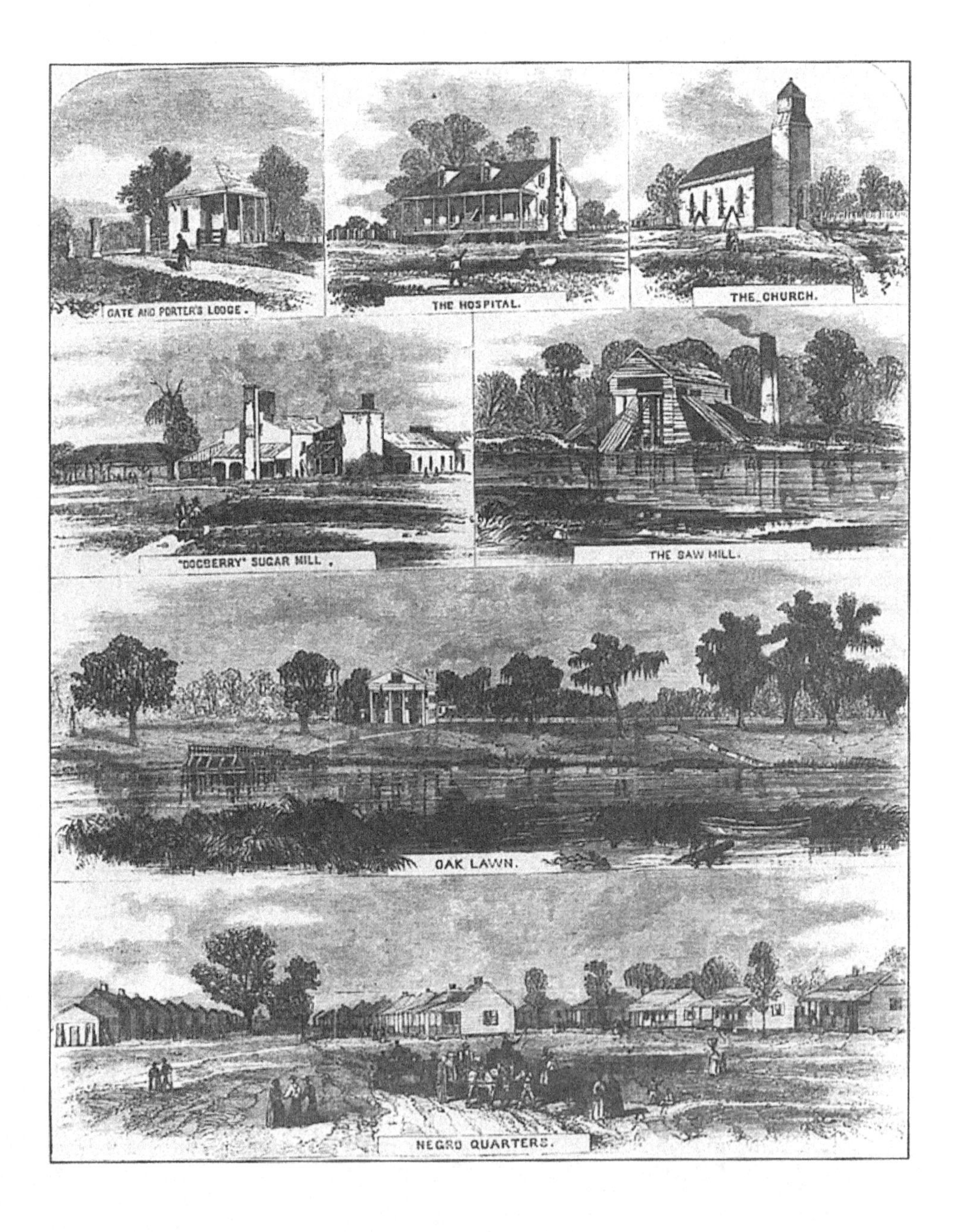

**Figure 4.1**
Even before the Industrial Revolution reached the South, when the area's economy depended almost completely on agriculture, mills using enslaved workers were hardly unknown on southern plantations. These scenes from the Oak Lawn plantation in Louisiana show a saw mill to cut lumber and a sugar mill to grind cane, both apparently powered by steam engines. Reprinted from *Frank Leslie's Illustrated Newspaper*, February 6, 1864, in Joe William Trotter, Jr., *The African American Experience*. (Boston: Houghton Mifflin Co., 2001), 173.

## Slave Labor as Reported in *Nile's Weekly Register* (1849) and *DeBow's Southern and Western Review* (1851)

Although it did not happen as often as enthusiasts hoped, there was a certain logic to the persistent southern belief that cotton ought to be manufactured into cloth close to where it was grown. A powerful cultural distaste for industry, reluctance to take capital out of agriculture in order to build factories, and fear of taking large number of enslaved Africans off the land and placing them in industrial settings, all kept the number of antebellum textile mills relatively small. Nevertheless, a few, like the much commented-upon Saluda Factory in South Carolina, and the unnamed one in Mississippi also discussed here, experimented with a possible industrial future.

### Nile's Weekly Register

#### Slave Labor in Factories

*From the Charleston Mercury* :—Much diversity of sentiment has heretofore prevailed relative to the availability of slave labor for manufacturing purposes. While some have contended that the negro could not be advantageously employed in manufactories, and that even if otherwise there would be manifest impropriety in congregating them in the numbers necessary for such purposes; others have maintained that there is no department of labor in which they can be more profitably or safely engaged. In such a conflict of opinion experience is the only safe arbiter, and as having a most important bearing upon the decision of this question, we copy from the Columbia Telegraph the statement of the Superintendent of the Saluda Factory. Mr. Graves is from New England, where he has had much experience in superintending operatives, and is familiar with the workings and discipline of the factories in that section of the Union. He came to the South with opinions made up as to the in capacity of the negro for Factory labor, and yet, after ample time has been taken to thoroughly test the question, he candidly avows that his former opinions were erroneous, and declares his conviction that slave labor is just as suitable for manufacturing purposes as the white labor of the North. Under all the circumstances of the case, we consider Mr. Graves' opinion as entitled to the highest respect, and it will necessarily have great influence upon minds which have heretofore entertained doubts upon this important subject:

Reprinted from: *Nile's Weekly Register* 75 (1849): 344; and excerpted from: *DeBow's Southern and Western Review*, 11 (September, 1851), 319–320, 433.

Mr. Editor—Dear Sir:
As the profitable employment of labor is engrossing the public mind at the present time, I cheerfully comply with your request, to furnish a statement of our experience in the employment of blacks, in the manufactory of Cotton goods.

Previous to my coming to this State, a little more than a year since, I had always supposed that blacks could not be employed to advantage in that department of labor. This impression was created, not by personal observations, but by the constant representation of their extreme indolence, carelessness, and utter want of ingenuity.

Upon my arrival at Saluda Factory, I found in the employ of that Company several black hands, and although a vote had been passed by the Stockholders to dismiss them, and to employ exclusively white hands—yet it was necessary to retain the blacks until the time for which they were engaged, should expire. This gave me an opportunity to notice their habits and to test their efficiency as operatives in a Cotton Factory. Their activity and promptness soon attracted my notice, and I watched with great interest and some curiosity, the progress of affairs, until the close of their term of service. At the expiration of that time, my former impressions had entirely given place to the conviction, that under all the circumstances connected with the Mill, it would be decidedly for the interest of the Stockholders still to retain in their employ a large proportion of black hands. And as the immediate cause for the passage of the above vote seemed to be removed, they acceded to the proposition.

To effect this it became necessary to employ several new hands, nearly all of whom had never before seen a Cotton Factory. They were put to the work as new hands, receiving no greater facilities for learning and performing their duties than is always allowed to such hands; and I have never seen an equal number of entirely new hands become efficient operatives in less time.

I believe that an equal number of persons may be taken from the farming districts of any of the Northern States, with the same discriminations as to native talent, and put to the same kind of work, and they will not become more efficient in the same length of time.

It is true that it requires skill and intelligence to manage cotton machinery to advantage; so it requires skill and intelligence to manage a farm or plantation to advantage. It does not follow, that because the person who works with the hoe, does not understand why one kind of compost is put in one place and a different kind in another, that therefore he cannot do justice to the plants with his hoe—neither does it follow, that because the operative is not versed in the sciences or skilled in the mechanic arts, that therefore he cannot be efficient at the spinning frame or the loom, as an operative.

I know very well that, in the selection of hands for the mills at the North, preference is always given to those who have enjoyed the advantages of intellectual culture; the entire want of which would be considered almost sufficient to disqualify the applicant for any service in the mill. But that deficiency in the white population of Massachusetts is an index to a very different state of things from that which the same deficiency denotes in the blacks of South Carolina.

In the former State there is a school brought within reach of every man's door, and he is permitted, nay *entreated*, to send his children to school and have them educated "without money and without price." If, therefore, such opportunities are allowed to pass unimproved, it is not difficult to divine in what habits such persons must have been drilled. The same deficiency intentionally universal among the blacks, bears no such evidence of indolence and recklessness of valuable acquisition. They are early trained to habits of industry and patient endurance, and by the concentration of all their faculties to the few departments of human acquisition to which they are necessarily restricted, their imitative faculties become cultivated to a very high degree, their muscles become trained and made obedient to the will, so that whatever they see done they are very quick in learning to do, without entering into any philosophical inquiry as to the *method* of doing it.

Our carding and spinning rooms are furnished with black hands almost entirely, and they perform their duties as promptly and as well as any hands I have ever seen.

We have thirty-eight cards and about five thousand spindles; we are making yarns of all numbers, from five to twenty. We have also introduced colored work into the Mill; and although our arrangements for the colored work are not quite completed, causing a small loss in the amount of work, yet we are manufacturing over twelve thousand pounds of Cotton per week—the work for the last week being twelve thousand one hundred and forty pounds.

Whether it be the true policy of the South to employ blacks in that department of labor, or whether there is any real danger to be apprehended from the influence of sensible men from the North to learn them, or whether it is advisable for the South to enter into that department of labor at all or not, there can be *no doubt as to the capacity or availability of the blacks in becoming efficient operatives*, or of the ultimate success of the working class of the white population (if they persevere) in becoming successful manufacturers.

J. Graves

---

### The Future of the South

*The Boston Atlas says* :

We fully agree with the statement which is often made, that so far as natural advantages are concerned, the South altogether leads the North in facilities for manufacturing, particularly in the manufacture of cotton. She has water power in abundance. She has coal and iron in inexhaustible supplies, of which the New England States have none; and more than all, she possesses the soil and climate on which to grow the raw material, and which no law, no capital, no enterprise, can take from her. In this last particular, the cotton growing States need fear no competition. Not all the free trade laws in the world, or all the protective tariffs that ever filled the pages of a statute book, can transfer the immense business of cotton growing from the South to the North. It remains there, fixed by the immutable laws of Providence. Possessing all these advantages, what is to hinder the South from outstripping the North in the manufacture of cotton? Nothing but the very thing which our South Carolina friend is so anxious to preserve and perpetuate, slavery.

*To which the Augusta Ga. Sentinel responds, as follows:*

The holders of slaves owe it to themselves to demonstrate, in a large way, that cotton can be picked, carded, spun, and woven, as well as grown at the South. Nothing short of this will stop the ceaseless reproaches and unjust imputations cast upon the relation of master and servant, as it exists in this quarter of the Union. It is the duty of all cotton planters to take hold of this great question of manufacturing and mechanical industry in good earnest. Of all men, you are most deeply interested in creating a steady home market for your great staple. Of all men, you are most to be benefitted by proving that slave labor in Georgia is as profitable to you, and as useful to the world, as free labor is at the North, or can be at the

South. The whole matter will turn in the end on the one pivot of dollars and cents. Slavery was abolished in New York because experience proved that the relation of master and slave was not profitable to the master. The people of the non-slaveholding States firmly believe that institution is unprofitable at the South—that every planting State would be much better off if its citizens would emancipate their servants. This is also the deliberate opinion of ninety nine in every one hundred of the hundreds of thousands of emigrants from Europe, who annually flock to this country, remain permanently, and become a portion of its sovereign rulers.

We must show by visible results that slavery is not incompatible with improvement of the soil; is not inimical to common schools and a high standard of general intelligence; and is not hostile to the most successful manufacturing, mechanical and commercial industry. We can influence and control public opinion on all these points if we will only set ourselves properly and steadily at work to attain the objects indicated. Our sectional movements, our empty resolutions and "committees of safety," are taken by the civilized world as a confession of weakness; a consciousness of wrong which can not endure the searching light of truth and a free discussion.

So far is slavery from being naturally opposed to all progress and improvement in rural and mechanical arts, in internal trade and foreign commerce, in popular education and moral instruction, that it can easily be made auxilliary to all these important ends. It is the perfection of human wisdom to make the best possible use of all the means which a good Providence has placed at our disposal. To whom much is given, much will also be required. Because God has given us much, it will not do to say in practice that we need do nothing for ourselves. Our abundant means for labor, our great advantages of climate, soil and water power, demand the most skilful use, the most profitable employment.

---

***DeBow's Southern and Western Review***

## Slave Power in Cotton Factories

A correspondent of the New-York Herald having visited the Saluda Factory, near Columbia, S. C., thus comments upon the use of slaves for manufactories:

The factory in question ($100,000 capital) employs 98 operatives, or 128 including children. They are all slaves; and a large proportion of them are owned by the company. The mill runs 5,000 spindles, and 120 looms. The fabrics manufactured are heavy brown shirting and Southern stripe, a coarse kind of colored goods for house servants. The superintendent is decidedly of the opinion that slave labor is cheaper for cotton manufacture than free white labor. The average cost per annum of those employed in this mill, he says, does not exceed $75. Slaves not sufficiently strong to work in the cotton fields, can attend to the looms and spindles in the cotton mills; and most of the girls in this establishment would not be suited for plantation work. We dislike the idea of drawing a comparison between the labor of the fair and virtuous daughters of the North and that of the blacks of the South, in the cotton mills. It is unpleasant to put them on the same footing, even in the cotton mills, though one mill may be, in Massachusetts, exclusively occupied by the amiable, industrious, intelligent, and educated daughters of the old Bay State, and the other may be, in South Carolina, worked by negro slaves. We regret it; we have that sort of respect for the sex of our own race, which makes it painful to bring them to the same level with the colored races, though both may be employed in the same service. At the best, the work in a cotton mill is consumptive of lungs as well as cotton. We have been through the mills of Lowell, and other places in the North: the general appearance of the female operatives is neat and cleanly; but their prevailing complexion is an unhealthy pallor. Not many die at the mills, because they are young, and when they fall sick, they, if possible, return home. But the life of an operative in a cotton mill is a consumptive business at best.

Mr. Graves is of the opinion that the blacks can better endure the labor of the cotton mills than the whites. The slaves in this factory, male and female, appeared to be cheerful, well fed, and healthy. The mill has been worked by slave operatives (requiring only one white overseer,) for two years past, and the result, we are informed, is in favor of slave operatives:

| | |
|---|---|
| Average cost of a slave operative per annum | $75 00 |
| " " white operative, at least | 116 00 |
| Difference | $41 00 |

Or over thirty per cent. saved in the cost of labor alone....

**Cotton Factory in Mississippi**

Choctaw, Mi., June 4, 1850

Our mill is located ten miles south of Greensboro,' in a poor, healthy neighborhood; fine water, good society, churches, schools, &C. We have but one grog-shop within seven miles of us, and that will probably not last long.

Our building is made of wood, 108 feet long, 48 wide, and three stories high. We are now running about 800 spindles, 10 cards, 12 looms, and all the accompanying necessary machinery for spinning and weaving. Owing to the high price of cotton we have stopped our looms. We have 500 spindles and five cards more, not finished; we shall probably get them in operation for the next crop. We carry on a machine shop in which we make every variety of machinery for carding and spinning. Our looms are built by Messrs. Rogers, Kechum & Grovanon, of Patterson, N. J. They are heavy and substantial, and are built for making heavy Linsey and Osnaburgs, such as are most used in the South. I think that companies in this state intending to embark in the manufacturing business, would do well to call and see our machinery before buying elsewhere. We have just completed the finest flour mill in this state, or equal to any in the South. We will show flour with the St. Louis or any other mill North or South.

We use a large fine Semple Engine, made by Messrs. Thurston, Green & Co., Providence. It is admired by all visitors for its great capacity and simplicity. It is run by a negro engineer, who also serves as fireman, who had no acquaintance with engines until he took hold of this. We have a double cylinder wool card that cards the wool twice as well as most of the country cards that have only one, and will turn off two hundred pounds of rolls a day, for which we charge 8 c. a pound.

## *The History of the First Locomotives in America* (1874)

William H. Brown

No other icon of the industrial era can match the railroad for sheer power, drama, and economic importance. The South Carolina Railroad was not only the first in the South but the first to use an American-made locomotive, *The Best Friend of Charleston*. Before the pioneer generation of railroaders had completely died off, one chronicler captured the recollections of those early years, including profiles of the people who built and operated the young nation's first lines.

---

Next to Nicholas W. Darrell comes the veteran engineer, Mr. Henry G. Raworth, another employé of the South Carolina Railroad from its very beginning. When the Best Friend was blown up through the ignorance of the negro fireman, the wreck was taken by Mr. Julius D. Petsch to Mr. Dotterer's shop for repairs, as we mentioned in another chapter. Young Raworth, an apprentice of Mr. Dotterer, assisted Mr. Petsch in the work upon the engine, and when again ready for the road, and the name changed to the "Phœnix," Mr. Raworth ran it as engineer, and in that capacity has continued to serve the company up to the present time, and is now running an engine on the road, a period of consecutive service of over forty-two years. During all this time Mr. Raworth was never in the service of any other railroad, and never out of the service of the South Carolina Railroad, excepting only once during the Seminole War, when the Government applied to the South Carolina Railroad for an engineer to run the engine of a small steamboat engaged in transporting troops and supplies in the Everglades of Florida. On this duty Mr. Raworth was engaged ten months, then returned to his old position, resumed his engine, and is running now (August 1, 1873). In the fall of 1871, the author (when he visited South Carolina to examine the records of the old roads for statistics for his "History of the First Locomotives in America") had the pleasure of several interviews, and a ride upon the locomotive over his entire route, with Mr. Raworth, and received much valuable information as to the early history of the road. During the ride on the locomotive with Mr. Raworth, the author saw and conversed with another veteran in the service of the South Carolina Railroad, in the person of Mr. Raworth's old negro fireman. This faithful assistant has been Mr. Raworth's fireman on the locomotive successively from one to another, as occasion required a change, for a period of over nineteen years, and during all that time never quitted Mr. Raworth. Between these two, Mr. Raworth and his fireman, the most friendly understanding has existed.

---

Excerpted from: William H. Brown, *The History of the First Locomotives in America*, rev. ed. (New York: D. Appleton and Co., 1874), 159–162.

In one of Mr. Raworth's letters to the author, June 17, 1872, he states that his old negro fireman, "Adam Perry," was formerly the property of Major John Schmidt, of Barnwell District, South Carolina, who hired him to Mr. George B. Lythgoe, who was employed on the road as assistant civil engineer at that time; that Adam had been with him, in October coming, nineteen years; and in reference to his character would say that he was faithful, industrious, strictly temperate, and a most moral negro; always respectful to his superiors when a *slave*, and since a freeman, and has been working on the road in different positions thirty-seven years. Mr. Raworth also wrote that he had a white fireman for seventeen years, whose name was Thornton Randall, who died two years ago. When running on Aikin Hill, he had both firemen. The most perfect friendly relationship existed between these men. "You would" (he writes) "never hear an improper word from them; they were always kind to each other."

The president of the road informed the author, when in Charleston, that, during all the period of Mr. Raworth's service, as a locomotive-engineer, the engine was never in the shops an hour for repairs, excepting only when actually worn out from constant hard work, and some of its parts required renewing. Both Mr. Raworth and his firemen, white and black, have been total-abstinence men all their lives, and much, if not all, of this remarkable exemption from accident and disaster of all kinds incident to railroad running, may be attributed to that excellent trait in their characters.

During the author's ride on the locomotive last fall with Mr. Raworth, the veteran was well, hearty, and in the finest spirits, and in his own peculiar way he said that he had been running so long, a period now of nineteen consecutive years, over the same route and between the same two points, that he had become so familiar with every feature on it, that if the division boss removed one of the spikes, or put another in its place, he was sure to notice it. The same was the case with his old fireman. He thought he would be fit for service for ten years yet, and then the company would switch him off on some comfortable siding, put a shed over him, and take care of him the rest of his life, as they had done with old Darrell, another of their faithful servants, as the first and second locomotive-engineers in America.

Since the foregoing was prepared for our present volume, we received, on August 16th, from the railroad veteran, Mr. Raworth, the following letter, enclosing the photographs of himself and his old negro fireman:

Aiken, S. C., *August* 11, 1873.
Mr. Wm. H. Brown—
Dear Sir: Yours is received. I am very sorry I had to keep you so long for our photographs. My old fireman has been very sick, which is the reason why I could not send them earlier. Adam Perry has worked on the South Carolina Railroad thirty-seven years, and nineteen years with me as my fireman. Hoping the photographs will be in time for you,
I remain, very respectfully,
H. G. Raworth.

**Figure 4.2**
Portrait of Adam Perry.

## Advertisement in *The Liberator* Seeking Colored Inventors (1834)

A generation before the Civil War, the African-American newspaper *The Liberator* attempted to collect information about the work of black inventors. In a country increasingly basing its self-image and self-esteem on mechanical ingenuity, evidence of black inventors was an increasingly important mark of belonging and acceptance for African-Americans.

### Notice

Colored inventors of any art machine, manufacture or composition of matter, or any new or useful improvement of any art, machine, manufacture or composition of matter, not known or used before his application, are requested to make known their names and their respective inventions to the Editor of the Liberator, so far as they may deem it safe and proper to communicate the same. The objects of this notice are: —

1st. To collect proofs of colored talent and ingenuity in the United States.

2d. To aid colored inventors in obtaining their patents for valuable inventions.

Boston, Aug. 9th, 1834.

Reprinted from: "Notice," *The Liberator*, September 6, 1834.

## U.S. Patent to Norbert Rillieux for an "Improvement in Sugar-Works" (1843)

Although a resident of New Orleans during the era of slavery, Norbert Rillieux was a free person, having a quadroon (one-fourth "negro" blood) mother and a white engineer for a father. At the time sugar was a major, and probably the most industrialized, sector of Louisiana agriculture. Rillieux (1806–1894) used his status to make significant improvements in the way sugar was manufactured.

**United States Patent Office**

**Norbert Rillieux, of New Orleans, Louisiana, Assignor to Saml. V. Merrick and John H. Towne**
Improvement in Sugar-Works

Specification Forming Part of Letters Patent No. **3,237**, Dated August 26, 1843.

To all whom it may concern:

Be it known that I, Norbert Rillieux, of New Orleans, in the parish of Orleans and State of Louisiana, have invented certain improvements in the method of evaporating and concentrating saccharine juices and sirups in the manufacture of sugar, and which is applicable to the evaporation of other fluids; and I do hereby declare that the following is a full, clear, and exact description of said improvements.

My invention consists of four leading improvements on the methods known, viz:

The first improvement is in the manner of connecting a steam-engine with the evaporating pan or pans in such manner that the engine shall be operated by the steam in its passage to the evaporating pan or pans, and the flow of steam be so regulated by a weighted or other valve as to reach the said pan or pans at the temperature required for the process—that is to say, where the saccharine juice boiled—the steam at the same time having access to the pan or evaporator without passing through the engine by the said valve, which is weighted or otherwise regulated to insure the supply of steam to the said pan or evaporator at the required pressure.

The second improvement is for the combination of the vacuum-pan or evaporator, (known as the "Howard Saccharine Evaporator,") in which the sirups are evaporated *in vacuo*, with a pan or evaporator or boiler in which the saccharine juices are

Reprinted from: U.S. Patent No. 3,237, dated August 26, 1843, issued to Norbert Rillieux, of New Orleans, for an "Improvement in Sugar-Works."

prepared to be transferred to the vacuum-pan or evaporator, and which at the same time generates the vapor from the saccharine juices to supply the vacuum pan or pans with the required quantity of steam or vapor under sufficient pressure and of a temperature sufficiently high to produce ebullition in the vacuum-pan or evaporator.

The third improvement relates to an amelioration of that kind of evaporator known as the "Champenoise column," and consists of an outer envelope to this column, by which I am enabled to adapt it to the condition of my second improvement.

The fourth improvement is a method of regulating the concentration of the sirup by means of a differential thermometer, which indicates the degree of concentration of the sirup without being affected by any change in the pressure under which the sirup is evaporated or concentrated, and also for an arrangement of such thermometer by which its range of action up to a given point does not act on the regulator which governs the supply of the concentrated or non-concentrated saccharine juice, either of which will regulate the degree of concentration of the sirup.

Having described the nature of my improvements, I shall proceed to describe the construction of the apparatus and the operation thereof by reference to the accompanying drawings, which make a part of this specification, and in which the letters of reference indicate the same parts in all the sections of the first improvement.

The mechanism and apparatus under this head are represented in the accompanying drawings, in which Figure No. 4, Plate 1, is an elevation of the steam boiler or boilers, steam-engine, the evaporating-pans, and the connection of the three; and Fig. 1, Plate 4, is a section of the engine with the steam-pipes which form the connection between the boiler, the engine, and the evaporating-pans. The main steam-pipe B′ from the boilers A′ extends to the pan or evaporator A, and introduces the steam in the false bottom *d*, and at *a* there is a branch pipe conducting to the cyllinder C of the steam-engine, and another at *c*, which connects with the eduction-valve of the engine, and between these two branch pipes, which are the induction and eduction pipes, the main steam-pipe B′ is provided with a loaded throttle or other valve, as at *b*, Fig. 1, Plate 4, and *v*, Fig. 4, Plate 1. By this arrangement it will be evident that the engine can only be worked by the pressure of the steam over and above that required for the pan or evaporator A, and as the steam cannot pass the weighted or throttle or other regulating valve, except it be of a higher pressure than is required to move the steam-engine, it passes through the pipe *a* to the engine, and from thence through the eduction-pipe *c* to the false bottom *d* of the pan, and all the additional steam that is required beyond that which passes through the engine goes directly through the eduction-pipe *c* and regulating-valves to the evaporating-pan. The difference between the pressure in the boilers A′ and the false bottom *d* will be the effective pressure exerted on the piston of the engine. Thus all the steam generated in the boilers A′, except the small quantity condensed, will be conducted and give out its heat to the pan or evaporator A. It is of course to be remembered that the pressure of the steam in the false bottom *d*, or rather its temperature, is to be regulated in the usual way. Should it be desired at any time to conduct the steam directly to the pan or evaporator A without working the engine, the valve *b*′ is closed, and the steam passes along the main pipe B′,

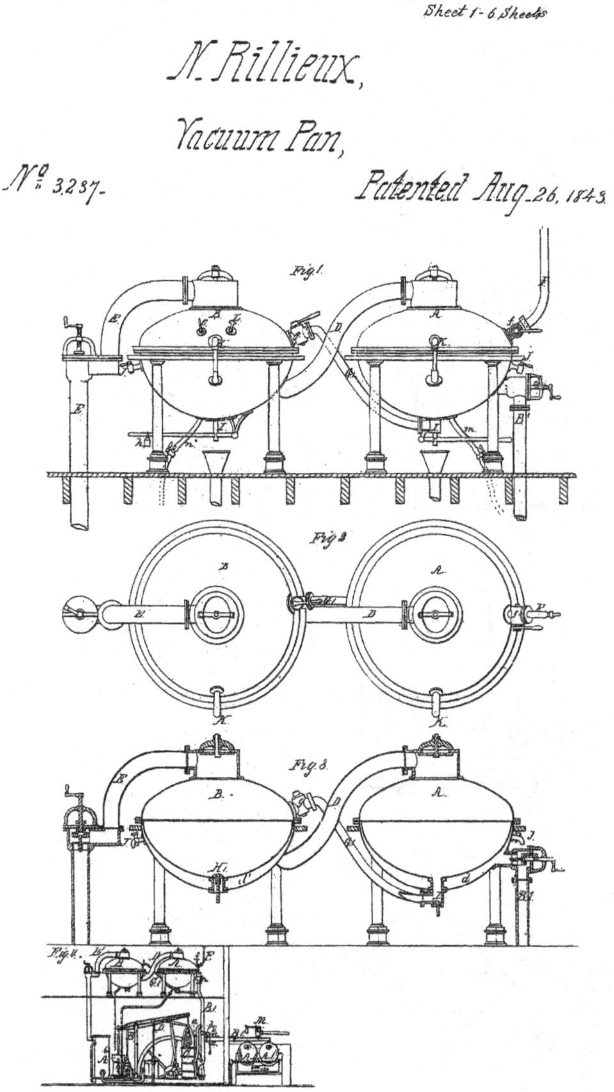

(a)

**Figures 4.3a–c**
Patent drawings for Norbert Rillieux's Improvement in Sugar-Works.

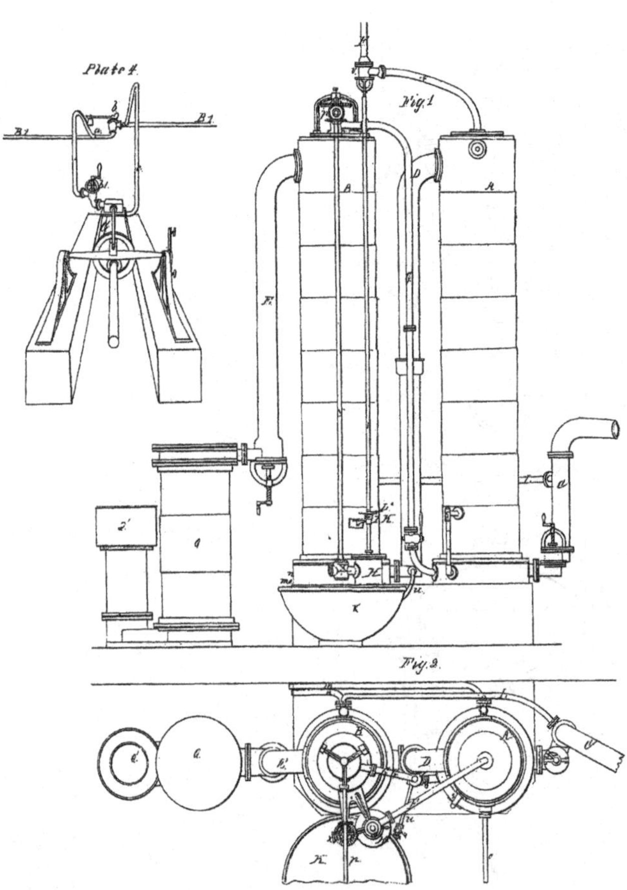


(b)

**Figures 4.3a–c**
(continued)

Sheet 6–6 Sheets.

N. Rillieux,

Vacuum Pan,

No. 3,237. Patented Aug. 26, 1843.

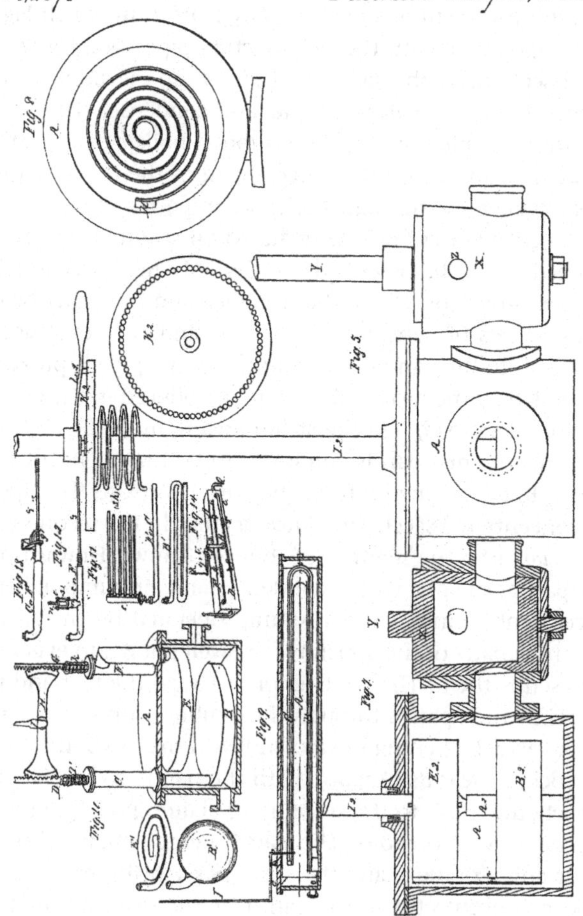

(c)

**Figures 4.3a–c**
(continued)

its pressure or temperature being regulated in any of the well-known methods. This arrangement is applicable to all kinds of boilers and evaporators. The throttle-valve in this arrangement can be connected with the governor of the steam-engine in manner well-known to every engineer, and need not therefore to be described; but for this arrangement, or something substantially the same, the pressure of steam would be the same on both sides of the piston, which would prevent the operation of the engine of the second improvement. This improvement is represented by Plate 1, in which Fig. 1 is an elevation of the pans or evaporators, with their connecting-pipes, &c. Fig. 2 is a plan, and Fig. 3 a longitudinal section, of the same; and Fig. 4, an elevation of the whole apparatus connected with the steam-boilers, engine, and condenser. The pan A or evaporator is constructed with a double bottom, or by tubular boilers or any other form of vacuum sugar-pan, leaving a space, *d*, into which steam of the required pressure is introduced from a steam-generator through the steam-pipe B′, there being a stop-cock or other valve, *v*′, to regulate or cut off the steam when required. The peculiar construction of this pan need not to be described, as it is only necessary to state that it should have two chambers—one for the saccharine juice and the other below it for the steam—the two pans or sets of pans, which may be similarly constructed. The saccharine juice is introduced into the pan or evaporator A by the pipe F, provided with a stop-cock, *f*, and there boiled by the heat of the steam introduced in the double bottom *d* or other form, as before described, and the steam arising from the evaporation of the saccharine juice in this the first pan is employed to boil the concentrated juice or sirup in the second pan, B, by passing off from the pan A through the pipe D to the double bottom *d*′. The concentrated saccharine juice or sirup in the second pan, B, is boiled or evaporated *in vacuo*, after the manner of the well-known Howard process, by connecting the upper part of the pan with a condenser and air-pump, E being the connection-pipe, G the condenser, and H the air-pump operated by the steam-engine or other power, and as these parts of the apparatus are well known to every engineer, it is not necessary to describe them. The connection between the first and the second pans, A B, is by a pipe, G′, which passes through the double bottom of the former, where it is provided with a valve, I, opening inward and with the top of the latter, it being provided with a stop-cock, *g*, near its junction with the pan B, to regulate the supply of the concentrated saccharine juice to the second or vacuum pan. The operation of this part of the operation is very obvious. The pan B being exhausted by its connection with the condenser and air-pump, and the stop-cock *g* being opened, the saccharine juice, which has been concentrated in the pan A by the steam in the false bottom *d*, is forced up the pipe G′ into the pan B by the elastic force of the steam generated from the saccharine juice, and after the sirup has been sufficiently reduced in the second or vacuum pan, B, it is drawn off through a pipe, H′, which is provided with a weighted valve, I, opening inward, and having a lever and weight, *n*, the stem of the valve passing down through a pipe, which connects the double bottom. These pans are provided with man-holes, a glass gage or indicating-tube, K, to indicate the level of the juice, blow-cocks J J, to expel the air from the double bottom, thermometer and ebullition-gage O L, and pipes *m n*, to discharge the water produced by the condensation of the steam in double bottoms or steam-chambers *d d'*.

Another method of applying this my second improvement is represented in the same plate, in which Fig. 5 is an elevation, and Fig. 6 a plan. In this modification the saccharine juice, instead of being concentrated by steam of a pressure equal to or greater than the atmosphere, which at the same time generates the steam from the saccharine juice to boil the sirup in the vacuum-pan, is concentrated in a boiler heated by fuel in the usual manner of a steam-boiler. By reference to the drawings it will be seen that all the connections in this modification are substantially similar to the first, except in form and the application or substitution of a common steam-boiler for the first pan, A, in the first-dedescribed modification. The boiler A and vacuum-pan B, under this modification, are represented as cylindrical; but under either of the modifications any form can be adopted in the judgment or fancy of the constructer. The supply pipe F is provided with a float and valve, *i i*, to regulate the supply of saccharine juice. The float *i* rests on the surface of the saccharine juice, and is attached to the lever of the valve *i*, which governs the aperture of the feed-pipe, so that when the saccharine juice rises too high the float is elevated and closes the aperture, which stops the supply, and when it sinks too low the float, following it, opens the valve, and thus regulates the supply.

The degree of concentration of the sirup in the pan B may be regulated by an apparatus attached to the pipe G, that conveys the concentrated saccharine juice from the boiler or pan A to the pan B, which apparatus acts by the specific gravity of the sirup in the following manner, viz: The reduced sirup is regularly discharged from the pan B by a pipe, H, which leads and discharges into a vessel, I, provided with what is well known as a "level" tube, through which the sirup is discharged into a double-bottomed pan, K, after it reaches a certain height in the vessel. This double-bottomed pan K is heated by vapor from the boiler A, and in it the sirup is further reduced and prepared for casting into form. The vessel I is hung by journals to a bent lever, *k*, and by rods or chains to the lever of a valve, *l*, in the conducting-pipe G', that supplies the pan B with the concentrated saccharine juice. The weight on the bent lever is so regulated with reference to the specific gravity of the quantity of sirup which the vessel I can contain and to its connection with the feed-valve in the pipe G' as to open the valve *l* when the specific gravity of the sirup becomes too great and to close it when its specific gravity is not sufficient. In this way it is evident that the degree of concentration of the sirup can be regulated with the most perfect accuracy. This method of regulating the supply of concentrated saccharine juice and of insuring the discharge of the sirup at the point of concentration required I intend to secure by a separate patent, as I shall claim under this patent another method of effecting this object.

Of the third improvement: This improvement is represented in Plate 2 of the accompanying drawings, in which Fig. 1 is a longitudinal elevation; Fig. 2, a plan; Fig. 3, a transverse elevation, representing the condenser in section; Fig. 4, a longitudinal section of the vacuum-evaporator; Fig. 5, the same of the pressure-evaporator, and Figs. 6, 7, 8, 9, and 10 separate parts in section. As the two evaporators are similarly constructed, it will only be necessary to describe the construction of one of them and the manner of connecting the two.

The apparatus known as the "Champenoise column," on which my said third improvement is based, consists of a vertical metallic column, A′, the upper end of which is domeshaped or semi-spherical. Within this column there is a steam-pipe, D′, extending from the base of the column to within a short distance of the top, through which steam is admitted to the inside of the column to heat it to the required temperature. The saccharine juice or sirup is discharged on the semi-spherical top or dome of this column, and in passing down is evaporated; but by this arrangement it is evident that the evaporation must take place under the pressure of the atmosphere, and that thus constructed it is not applicable to the Howard process of evaporation *in vacuo*, which is highly important in the concentration of sirups. I have therefore so improved this apparatus as to bring it within the conditions of the Howard process; and my second recited improvement—viz., the employment of an evaporator which works at or above the pressure of the atmosphere, in combination with a second, which evaporates *in vacuo*, the steam for heating the second being supplied from the first by the evaporation of the saccharine juice, and hence I have to employ two such columns. The column is represented at A′, Fig. 5, and the inner steam-pipe which it envelops is represented at D′, Fig. 4.

My improvement consists simply in enveloping this column with an outer casing, as at A B. The two evaporators being thus constructed and enveloped by an outer casing, it remains simply to describe the manner of connecting them with each other and with the boiler and condenser, and also the manner of supplying the saccharine juice to the first and the concentrated juice to the second. The main steam-pipe C communicates with the vertical steam-pipe D′ within the first column, A, through a passage, R, in the base; and the saccharine juice is introduced and discharged onto the top or dome of this column A′ by the pipe F, which is connected with a bifurcated pipe, *u*, that turns in a collar or stuffing-box in the upper end of the outer casing, and is made to rotate, to insure the regular discharge of the juice, by miter-wheels on the shaft of the bifurcated pipe, and a horizontal shaft, *o*, that receives motion from the steam-engine or other first mover, the shaft and pipe being provided with stuffing-boxes to prevent the escape of steam. The saccharine juice, being thus equally distributed on the dome of the column, runs down to the bottom, and, in passing over the surface heated by steam, as before described, is sufficiently evaporated or concentrated to be transferred to the other column, where it is further evaporated *in vacuo*. It is then forced out by the pressure of the steam generated from the juice in concentrating it through the passage *w* in the base of the column, (see Fig. 8,) which communicates with the top of the vacuum-evaporator B by means of the connecting-pipe G; and it is there discharged on the top or dome of the column B′ in the same manner and by the same means as the saccharine juice is delivered on the first column, except that in the latter the miter-wheels, &c., to rotate the discharge-pipe are outside of the surrounding case at *p*. The steam or vapor from the saccharine juice in the evaporator A passes off through the pipe D, enters the base of column B′ at R′, up the pipe D′, and is discharged at the required temperature near the upper part of the column. The upper part of the evaporator B communicates by the pipe E with the condenser Q, of any construction; but the one represented in the drawings is made after the manner of the well-known Hale condenser, and therefore needs not to be described. It is of course understood that the condenser is provided with an exhausting or air pump. The evaporators A and B may be provided with a glass gage, S, to indicate the level of the saccharine juice and sirup, and with a thermometer and steam-gage. The water or condensed vapor is discharged from the base of column A′ through an aperture, *r*, (represented by dotted lines in Fig. 8,) and from the column B′ through the aperture *v*, Figs. 6, 7, and 9, each of which communicates with its appropriate pipe, *m* and *n*. The sirup passes out from the evaporator B, through the aperture *w*, into the vessel H, that contains a differential thermometer to regulate the bake or concentration of the sirup, and from the vessel it is discharged into the double-bottom pan K, already described, by a rotating cock, X, known as the "rotating feeding-cock." This cock has a chamber or chambers in it, which, as it rotates, is alternately presented to the pipe leading from the vessel H and the pipe discharging into the pan K; and in this manner, for each chamber in the cock, at each rotation, a measure of sirup is discharged. Motion is given to this cock by the vertical arbor Y, which is geared with the horizontal shaft *p*, that communicates motion to the discharge-pipe *u* in the evaporator B. A pipe, L, extends from the main steam-pipe C to each of the evaporators A and B, for the purpose of admitting steam to cleanse the surface of the columns A′ and B′.

Of the fourth improvement: The manner of constructing the differential thermometer to govern the bake or concentration of the sirup under the head of my fourth improvement is represented under its various modifications in Plate 3, and the manner of applying it to the apparatus is represented in Figs. 1, 2, 3, and 6 of Plate 2, referred to in the description of my third improvement; but it is to be understood that this manner of applying it is only given as an exemplification, for it can be variously applied without changing the character of this invention—as, for instance, it is immaterial whether the application of it be made to govern the supply of the saccharine juice to the first evaporator, A, or to the second, B, from the first, as the object is to regulate the concentration of the sirup in the last evaporator B, and it is immaterial whether this is effected by one or the other. When applied to govern the supply of juice to the first evaporator, then the sirup should be regularly discharged from the last, for when the regulator indicates the concentration to be too great, the valve is opened, which admits a greater supply of saccharine juice to the first, which is forced by the pressure of the steam into the second the moment it has reached the bottom of the column, and when applied to regulate the discharge from the first to the last the discharge from the last should be regular, and the supply of the saccharine juice to the first should be regulated by a self-feeding apparatus or by hand, so as to make the supply correspond with the transfer from the first to the second. The object to be obtained is to increase the supply from the first to the last evaporator when the thermometer indicates the concentration to be too great, and to reduce it when it indicates insufficient concentration. In the vessel H, Plate 2, through which the sirup passes after it has been discharged from the second evaporator, B, is placed a differential thermometer, constructed in the manner represented by Figs. 5, 6, 7, and 8, Plate 3. A stem, $A^2$, projects from the bottom of this vessel H, to which is attached a metallic volute thermometer, $B^2$, composed of two plates of metal of different dilatation, soldered or otherwise firmly fastened together along their whole length. To the outer end of this volute is attached a bar of metal, $H^2$, and to this bar is also attached the outer end of another and similar metallic thermometer, $E^2$, the inner end of which is attached to a vertical spindle, $I^2$. The lower end of this spindle rests on the standard $A^2$, and its upper end is provided with an index-plate, $K^2$, with holes in its upper surface, and it is there connected with the spindle of the regulating or throttle valve V in the supply-pipe F, the lower end of the valve-spindle having a lever, and pin $L^2$ to take into the holes or the plate $K^2$, and thus establish the connection between the thermometer and regulating-valve, which can, by the index-holes and pin, be adjusted to any range of temperature required for the concentration of the sirup. The lower half, $B^2$, of this thermometer is always immersed in the sirup as it is discharged from the evaporator B, and the upper half, $E^2$, is acted upon only by the vapor arising from the said sirup.

It is a well known law that as the density of a fluid increases so does the difference between its temperature and that of the vapor arising therefrom also increases, and this difference at any given temperature and density, under all pressures, is a constant number; but the temperature of the two varies with the pressure, and as the sirup is concentrated in a vessel connected with a condenser exhausted by an air-pump, the pressure will necessarily change with the change of pressures, and this change takes place without affecting the degree of concentration of the sirup, but does change its temperature, so that simple thermometers could not effect the object in view, which is to regulate the degree of concentration; but as the difference between the temperature of the sirup and its vapor is a constant number at all degrees of concentration, running through every degree of pressure, it is evident that the range of pressure in the condenser or any vessel in connection therewith will not affect the differential thermometer, which indicates only a change in the difference of the temperature between the two, which change in the said difference also indicates a change in the concentration of the sirup. It is therefore immaterial in what manner these differential thermometers are constructed or in what manner they are applied to regulate the supply of saccharine juice or the discharge of the concentrated juice or sirup. Figs. 9, 10, 11, 12, and 13 represent different methods of constructing differential thermometers and different methods of applying them to this purpose. They may operate either by the expansion of metals or fluids. If fluids are employed, a greater range may be given to the follower or piston by constructing the pipes in any of the forms represented in Fig. 11.

As the thermometer is attached to a regulating-valve, it is important so to construct it that all its range up to a certain point should not act on the valve. For this purpose I construct them in the manner represented in Figs. 12, 13, and 14. The fluid or semi-fluid of which the thermometers are composed is put in the cylinders C F C F, and to each of these is fitted a piston on which the expanding fluid is to act. In the modification, Fig. 12, the cylinder C F is provided with a vessel, *m*, into which is fitted a follower, *n*, pressed down by a spiral or other spring, S′, on its stem, and as it requires less force to contract this spring than it does to operate the regulating-valve, all that part of the dilatation which takes place before the valve should be acted upon acts on the follower *m*, and when that has reached the limit of its range, then of course the valve is acted upon. In the modification, Fig. 13, instead of the additional vessel *m*, the piston *m′* is attached to the connecting-rod *g* by a spring, which yields to all the dilatation that takes place before the instrument should act on the valve.

Fig. 14 represents the manner in which the piston-rods *g* act upon the cog-wheel *r*, attached to the differential lever *s* on the spindle *u* of the arm I, that communicates by the rod X with the regulating-valve.

It is to be remarked that the rods *g g* are to be provided with springs to force back the pistons when the fluid or semi-fluid contracts, and that these springs are to be of such force as not to be acted upon until after the small springs *s′* are entirely contracted.

Having fully described the principle or character of my improvements and the manner of constructing the apparatus and of operating with the same, I shall proceed to point out what I claim as my invention and desire to secure by Letters Patent.

I am aware that the escape steam from a steam-engine has been variously applied to economize fuel, and therefore it is to be understood that I do not claim this as of my invention under the head of my first improvement.

I am also aware that the vapor arising from a fluid submitted to the action of heat for evaporation or concentration has been conducted to other vessels to assist in heating them, and therefore it is to be understood that I do not, under the head of my second improvement, claim this as of my invention; and I am also aware that a single thermometer has been applied to regulate the supply of water to saline baths whenever, by concentration, it exceeds in its boiling temperature the desired degree of temperature, as this has been long since proposed by Dr. Ure; and I do not, therefore, under the head of my fourth improvement, claim as my invention the application of a thermometer to regulate the supply of liquids to evaporating-vessels; but

What I do claim as my invention, and desire to secure by Letters Patent, is as follows, viz:

1. Under the head of my first improvement, I claim the employment of a weighted throttle or other regulating valve in the main steam-pipe leading from the boiler to the evaporating pan or pans and the steam-engine, which valve shall be situated between the induction-valve of the engine and the evaporating pan or pans, for the purpose and in the manner described.
2. Under the head of my second improvement, I claim a vacuum pan or pans—that is to say, an evaporating pan or pans connected with a condenser—in combination with an evaporating pan or pans, or boiler, in which the saccharine juice or other fluid is

evaporated under a pressure lower, equal to, or greater than the atmosphere, which last-mentioned pan or pans, or boiler, prepares the saccharine juice, &c., from the vacuum pan or pans, and at the same time supplies the necessary vapor from the saccharine juice, &c., to complete the evaporation or concentration of the sirup, &c., in the vacuum pan or pans, as fully described above.

3. Under the head of my third improvement, I claim surrounding the evaporating-column, known as the "Champenoise" column, with an outer column or jacket, by which I am enabled to adapt it to the condition of my secondrecited improvement, as described.

4. Under the head of my fourth improvement, I claim the employment of a differential thermometer to regulate the concentration of the sirup, in the manner substantially as herein described.

5. The so constructing the differential thermometer that all the range of its action up to the point desired shall not act on the regulating-valve, as described.

N. Rillieux [L. S.]

Witnesses:

Geo. Griscom

Joseph Greer

## U.S. Patent to Norbert Rillieux for an "Improvement in Sugar-Making" (1846)

Rillieux's continued work resulted in another patent, for a complementary device, three years later. It is significant that that study and improvement of the important sugar industry was a continuing object of his attention.

**United States Patent Office**

**Norbert Rillieux, of New Orleans, Louisiana**
Improvement in Sugar-Making

Specification Forming Part of Letters Patent No. **4,879,** Dated December 10, 1846.

To all whom it may concern:

Be it known that I, Norbert Rillieux, of New Orleans, in the parish of Orleans and State of Louisiana, have invented new and useful Improvements in the Method of Heating, Evaporating, and Cooling Liquids, especially intended for the manufacture of sugar; and I do hereby declare that the following is a full, clear, and exact description of the principle or character which distinguishes them from all other things before known, and of the manner of making, constructing, and using the same, reference being had to the accompanying drawings, making part of this specification. . . .

My invention consists, first, of a heater for clarifying saccharine juices preparatory to the evaporating process, but which may be employed simply for heating the juice preparatory to clarifying; second, of a cooler employed in connection with the vacuum-pans or evaporators or boiling apparatus, by means of which the saccharine juices are cooled by a current of air that they may be employed as a means of condensation for the vacuum-pans, at the same time preparing them by partial evaporation for the evaporating-pans; and, third, of an arrangement of vacuum-pans or evaporators. . . .

It will be obvious that this boiling or evaporating apparatus can be employed in connection with my improved heater and cooler by adopting the connections pointed out in the description of the entire apparatus.

Having thus pointed out the principle or character of my improvements and the manner of constructing and applying the same, what I claim as my invention, and desire to secure by Letters Patent, is—

Reprinted from: U.S. Patent No. 4,879, dated December 10, 1846, issued to Norbert Rillieux, of New Orleans, for an "Improvement in Sugar-Making."

1. The method of heating the saccharine juice in a heater preparatory to its introduction in the evaporating-pans, by means of the waste hot water or escape steam from the evaporating-pans, substantially as described.
2. The method of clarifying saccharine juice by heating it in a heater provided with a spout for the discharge of the impurities in the form of scum, and a pipe for drawing off the clear liquid, the said pipe being so arranged as to receive the liquid from the heater below the level of the spout which discharges the scum, and then bending up above the said spout to cause the liquid in the heater to rise sufficiently high to discharge the scum, substantially as described.
3. The method of cooling and partially evaporating saccharine juice or other liquids by discharging the same in the form of spray or drops in a chamber, where it meets with a current of air, substantially as described; and this I also claim in combination with a condenser, substantially as herein described, whereby the liquid intended to be concentrated is prepared for the evaporating-pans and used as a means of condensing the vapor from the pans in which it is to be concentrated, or by means of which the water used for the condensing jet is recooled, substantially as described.
4. The method, substantially as described, of combining a vacuum striking-pan with a series of evaporating-pans, the last of which is independent of the striking-pan, and the last of the series of evaporating-pans can be in connection with the condenser and work independently of each other, that either the striking-pan or the series of evaporating-pans can be worked without the other, as described.

N. Rillieux

Witnesses:

Chs. M. Keller

Ch. L. Fleischmann, Jr.

## The Confederate Patent Act (1861)

During the antebellum period, the U.S. Patent Office had refused to recognize inventions made by enslaved Africans, even when their owners wished to take out patents in their own names. After the outbreak of the Civil War, however, the new Confederate government allowed owners to claim patent rights for the discoveries of their enslaved laborers.

### Sec. 50

*And be it further enacted,* That in case the original inventor or discoverer of the art, machine or improvement for which a patent is solicited is a slave, the master of such slave may take an oath that the said slave was the original inventor; and on complying with the requisites of the law, shall receive a patent for said discovery or invention, and have all the rights to which a patentee is entitled by law.

### Sec. 51

*And be it further enacted,* That all patents issued by the government of the United States, in favor of citizens or subjects of foreign countries, prior to the eighth day of February last, shall have the same force and effect in these Confederate States as if issued under the authority of these States: *Provided,* That this section shall not take effect in favor of any alien enemy, holder or assignee of any such patent as aforesaid.

### Sec. 52

*And be it further enacted,* That this act shall take effect and be in force from and after its passage.
Approved May 21, 1861.

The Confederate Patent Act, Section 50, May 21, 1861.

# III WAR, RECONSTRUCTION, AND SEGREGATION

The stew of factors that brought about the American Civil War included, preeminently, the growing success of the northern abolition movement, led by such figures as Frederick Douglass, and the widening disparity between the economic systems of the North and South. As the rest of the nation embraced and contributed to the Industrial Revolution that had been imported from Great Britain, the South seemed more and more committed to a system of large-scale commercial agriculture resting upon the festering base of enslaved labor.

When the war finally came, after years of threats and near misses, the lack of industrialization in the South turned out to be the Achilles heel of its economy. The North marshaled its vast productive plants to produce guns, uniforms, iron, wagons, foodstuffs, and all the other materials of war and excelled in the adaptation of other new technologies to the needs of combat. The railroads, observation balloons, Gatling guns, the rifled cannon, and the ironclad ships made this, in some ways, the first modern war, and it attracted members of the general staffs of all the major European powers. In the end, as in the world wars of the twentieth century, the ability to mobilize technology and increase production proved decisive. The Confederacy took the sensible step of allowing, for the first time, the recognition of any inventions made by enslaved laborers, but the resulting patents were to the property of the slaves' masters. It was clearly much too little and far too late.

After the war, the entire physical infrastructure of the South required rebuilding, and a host of neglected social programs—such as compulsory free schools—needed to be addressed. Black legislators and their northern allies made a good start on these projects, but the failure to provide economic security for the newly emancipated masses of black Americans, and the North's cynical abandonment of Reconstruction, allowed the reinstitution of racist laws and customs designed to, among other things, force former enslaved laborers to remain an exploited economic and social class. Efforts by some southern advocates and northern investors to create a "New South," at last committed to industrial development and the elaboration of an engineering infrastructure to support it, were only partially successful. In the South as elsewhere, however, African-Americans continued to participate in the technological life of the nation even in the face of concerted efforts to keep them from doing so on anything approaching equal terms.

# 5 Finding a Place in the Industrial Age

**Mechanism and Art (1873)**

*Using Jokes to Invent the Expert*

**John Henry, the Steel Driving Man (1929)**

**The Eclipse Clothes Wringer (c. 1880s)**

**Woman with Bicycle (c. 1890s)**

***The Taint of the Bicycle* (1902)**
W. F. Fonvielle

## Mechanism and Art (1873)

In the wake of the Gold Rush, San Francisco emerged as a city where white labor was relatively well organized and where they viewed "people of color," particularly Chinese immigrants, as dangerous competition for their jobs. The city had an old and well-established African-American community, however, and one of their newspapers urged black youth to seek industrial and artistic employment despite white opposition.

We have repeatedly urged upon our our readers the importance of learning the mechanical arts and have endeavored to impress upon parents the necessity of bringing their children up to learn a trade or profession. We witness with sorrow our young man growing up to manhood who have no higher aspiration of business than becoming waiters or laborers in stores. Not that we condemn any honest occupation; labor of all kinds is honorable; but a mechanic or an artist can always command more respect, and receive higher remuneration than an ordinary laborer.

We have been led to these reflections, and to allude again to this subject, because we have recently had an application from a large mechanical establishment for a colored boy as an apprentice. Here is a chance but seldom offered, but after some weeks search and inquiries we have been unable to procure a boy. Some parents object because their sons can earn more at some servile occupation, such as waiter or jobbing than he can while learning his trade. Others have not sufficient control over their children, but give them too much their own way; and we suppose the excitement and variety of jobbing and the change of scene which a flunkey on a steamship experiences have their attractions over the monotony of a mechanic's workshop or artist's studio.

There are often opportunities like above. The prejudice against colored mechanics and artizans is wearing away; journeymen are now more willing to work in the shop with colored men than formerly, and the time will shortly arrive when a colored mechanic if proficient, can obtain employment as readily as a white man. It is necessary, however, that a colored workman should be equal in ability to any others, and if possible, superior, for it is only by superiority that we can conquer prejudices.

There are many other opportunities now offering for colored boys to learn trades. In the works of the Central Pacific Railroad at Sacramento almost every branch of mechanical science is carried on. There are carpenters, joiners, painters, upholsterers, iron workers, brass founders and artificers in wood and metal of every description—there colored apprentices of ordinary capacity will at any time be taken, and not only learned any trade they may desire, but receive board wages while learning.

Reprinted from: "Mechanism and Art," *Elevator* (San Francisco), September 6, 1873.

There are now several colored journeymen employed in the machine shop and other works of that company. How many colored boys have availed themselves of the opportunities offered we have not learned, but we have good authority for saying that the company will take colored apprentices whenever application is made.

It is not only the mechanical branch to which we would call the attention of the rising generation but we would have them excel also in the fine arts. The world of arts is cosmopolite—it is universal. It is confined to no race, nation or color. The schools of Music, Painting, Sculpture are open to all who possess inclination for the study and practice of those arts. Italy, Germany, France no longer claim a monopoly in the fine arts. While the facilities for study and practice are greater in those countries than elsewhere, and students can find associations more congenial to their tastes and habits, it is not to the natives of those countries alone, to whom the development of those arts is now particularly confined, as in former days. England and America are filling the Academies, the Conservatories and the Studios with pupils and professors. The Saxon, the Celt, and even the descendant of the Indian and the African are crowding out the Latin and the Teuton in Rome, Naples, Düsseldorf, Dresden, Vienna and Paris.

We have now in this city an evidence of the progress which can be made in the highest and most difficult of the fine arts in the person of Miss EDMONIA LEWIS; the celebrated Sculptress. This young lady having an intuitive feeling that she could accomplish certain things, applied herself to the task, and finally by perseverance and study has become equal to the great artists of ancient or modern times, and the emanations of her mind and productions of her chisel will compare favorably with the works of Praxilities, Canova, Powers or Miss Harriet Hosmer.

What Edmonia Lewis has done, others may likewise do. If all cannot arise to the same eminence, and occupy as high a niche in the temple of fame, every one can learn some useful trade or profession.

## *Using Jokes to Invent the Expert*

With the new electrical industry after the war, those who worked with this new technology were anxious to establish themselves as acknowledged experts on the subject. One common way of differentiating themselves was to make jokes about the Other—farmers, women, immigrants, African-Americans—who could be marginalized as technologically ignorant and even, in an important sense, not yet modern. One such racist joke was quoted by the communications scholar Carolyn Marvin in her effort to trace the activity she called "inventing the expert."

The *Electrical Review* reported that Harry B. Cox of Cincinnati and his brother, an Episcopal clergyman "having charge of the city missions," had made "amusing" tests with an electric speaking trumpet devised by Harry Cox.

---

They experimented upon an old darky, and completely frustrated the old fellow, who was walking up the road. Using the bell end of the horn, they began talking to the colored man as he walked along.

The peculiarity of sound transmitted by the trumpet is that, to the person hearing it, it appears to come from some one near him, and not from a distance. The old darky hearing the voice was at first annoyed, then puzzled, and finally so badly frightened that he started up the road on the dead run, no doubt attributing his adventure to some supernatural agency.

---

Excerpted from: *Electrical Review*, quoted in Carolyn Marvin, *When Old Technologies Were New* (New York: Oxford University Press, 1988), 33. Reprinted by courtesy of *Electrical Review* and Highbury House.

## John Henry, the Steel Driving Man

Of the American folk heroes, all but one are white males. The exception is John Henry, the "steel-driving man." Like many folk stories, this one tells of a skilled worker who pits himself against a new industrial technology and wins, but in that victory loses his life. Many different versions of the tale exist, and some scholars claim that it may have some basis in actual events: the driving of a tunnel through the mountains of West Virginia for the Baltimore & Ohio Railroad in the 1880s. This version is the first known published text.

John Henry was a railroad man,
He worked from six 'till five,
"Raise 'em up bullies and let 'em drop down,
I'll beat you to the bottom or die."

John Henry said to his captain:
"You are nothing but a common man,
Before that steam drill shall beat me down,
I'll die with my hammer in my hand."

John Henry said to the Shakers:
"You must listen to my call,
Before that steam drill shall beat me down,
I'll jar these mountains till they fall."

John Henry's captain said to him:
"I believe these mountains are caving in."
John Henry said to his captain: "Oh Lord!"
"That's my hammer you hear in the wind."

John Henry he said to his captain:
"Your money is getting mighty slim,
When I hammer through this old mountain,
Oh Captain will you walk in?"

John Henry's captain came to him
With fifty dollars in his hand,
He laid his hand on his shoulder and said:
"This belongs to a steel driving man."

John Henry was hammering on the right side,
The big steam drill on the left,
Before that steam drill could beat him down,
He hammered his fool self to death.

They carried John Henry to the mountains,
From his shoulder his hammer would ring,
She caught on fire by a little blue blaze
I believe these old mountains are caving in.

John Henry was lying on his death bed,
He turned over on his side,
And these were the last words John Henry said
"Bring me a cool drink of water before I die."

John Henry had a little woman,
Her name was Pollie Ann,
He hugged and kissed her just before he died,
Saying, "Pollie, do the very best you can."

John Henry's woman heard he was dead,
She could not rest on her bed,
She got up at midnight, caught that No. 4 train,
"I am going where John Henry fell dead."

They carried John Henry to that new burying ground
His wife all dressed in blue,
She laid here hand on John Henry's cold face,
"John Henry I've been true to you."

W. T. Blankenship

## The Eclipse Clothes Wringer (c. 1880s)

**Figure 5.1**
When African-American women did laundry work for white families, they either took it home or completed in the employers' homes. Like other workers, most launderers worked with hand tools, though some employers invested in machinery. These machines, such as the Eclipse Clothes Wringer, were marketed as labor-savers that could make this difficult work easier. Reprinted from "The Eclipse Clothes Wringer," c. 1880s, #60, Warshaw Collection of Business Americana Series: Laundry Machinery and Accessories, Box 1, Folder 1/43, NMAH Archives, Smithsonian Institution, Washington, D.C.

## Woman with Bicycle (c. 1890s)

**Figure 5.2**

Bicycles, which became wildly popular during the 1890s, were the first transportation devices providing truly individual mobility, and at a price that many Americans could afford. Women's use of bicycles was of particular importance because their mastery of this machine, the physical exertion it demanded, and the freedom of movement it provided made women's bicycle riding a subversive attack upon Victorian notions of femininity. Even though many women did not own bicycles, its powerful message of adventure and liberation made the bicycle an appealing icon of self-assertiveness and self-confidence. This young woman had her picture taken, probably in an Atlanta photography studio, with a bicycle that was both a prop and a promise. Reprinted by courtesy of the Atlanta History Center, KHN Collection.

## *The Taint of the Bicycle* (1902)

W. F. Fonvielle

Female agency, in the form of access to the freedom the bicycle allowed, was not endorsed by everyone. Like the telephone, and countless other technologies, the bicycle made possible a refiguring of power relationships between the classes, between the sexes, as well as between parents and children. African-American author W. F. Fonvielle offered a pessimistic view of the way in which established hierarchies and values might be threatened.

---

A Northern manufacturing concern constructed one hundred safety bicyles, and put them on the market at a selling price of $100 apiece. The rich folk gobbled them up at that price, and incidentally remarked: How cheap. The bicycle makers were smilingly satisfied, and on the best kind of terms with themselves and everybody else; being thus, they straightway attempted to put a thousand of the same kind before the public. The people who have the honor of making up the public were getting mad, so they turned in and purchased them before the workmen could turn them out. Rich people bought most of these; yet some who were not listed as wealthy bought a few of them at $100 apiece. Other people went into the manufacture and sale of "bikes," "wheels" and "safeties." While the artisans toiled night and day to supply the demand. Then came names, trade marks and lamps, bells, pumps, whistles and confusion. . . .

Then, sad regret—woman went "a coasting," and "a scorching." For a time she rode her brother's bicycle—when he would let her; she rode her son's "bike," if she were larger and stronger than he; on her lovers "safety," if he loved her better than she loved him; and finally on her husband's "wheel," if he were a hen-pecked husband. She rode; and what is worse, she rode just as the men ride. But some good people still exist. Some who persistently and stubbornly refuse to fall down and worship two wheels, a handle bar and a chain. So these good people protested, using the words, "shame," "awful," "indecent," et cetera. Public sentiment was so strong that all got down, save the more brazen among them, till the manufacturing folk could make her one; which they proceeded to do without loss of time. Then with stiff hat, laundried shirt, high collar, four-in-hand tie, low cut vest, cuttaway coat, plus a substitution for pantaloons, she rode and pretended to be happy in her new role. Perhaps she was happy. . . .

Ere long this one step led to something else. It bore fruit quickly. The churches were thrown open, raffles, ticket selling and buying took place in the houses dedicated to the worship of "Our Father in Heaven." The purpose being to give a bicycle to the

---

Excerpted from: *The Taint of the Bicycle* (n.p., 1902), 8, 10–11, 14–22.

most popular minister in town, or the most popular young woman. Then the most popular Grand Worthy boss of the local societies, and the Secretary of the Burying Society had to have a bicycle as he afterwards explained, that he might get around to a dead member's house in a hurry. Girls of tender ages and pure lives, the young flowers in the garden of our civilization, went forth armed with credentials—tickets and punch cards—holding up men on the streets and enticing other people to give in order that they might win in a "beauty contest." When these contests were decided, old people and young people, men whose breaths gave good evidence that it had been quite a while since they signed the temperance pledge, and women of questionable character—old sinners jostled each other good naturedly and voted together for their favorites. A young woman got the prize bicycle, but no one knows better than she does, that she is not the most popular young woman in her community.

Then the cooks, washerwomen, nurse girls, the farmers, boot blacks and street loafers took to the wheel. It may never be known as to how they obtained them. Often it has been noted and commented upon that the bicycle which the cook owned and rode, was the same make and cost the same money as did the one which the mistress of the house rode, yet the cook received but $1.25 for her week's labor. Dressed in a homespun apron, bare-headed and barefooted, I have seen them enjoying themselves upon the thoroughfares while they rode a wheel which cost seventy-five dollars.

The nurse girl pushed the carriage all the week for fifty cents, yet after sundown she rode a new bicycle in company with her friends and it belonged to her. The washer woman, with six children and an indolent husband on her hands, standing over the wash tub all day, rinsing, starching, rubbing and ironing two days more, drew $1.15 on Saturday afternoon for the week's work, yet Sunday afternoon, if you chanced to stroll in the park, it is highly probable that she would be among the first persons to kindly request you to step to one side in order that she and her wheel might pass.

The farmer boy working all the year for $19.25, his board and clothes, rides into town on his bicycle on Saturday afternoon, peddling with brogan shoes, minus socks, dressed in jeans pantaloons, hickory shirt and cotton hat. The bicycle he rode was his and he enjoyed himself. The boot-blacks own the best make of wheels and ride like princes. Six months ago a boy who sold lemonade at a penny a glass owned two wheels, one of which he offered to sell. The street loafers own but few wheels, but they ride; they rent and pay twenty-five cents per hour for the privilege and the pleasure of the thing.

Where do these people get the money to buy these bicycles?

In the town of K———lived a husband and wife whose chief treasure was a daughter, the sweetest little damsel in the town. Idolized by her parents, greatly beloved by her friends, respected by the community in which she lived, it was a pleasure to know her. Gentle, yet shy as an untamed bird, perfect in figure, beautiful in facial expression, with the velvet violet eyes—eyes which told in their eloquence the simple, unaffected story of Sadie Southwell, the brightest member of her class in the home school until she had passed her fifteenth summer, when she was sent from home to a distant town to attend a more pretentious school. She was gladly received at the Seminary, and as at home, soon became a prime favorite with the young people with whom she came in contact, and likewise with the authorities. Graduating at the head of her class in June three years after matriculation, she left receiving the benedictions of the school, the kisses of her class-mates, flowers from her friends, and good wishes from everybody. But not before she had given her heart to one who had beheld her passing in the throng. A great strong man of fine attainments, worthy in every way to become the life companion of this dear girl. Scarcely eighteen, she was married during the first bright days of October. Surrounded by those who loved her best, together with music, flowers, the sweet, sad peal of the wedding bells, she found herself happy. Her husband took her away from the old folks and the old home—associations which she had loved for eighteen years—a thousand miles away. They were both very happy, for here, as at other places, the young wife soon became a favorite and was foremost in doing acts of kindness.

The people among whom they moved, made up as they were of preachers, lawyers, physicians, writers, merchants, teachers etc., 'tis true that most of these people possessed and rode bicycles. Heretofore this young woman had been content, she had never cared for such things, but in an unlooked for moment, she too found herself longing to possess one. She was too modest, too good, too thoughtful to tease, worry or beg her husband into purchasing a bicyle, but her young lord read the request in the liquid, pleading, violet eyes. Who could withstand their beseeching? When the winter had scarcely gone, when the violets had returned, before the May roses had fully blown, he bought her one. Her happiness was unbounded—complete. This good man was a minister of the gospel, pastor of one of the largest local churches, he had no love for the bicycle, and often remarked that he could not afford to ride one. But he gave the young wife one, and was happy in knowing that she was happy. The wife soon learned to ride, and joined bicycle parties of different sizes, shapes and sexes, having a thoroughly good time. A bicycle club was formed, she joined that, and the summer was one long day of wheeling and pleasure. Autumn came and went, and then came the ice and the snow, but hardly had the grass begun to spring ere the bicyclists went forth again. Busy with his parish work, the young, patient husband never inquired as to when and how the young wife went and came. It was enough to know that she was with her friends and enjoying herself.

But one day—how well I remember it—it was the 30th of May. Like Byron, I want to be particular about dates. Decoration Day. One sad day when fair women and lovely children and strong men with band and horse and sabre and cannon, and musket and bayonet and trailing banner had gone forth to place flowers—roses, the laurel, the bay and the pine upon the mounds of the Nation's honored dead. It was a lovely day, as nearly perfect as a day in May can be. The young wife returned home alone; not riding, but walking, leading her bicycle. Bursting into the hall wild-eyed, nervous, pale with ashen lips, she left her bicycle in the hall and sought her husband. Alarmed at her appearance, he questioned her. Then falling into a chair, bursting into tears, between sobs and much incoherent speech, she made a confession—told him all.

Overcome with what he had just learned, the good man sat blinded in his stony grief. Then raising his arms above his head, there issued a cry from his lips not unlike that of some wild beast, wounded at bay, still defending the cub which the hunter is seeking to despoil her of. That is all. They live together in the same house, in the same city; but there is no happiness now for either of them; and while they do not discuss it, the little wife grows thinner day by day. She smiles sometimes—such a sad, sweet smile, but her heart is not in it. Her beautiful eyes have lost their fire and their lustre. And the husband generous, good, once strong, now a wreck, sitting all the day long, all the night long in his study, calm, unmoved, dazed—looking back upon his short married life—short, sweet days of the past, into the wide unknown expanse of the future, and wonders what has it now in store for him. And the neighbors, friends, guessing what the hideous skeleton is in the closet of that home, generously remain away. They do not intrude their presence upon them. And the little modest wife, once the joy and pride of all who knew her, never leaves her home now—growing more pale, more thin, the roses have long since left her cheeks, her hands almost transparent she sits and thinks all alone. So when the leaves had begun to turn, when there was a mist in the sky, when the first frosts had lain their icy fingers upon the forests and turned them into gardens of riotous colors—when time stood on the rim, the edge, the line between autumn and winter—one sad day in the lonesome October the 16th the little wife died. Her punishment, like Cain's had been greater than she could bear.

# IV THE PROGRESSIVE ERA

The years around the turn of the twentieth century witnessed what some have called the Second Industrial Revolution. Startling new applications of electricity and the beginnings of a conservation movement, a vast growth in the number of collegiately trained engineers, the rise of industrial research and mass production, the birth of home economics and scientific management, all marked this as a period of unprecedented change in the nation's technology.

At the same time, the country was going though one of its periodic times of reform. The growth of industrial cities, increasing immigration (especially from eastern and southern Europe), the development of an industrial proletariat, the alarming disappearance of once-abundant natural resources, the closing of the frontier, and the perceived crisis in rural life all were national problems that cried out for solution. One avenue of reform was through radical political action; socialism was an increasingly attractive option for many Americans, and the third parties of the Populists and Progressives made fundamental contributions to social change and economic reform. More popular with conservatives in general and large corporations in particular was a new recourse to science and technology; an appeal, as it were, from technology drunk to technology sober.

Both the industrialization and the reform impulse of the period reached a climax in the Great War. From August 1914, when war broke out in Europe, through April 1917, when the United States officially became a combatant, and until November 1918, when the Armistice was signed, American science and technology acted through the laboratories and workshops of the country to give full-scale expression to the engineering ideals of mass production, scientific management, and industrial research that had been unfolding for nearly a quarter of a century.

African-Americans were by no means immune from these developments. The great debate at the turn of the century over the proper role of industrial education for young people—the debate personalized by Booker T. Washington and W. E. B. Dubois—both looked back to a time when black artisans had done the technological work of the South, and forward to the emerging possibilities of the second Industrial Revolution. The century began with young people at Tuskegee Institute learning a range of trades from the traditional to the modern. Soon, World War I drew a Great Migration to staff the wartime factories of the industrial northern heartland. Engineers and foundry workers, inventors and entrepreneurs, African-Americans struggled to find a place in the Second Insutrial Revolution.

# 6 Training for the Industrial Age

## Comments on the Advisibility of Instructing Engineering Students in the History of the Engineering Profession (1903)

L. S. Randolph

The explosive demand for engineers in the late nineteenth century, and the increasing number of people training to meet that need, created a profession that was rapidly becoming the largest in the nation. At the same time, technology was changing so rapidly that new fields of engineering were being created and old ones redefined. Having grown out of the skilled craft of surveyor and machinist, electrician and chemist, the new engineers experienced a great unease about their role and place in modern society; they were, it seemed to them, poorly defined and underappreciated by the public. Not surprisingly, threats to their status were defined in terms of class and gender, as well as, of course, race. The following remarks were made in response to an argument that courses in the history of technology would give engineering students pride and confidence in their profession.

---

The writer is heartily in accord with Mr. Waddell in his ideas, and believes that such a history as he advocates will be extremely valuable. One of the great values of history and the study thereof, is that it gives a man more definite conception than he will otherwise get, and there is no doubt in the writer's mind that definite conception in regard to what constitutes an engineer and what an engineer should be, are not only badly needed among the people generally but also among those whose duty it is to train and teach engineers.

The multiplicity of scholastic degrees with the word "engineer" attached, and the planning of many of the courses of instruction in engineering, are proof positive of the absolute necessity for some more thorough knowledge of the work of an engineer. The civil engineer retains more than any other class his high position. By civil engineer is meant all of those included in the broad term of civil engineers, with the exception of what are sometimes called dynamic engineers. In one or two of the old schools of engineering, which for thirty or forty years have been turning out civil engineers and which therefore are thoroughly acquainted with the history of the same, the course of instruction is well fitted to the work to be accomplished and the professional standing of those practising the profession is more thoroughly recognized.

The danger comes from those who, having for their own work nothing of the training and preparation which is given to the civil engineer, seize upon the title of such high renown and boldly appropriate to themselves the accompanying unearned

---

Reprinted from: Comments by L. S. Randolph on J. A. L. Waddell, "The Advisibility of Instructing Engineering Students in the History of the Engineering Profession," Society for the Promotion of Engineering Education, *Proceedings* 11 (1903): 201–202.

honors and emoluments. We have the man who fires the boiler and pulls the throttle dubbed a locomotive or stationary engineer; we have the woman who fires the stove and cooks the dinner dubbed the domestic engineer, and it will not be long before the barefooted African, who pounds the mud into the brick molds, will be calling himself a ceramic engineer. Those of the teaching profession have seen how this thing goes and are familiar with the tonsorial artists who are called professors, and the dancing masters who have the same high sounding title.

In order to place the matter briefly, it can be said that the profession of engineering is to-day at the parting of the ways. If the engineer is to fulfill the definition which is so generally accepted, "as one who applies the discoveries of the scientists to the structural needs of mankind," something must be done at once, and there are but two ways of handling the question. One is boldly to adopt a new title and leave the title of engineer, which is rapidly falling into disrepute, to the locomotive engineers, the domestic engineers, the sugar engineers, etc., who have caused its fall; or to begin a campaign of education which shall bring clearly before the public, before our college presidents and boards of control, what the engineer is, or rather what he should be. I know of no better place for this to be done than in this Society, and of no better method than the preparation of a history of engineering as outlined by the author of the paper.

## Industrial Education; Will It Solve the Negro Problem, II (1904)

Booker T. Washington

One of the great debates within the African-American community at the turn of the century was regarding the proper role of education in "uplifting the race." Perhaps the best know African-American leader, at least among the white establishment, was Booker T. Washington, a passionate advocate of what he called "industrial education." While not denying the importance of "higher" education, he called for the training of black youth in the practical technical skills of the day. This article was one of a series on the subject, published by a prominent African-American magazine.

Since the war no one object has been more misunderstood than that of the object and value of industrial education for the Negro. To begin with, it must be borne in mind that the condition that existed in the South immediately after the war, and that now exists, is a peculiar one, without a parallel in history. This being true, it seems to me that the wise and honest thing is to make a study of the actual condition and environment of the Negro, and do that which is best for him, regardless of whether the same thing has been done for another race in exactly the same way. There are those among our friends of the white race, and those among my own race, who assert with a good deal of earnestness, that there is no difference between the white man and the black man in this country. This sounds very pleasant and tickles the fancy but when we apply the test of hard, cold logic to it, we must acknowledge that there is a difference; not an inherent one, not a racial one, but a difference growing out of unequal opportunities in the past.

If I might be permitted to even seem to criticise some of the educational work that has been done in the South, I would say that the weak point has been in a failure to recognize this difference.

Negro education, immediately after the war, in most cases, was begun too nearly at the point where New England education had ended. Let me illustrate: One of the saddest sights I ever saw was the placing of a $300 rosewood piano in a country school in the South that was located in the midst of the "Black Belt." Am I arguing against the teaching of instrumental music to the Negroes in that community? Not at all; only I should have deferred those music lessons about twenty-five years. There are numbers of such pianos in thousands of New England homes, but behind the piano in the New England home, there were one hundred years of toil, sacrifice and economy; there was the small manufacturing industry, started several years ago by hand power,

Reprinted from: Booker T. Washington, "Industrial Education; Will It Solve the Negro Problem, II," *The Colored American Magazine* 7 (February 1904): 87–92.

now grown into a great business; there was the ownership in land, a comfortable home free from debt, a bank account. In this "Black Belt" community where this piano went, four-fifths of the people owned no land, many lived in rented one-room cabins, many were in debt for food supplies, many mortgaged their crops for the food on which to live and not one had a bank account. In this case, how much wiser it would have been to have taught the girls in this community how to do their own sewing, how to cook intelligently and economically, house-keeping, something of dairying and horticulture; the boys something of farming in connection with their common school education, instead of awakening in these people a desire for a musical instrument, which resulted in their parents going in debt for a third-rate piano or organ before a home was purchased. These industrial lessons should have awakened in this community a desire for homes, and would have given the people the ability to free themselves from industrial slavery, to the extent that most of them would have soon purchased homes. After the home and the necessaries of life were supplied, could come the piano; one piano lesson in a home is worth twenty in a rented log cabin.

Only a few days ago I saw a colored minister preparing his Sunday sermon just as the New England minister prepares his sermon. But this colored minister was in a broken down, leaky, rented log cabin, with weeds in the yard, surrounded by evidences of poverty, filth and want of thrift. This minister had spent some time in school studying theology. How much better would it have been to have had this minister taught the dignity of labor, theoretical and practical farming in connection with his theology, so that he could have added to his meagre salary, and set an example to his people in the matter of living in a decent house and correct farming—in a word, this minister should have been taught that his condition, and that of his people, was not that of a New England community, and he should have been so trained as to meet the actual needs and condition of the colored people in this community.

God, for 250 years, was preparing the way for the redemption of the Negro through industrial development. First, he made the Southern white man do business with the Negro for 250 years in a way that no one else has done business with him. If a Southern white man wanted a house or a bridge built, he consulted a Negro mechanic about the plan, about the building of the house or a bridge. If he wanted a suit of clothes or a pair of shoes made, it was the Negro tailor or shoemaker that he talked to. Secondly, every large slave plantation in the South was in a limited sense, an industrial school. On these plantations there were scores of young colored men and women who were constantly being trained, not alone as common farmers, but as carpenters, blacksmiths, wheel-wrights, plasterers, brickmasons, engineers, bridge-builders, cooks, dressmakers, eers, bridge-builders, cooks, dressmakers, etc., more in one county than now in the whole city of Atlanta. I would be the last to apologize for the curse of slavery, but I am simply stating facts. This training was crude and was given for selfish purposes, and did not answer the highest purpose, because there was an absence of literary training in connection with that of the hand. Nevertheless, this business contact with the Southern white man and the industrial training received on these plantations, put us at the close of the war into possession of all the common and skilled labor in the

South. For nearly twenty years after the war, except in one or two cases, the value of industrial training given by the Negroes' former masters on the plantations and elsewhere, was overlooked. Negro men and women were educated in literature, mathematics and the sciences, with no thought of what had taken place on these plantations for two and a half centuries. After twenty years, those who were trained as mechanics, etc., during slavery, began to disappear by death, and gradually we awoke to the fact that we had no one to take their places. We had trained scores of young men in Greek, but few in carpentry, or mechanical or architectural drawing; we had trained many in Latin, but almost none as engineers, bridge-builders and machinists. Numbers were taken from the farm and educated, but were educated in everything except agriculture; hence they had no sympathy with farm life and did not return to it.

The place made vacant by old Uncle Jim, who was trained as a carpenter during slavery, and who, since the war, had been the leading contractor and builder in the Southern towns, had to be filled. No young colored carpenter capable of filling Uncle Jim's place could be found. The result was that his place was filled by a white mechanic from the North, or from Europe or from elsewhere. What is true of carpentry and house-building in this case, is true, in a degree, of every line of skilled labor, and is becoming true of common labor. I do not mean to say that all of the skilled labor has been taken out of the Negro's hands, but I do mean to say that in no part of the South is he so strong in the matter of skilled labor as he was twenty years ago, except, possibly, in the country districts and the smaller towns. In the more Northern of the Southern cities, such as Richmond and Baltimore, the change is most apparent, and it is being felt in every Southern city. Wherever the Negro has lost ground industrially in the South, it is not because there is a prejudice against him as a skilled laborer on the part of the native Southern white man, for the Southern white man generally prefers to do business with the Negro mechanic, rather than with the white one; for he is accustomed to doing business with the Negro in this respect. There is almost no prejudices against the Negro in the South in matters of business, so far as the native whites are concerned, and here is the entering wedge for the solution of the race problem. Where the white mechanic or the factory operative gets a hold, the trades union soon follows and the Negro is crowded to the wall.

But what is the remedy for this condition? First, it is most important that the Negro and our white friends honestly face the facts as they are, otherwise the time will not be far distant when the Negro in the South will be crowded to the ragged edge of industrial life, as he is in the North. There is still time to repair the damage and to reclaim what we have lost.

I stated in the beginning that the industrial education for the Negro had been misunderstood. This has been chiefly because some have gotten the idea that industrial development was opposed to the Negro's higher mental development. This has little or nothing to do with the subject under discussion; and we should no longer permit such an idea to aid in depriving the Negro of the legacy in the form of skilled labor, that was purchased by his forefathers at the price of 250 years in slavery. I would say to the black boy what I would say to the white boy; get all the mental development that your time and pocketbook will afford—the more the better, but the time has come when a larger proportion, not all, for we need professional men and women, of the educated colored men and women, should give themselves to industrial or business life. The professional class will be helped in proportion as the rank and file have an industrial foundation so that they can pay for professional services. Whether they receive the training of the hand while pursuing their academic training or after the academic training is finished, or whether they will get their literary training in an industrial school or college, is a question which each individual must decide for himself, but no matter how or where educated, the educated men and women must come to the rescue of the race in the effort to get and hold its industrial footing. I would not have the standard of mental development lowered one whit, for with the Negro, as with all races, mental strength is the basis of all progress, but I would have a larger proportion of this mental strength reach the Negro's actual needs through the medium of the hand. Just now the need is not so much for common carpenters, brick-masons, farmers and laundry-women, as for industrial leaders; men who, in addition to their practical knowledge, can draw plans, make estimates, take contracts; those who understand the latest methods of truck-gardening and the science underlying practical agriculture; those who understand machinery to the extent that they can operate steam and electric laundries, so that our women can hold on to the laundry work in the South, that is so fast drifting into the hands of others in the large cities and towns.

It is possible for a race or an individual to have mental development and yet be so handicapped by custom, prejudice and lack of employment, as to dwarf and discourage the whole life, and this is the condition that prevails among my race in most of the large cities of the North, and it is to prevent this same condition in the South that I plead with all the earnestness of my heart. Mental development alone will not give us what we want, but mental development tied to hand and heart training, will be the salvation of the Negro.

In many respects, the next twenty years are going to be the most serious in the history of the race. Within this period it will be largely decided whether the Negro is going to be able to retain the hold which he now has upon the industries of the South, or whether his place will be filled by white people from a distance. The only way that we can prevent the industries slipping from the Negro in all parts of the South, as they have already in certain parts of the South, is for all the educators, ministers and friends of the Negro to unite to push forward, in a whole-souled manner, the industrial or business development of the Negro, either in school or out of school, or both. Four times as many young men and women of my race should be receiving industrial training. Just now the Negro is in the position to feel and appreciate the need of this in a way that no one else can. No one can fully appreciate what I am saying who has not walked the streets of a Northern city day after day, seeking employment, only to find every door closed against him on account of his color, except along certain lines of menial service. It is to prevent the same thing taking place in the South that I plead. We may argue that mental development will take care of this. Mental development is a good thing. Gold is also a good thing, but gold is worthless without opportunity to make it touch the world of trade. Education increases an individual's wants many fold. It is cruel in many cases to increase the wants of the black youth by mental development alone, without at the same time increasing his ability to supply these increased wants along the lines at which he can find employment.

I repeat that the value and object of industrial education has been misunderstood by many. Many have had the thought that industrial training was meant to make the Negro work, much as he worked during the days of slavery. This is far from my idea of it. If this training has any value for the Negro, as it has for the white man, it consists in teaching the Negro how rather not to work, but how to make the forces of nature—air, water, horse-power, steam and electric power work for him, how to lift labor up out of toil and drudgery, into that which is dignified and beautiful. The Negro in the South works, and he works hard; but his lack of skill, coupled with ignorance, causes him to do his work in the most costly and shiftless manner, and this keeps him near the bottom of the ladder in the business world. I repeat that industrial education teaches the Negro how not to work. Let him who doubts this, contrast the Negro in the South, toiling through a field of oats with an old-fashioned reaper, with the white man on a modern farm in the West, sitting upon a modern "harvester" behind two spirited horses, with an umbrella over him, using a machine that cuts and binds the oats at the same time—doing four times as much work as the black man with one half the labor. Let us give the black man so much skill and brains that he can cut oats like the white man, then he can compete with him. The Negro works in cotton, and has no trouble so long as his labor is confined to the lower forms of work—the planting, the picking and the ginning; but when the Negro attempts to follow the bale of cotton up through the higher stages; through the mill where it is made into the finer fabrics, where the larger profit appears he is told that he is not wanted. The Negro can work in wood and iron, and no one objects, so long as he confines his work to the felling of trees and the sawing of boards, to the digging of iron ore and the making of pig iron; but when the Negro attempts to follow his tree into the factory, where it is made into chairs and desks and railway coaches; or when he attempts to follow the pig iron into the factory, where it is made into knife blades and watch springs, the Negro's trouble begins. And what is the objection? Simply that Negro lacks skill, coupled with brains, to the extent that he can compete with the white man, or that when white men refuse to work with colored men, enough skilled and educated colored men cannot be found able to superintend and manage every part of any large industry, and hence, for these reasons, we are constantly being barred out. The Negro must become, in a larger measure, an intelligent producer, as well as consumer. There should be more vital connection between the Negro's educated brain and his opportunity of earning his daily living. Without more attention being given to industrial development, we are likely to have an over-production of educated politicians—men who are bent on living by their wits. As we get farther away from the war period, the Negro will not find himself held to the Republican party by feelings of gratitude. He will feel himself free to vote for any party; and we are in danger of having the vote or "influence" of a large proportion of the educated black men in the market for the highest bidder, unless attention is given to the education of the hand, or to industrial development.

A very weak argument often used against pushing industrial training for the Negro, is that the Southern white man favors it and, therefore, it is not best for the Negro. Although I was born a slave, I am thankful that I am able to so far rid myself of prejudice as to be able to accept a good thing, whether it comes from a black man or from a white man, a Southern man or a Northern man. Industrial education will not only help the Negro in the matter of industrial development, but it will help in bringing about more satisfactory relations between him and the Southern white man. For the sake of the Negro and the Southern white man, there are many things in the relations of the two races that must soon be changed. We cannot depend wholly upon abuse or condemnation of the Southern white man to bring about these changes. Each race must be educated to see matters in a broad, high, generous, Christian spirit; we must bring the two races together, not estrange them. The Negro must live for all time by the side of the Southern white man. The man is unwise who does not cultivate in every manly way the friendship and good will of his next door neighbor, whether he is black or white. I repeat that industrial training will help cement the friendship of the two races. The history of the world proves that trade, commerce, is the forerunner of peace and civilization as between races and nations. We are interested in the political welfare of Cuba and the Sandwich Islands, because we have business interests with these islands. The Jew, that was once in about the same position that the Negro is to-day, has now complete recognition, because he has entwined himself about America in a business or industrial sense. Say or think what we will, it is the tangible or visible element that is going to tell largely during the next twenty years in the solution of the race problem. Every white man will respect the Negro who owns a two-story brick business block in the center of town and has $5000 in the bank. When a black man is the largest taxpayer and owns and cultivates the most successful farm in his county, his white neighbors will not object very long to his voting and to having his vote honestly counted. The black man, who is the largest contractor in his town and lives in a two-story brick house, is not very liable to be lynched. The black man that holds a mortgage on a white man's house, which he can foreclose at will, is not likely to be driven away from the ballot-box by the white man.

I know that what I have said will likely suggest the idea that I have put stress upon the lower things of life—the material; that I have overlooked the higher side, the ethical and religious. I do not overlook or undervalue the higher. All that I advocate in this article is not as an end, but as a means. I know as a race, we have got to be patient in the laying of a firm foundation, that our tendency is too often to get the shadow instead of the substance, the appearance rather than the reality. I believe further, that, in a large measure, he who would make the statesmen, the men of letters, the men for the professions for the Negro race of the future, must, to-day, in a large measure, make the intelligent artisans, the manufacturers, the contractors, the real estate dealers, the land-owners, the successful farmers, the merchants, those skilled in domestic economy. Further, I know that it is not an easy thing to make a good Christian of a hungry man. I mean that just in proportion as the race gets a proper industrial foundation—gets habits of industry, thrift, economy, land, homes, profitable work, in the same proportion will its moral and religious life be improved. I have written with a heart full of gratitude to all religious organizations and individuals for what they have done for us as a race, and I speak as plainly as I do because I feel that I have had opportunity, in a measure to come face to face with the enormous amount of work that must still be done by the generous men and women of this country before there will be in reality, as well as in name, high Christian civilization among both races in the South.

To accomplish this, every agency now at work in the South needs reinforcement.

## The Training of Negroes for Social Power (1904)

W. E. B. Dubois

Opposition to an overemphasis on industrial education, and any implication that black youth were not fit for collegiate education, popularly was associated especially with the Harvard-trained intellectual W. E. B. Dubois. Rather than emphasizing, like Washington, a concentration on manual training for the black masses, Dubois championed the higher education of what he called "The Talented Tenth," those young people who could act as leaders and educators of the masses.

The responsibility for their own social regeneration ought to be placed largely upon the shoulders of the Negro people. But such responsibility must carry with it a grant of power; responsibility without power is a mockery and a farce. If, therefore, the American people are sincerely anxious that the Negro shall put forth his best efforts to help himself, they must see to it that he is not deprived of the freedom and power to strive. The responsibility for dispelling their own ignorance implies that the power to overcome ignorance is to be placed in black men's hands; the lessening of poverty calls for the power of effective work, and one responsibility for lessening crime calls for control over social forces which produce crime.

Such social power means, assuredly, the growth of initiative among Negroes, the spread of independent thought, the expanding consciousness of manhood; and these things to-day are looked upon by many with apprehension and distrust, and there is systematic and determined effort to avoid this inevitable corollary of the fixing of social responsibility. Men openly declare their design to train these millions as a subject caste, as men to be thought for, but not to think; to be led, but not to lead themselves.

Those who advocate these things forget that such a solution flings them squarely on the other horn of the dilemma; such a subject child-race could never be held accountable for its own misdeeds and shortcomings; its ignorance would be part of the nation's design, its poverty would arise partly from the direct oppression of the strong and partly from thriftlessness which such oppression breeds; and, above all, its crime would be the legitimate child of that lack of self-respect which caste systems engender. Such a solution of the Negro problem is not one which the saner sense of the nation for a moment contemplates; it is utterly foreign to American institutions, and is unthinkable as a future for any self-respecting race of men. The sound afterthought of the American people must come to realize that the responsibility for dispelling igno-

Reprinted from: W. E. B. Dubois, "Industrial Education—Will It Solve the Negro Problem, VII. The Training of Negroes for Social Power," *The Colored American Magazine* 7 (May 1904): 333–339.

rance and poverty and uprooting crime among Negroes cannot be put upon their own shoulders unless they are given such independent leadership in intelligence, skill, and morality as will inevitably lead to an independent manhood which cannot and will not rest in bonds.

Let me illustrate my meaning particularly in the matter of educating Negro youth.

The Negro problem, it has often been said, is largely a problem of ignorance—not simply of illiteracy, but a deeper ignorance of the world and its ways, of the thought and experience of men; an ignorance of self and the possibilities of human souls. This can be gotten rid of only by training; and primarily such training must take the form of that sort of social leadership which we call education. To apply such leadership to themselves, and to profit by it, means that Negroes would have among themselves men of careful training and broad culture, as teachers and teachers of teachers. There are always periods of educational evolution when it is deemed proper for pupils in the fourth reader to teach those in the third. Such a method, wasteful and ineffective at all times, is peculiarly dangerous when ignorance is widespread and when there are few homes and public institutions to supplement the work of the school. It is, therefore, of crying necessity among Negroes that the heads of their educational system—the teachers in the normal schools, the heads of high schools, the principals of public systems, should be unusually well-trained men; men trained not simply in common-school branches, not simply in the technique of school management and normal methods, but trained beyond this, broadly and carefully, into the meaning of the age whose civilization it is their peculiar duty to interpret to the youth of a new race, to the minds of untrained people. Such educational leaders should be prepared by long and rigorous courses of study similar to those which the world over have been designed to strengthen the intellectual powers, fortify character, and facilitate the transmission from age to age of the stores of the world's knowledge.

Not all men—indeed, not the majority of men, only the exceptional few among American Negroes or among any other people—are adapted to this higher training, as, indeed, only the exceptional few are adapted to higher training in any line; but the significance of such men is not to be measured by their numbers, but rather by the numbers of their pupils and followers who are destined to see the world through their eyes, hear it through their trained ears, and speak to it through the music of their words.

Such men, teachers of teachers and leaders of the untaught, Atlanta University and similar colleges seek to train. We seek to do our work thoroughly and carefully. We have no predilections or prejudices as to particular studies or methods, but we do cling to those time-honored sorts of discipline which the experience of the world has long since proven to be of especial value. We sift as carefully as possible the student material which offers itself, and we try by every conscientious method to give to students who have character and ability such years of discipline as shall make them stronger, keener, and better for their peculiar mission. The history of civilization seems to prove that no group or nation which seeks advancement and true development can despise or neglect the power of well-trained minds; and this power of intellectual leadership must be given to the talented tenth among American Negroes before this race can seriously be asked to assume the responsibility of dispelling its own ignorance. Upon the foundation-stone of a few well-equipped Negro colleges of high and honest standards can be built a proper system of free common schools in the South for the masses of the Negro people; any attempt to found a system of public schools on anything less than this—no narrow ideals, limited or merely technical training—is to call blind leaders for the blind.

The very first step toward the settlement of the Negro problem is the spread of intelligence. The first step toward wider intelligence is a free public-school system; and the first and most important step toward a public-school system is the equipment and adequate support of a sufficient number of Negro colleges. These are first steps, and they involve great movements: first, the best of the existent colleges must not be abandoned to slow atrophy and death, as the tendency is to-day; secondly, systematic attempt must be made to organize secondary education. Below the colleges and connected with them must come the normal and high schools, judiciously distributed and carefully manned. In no essential particular should this system of common and secondary schools differ from educational systems the world over. Their chief function is the quickening and training of human intelligence; they can do much in the teaching of morals and manners incidentally, but they cannot and ought not to replace the home as the chief moral teacher; they can teach valuable lessons as to the meaning of work in the world, but they cannot replace technical schools and apprenticeship in actual life, which are the real schools of work. Manual training can and ought to be used in these schools, but as a means and not as an end—to quicken intelligence and self-knowledge and not to teach carpentry; just as arithmetic is used to train minds and not to make skilled accountants.

Whence, now, is the money coming for this educational system? For the common schools the support should come from local communities, the State governments, and the United States Government; for secondary education, support should come from local and State governments and private philanthropy; for the colleges, from private philanthropy and the United States Government. I make no apology for bringing the United States Government in thus conspicuously. The General Government must give aid to Southern education if illiteracy and ignorance are to cease threatening the very foundations of civilization within any reasonable time. Aid to common-school education could be appropriated to the different States on the basis of illiteracy. The fund could be administered by State officials, and the results and needs reported upon by United States educational inspectors under the Bureau of Education. The States could easily distribute the funds so as to encourage local taxation and enterprise and not result in pauperizing the communities. As to higher training, it must be remembered that the cost of a single battle-ship like the "Massachusetts" would endow all the distinctively college work necessary for Negroes during the next half-century; and it is without doubt true that the unpaid balance from bounties withheld from Negroes in the Civil War would, with interest, easily supply this sum.

But spread of intelligence alone will not solve the Negro problem. If this problem is largely a question of ignorance, it is also scarcely less a problem of poverty. If Negroes are to assume the responsibility of raising the standards of living among themselves, the power of intelligent work and leadership toward proper industrial ideals must be placed in their hands. Economic efficiency depends on intelligence, skill, and thrift. The public-school system is designed to furnish the necessary intelligence for the ordinary worker, the secondary school for the more gifted workers, and the college for the exceptional few. Technical knowledge and manual dexterity in learning branches of the world's work are taught by industrial and trade schools, and such schools are of prime importance in the training of colored children. Trade-teaching cannot be effectively combined with the work of the common schools because the primary curriculum is already too crowded, and thorough common-school training should precede trade-teaching. It is, however, quite possible to combine some of the work of the secondary schools with purely technical training, the necessary limitations being matters of time and cost: the question whether the boy can afford to stay in school long enough to add parts of a high-school course to the trade course, and particularly the question whether the school can afford or ought to afford to give trade-training to high-school students who do not intend to become artisans. A system of trade-schools, therefore, supported by State and private aid, should be added to the secondary school system.

An industrial school, however, does not merely teach technique. It is also a school—a center of moral influence and of mental discipline. As such it has peculiar problems in securing the proper teaching force. It demands broadly trained men: the teacher of carpentry must be more than a carpenter, and the teacher of the domestic arts more than a cook; for such teachers must instruct, not simply in manual dexterity, but in mental quickness and moral habits. In other words, they must be teachers as well as artisans. It thus happens that college-bred men and men from other higher schools have always been in demand in technical schools, and it has been the high privilege of Atlanta University to furnish during the thirty-six years of its existence a part of the teaching force of nearly every Negro industrial school in the United States, and to-day our graduates are teaching in more than twenty such institutions. The same might be said of Fisk University and other higher schools. If the college graduates were to-day withdrawn from the teaching force of the chief Negro industrial schools, nearly every one of them would have to close its doors. These facts are forgotten by such advocates of industrial training as oppose the higher schools. Strong as the argument for industrial schools is—and its strength is undeniable—its cogency simply increases the urgency of the plea for higher training-schools and colleges to furnish broadly educated teachers.

But intelligence and skill alone will not solve the Southern problem of poverty. With these must go that combination of homely habits and virtues which we may loosely call thrift. Something of thrift may be taught in school, more must be taught at home; but both these agencies are helpless when organized economic society denies to workers the just reward of thrift and efficiency. And this has been true of black laborers in the South from the time of slavery down through the scandal of the Freedmen's Bank to the peonage and crop-lien system of to-day. If the Southern Negro is shiftless, it is primarily because over large areas a shiftless Negro can get on in the world about as well as an industrious black man. This is not universally true in the South, but it is true to so large an extent as to discourage striving in precisely that class of Negroes who most need encouragement. What is the remedy? Intelligence—not simply the ability to read and write or to sew—but the intelligence of a society permeated by that larger vision of life and broader tolerance which are fostered by the college and university. Not that all men must be college-bred, but that some men, black and white, must be, to leaven the ideals of the lump. Can any serious student of the economic South doubt that this to-day is her crying need?

Ignorance and poverty are the vastest of the Negro problems. But to these later years have added a third—the problem of Negro crime. That a great problem of social morality must have become eventually the central problem of emancipation is as clear as day to any student of history. In its grosser form as a problem of serious crime it is already upon us. Of course it is false and silly to represent that white women in the South are in daily danger of black assaulters. On the contrary, white womanhood in the South is absolutely safe in the hands of ninety-five percent of the black men—ten times safer than black womanhood is in the hands of white men. Nevertheless, there is a large and dangerous class of Negro criminals, paupers, and outcasts. The existence and growth of such a class, far from causing surprise, should be recognized as the natural result of that social disease called the Negro problem; nearly every untoward circumstance known to human experience has united to increase Negro crime: the slavery of the past, the sudden emancipation, the narrowing of economic opportunity, the lawless environment of wide regions, the stifling of natural ambition, the curtailment of political privilege, the disregard of the sanctity of black men's homes, and, above all, a system of treatment for criminals calculated to breed crime far faster than all other available agencies could repress it. Such a combination of circumstances is as sure to increase the numbers of the vicious and outcast as the rain is to wet the earth. The phenomenon calls for no delicately drawn theories of race differences; it is a plain case of cause and effect.

But, plain as the causes may be, the results are just as deplorable, and repeatedly to-day the criticism is made that Negroes do not recognize sufficiently their responsibility in this matter. Such critics forget how little power to-day Negroes have over their own lower classes. Before the black murderer who strikes his victim to-day, the average black man stands far more helpless than the average white, and, too, suffers ten times more from the effects of the deed. The white man has political power, accumulated wealth, and knowledge of social forces; the black man is practically disfranchised, poor, and unable to discriminate between the criminal and the martyr. The Negro needs the defense of the ballot, the conserving power of property, and, above all, the ability to cope intelligently with such vast questions of social regeneration and moral reform as confront him. If social reform among Negroes be without organization or trained leadership from within, if the administration of law is always for the avenging of the white victim and seldom for the reformation of the black criminal, if ignorant black men misunderstand the functions of government because they have had no decent instruction, and intelligent black men are denied a voice in government because they are black—under such circumstances to hold Negroes responsible for the suppression of crime among themselves is the cruellest of mockeries.

On the other hand, a sincere desire among the American people to help the Negroes undertake their own social regeneration means, first, that the Negro be given the ballot on the same terms as other men, to protect him against injustice and to safeguard his interests in the administration of law; secondly, that through education and social organization he be trained to work, and save, and earn a decent living. But these are not all: wealth is not the only thing worth accumulating; experience and knowledge can be accumulated and handed down, and no people can be truly rich without them. Can the Negro do without these? Can this training in work and thrift be truly effective without the guidance of trained intelligence and deep knowledge—without that same efficiency which has enabled modern peoples to grapple so successfully with the problems of the Submerged Tenth? There must surely be among Negro leaders the philanthropic impulse, the uprightness of character and strength of purpose, but there must be more than these; philanthropy and purpose among blacks as well as among whites must be guided and curbed by knowledge and mental discipline—knowledge of the forces of civilization that make for survival, ability to organize and guide those forces, and realization of the true meaning of those broader ideals of human betterment which may in time bring heaven and earth a little nearer. This is social power—it is gotten in many ways—by experience, by social contact, by what we loosely call the chances of life. But the systematic method of acquiring and imparting it is by the training of youth to thought, power, and knowledge in the school and college. And that group of people whose mental grasp is by heredity weakest, and whose knowledge of the past is for historic reasons most imperfect, that group is the very one which needs above all, for the talented of its youth, this severe and careful course of training; especially if they are expected to take immediate part in modern competitive life, if they are to hasten the slower courses of human development, and if the responsibility for this is to be in their own hands.

Three things American slavery gave the Negro—the habit of work, the English language, and the Christian religion; but one priceless thing it debauched, destroyed, and took from him, and that was the organized home. For the sake of intelligence and thrift, for the sake of work and morality, this home-life must be restored and regenerated with newer ideals. How? The normal method would be by actual contact with a higher home-life among his neighbors, but this method the social separation of white and black precludes. A proposed method is by schools of domestic arts, but, valuable as these are, they are but subsidiary aids to the establishment of homes; for real homes are primarily centers of ideals and teaching and only incidentally centers of cooking. The restoration and raising of home ideals must, then, come from social life among

Negroes themselves; and does that social life need no leadership? It needs the best possible leadership of pure hearts and trained heads, the highest leadership of carefully trained men.

Such are the arguments for the Negro college, and such is the work that Atlanta University and a few similar institutions seek to do. We believe that a rationally arranged college course of study for men and women able to pursue it is the best and only method of putting into the world Negroes with ability to use the social forces of their race so as to stamp out crime, strengthen the home, eliminate degenerates, and inspire and encourage the higher tendencies of the race not only in thought and aspiration, but in every-day toil. And we believe this, not simply because we have argued that such training ought to have these effects, or merely because we hope for such results in some dim future, but because already for years we have seen in the work of our graduates precisely such results as I have mentioned: successful teachers of teachers, intelligent and upright ministers, skilled physicians, principals of industrial school, business men, and, above all, makers of model homes and leaders of social groups, out from which radiate subtle but tangible forces of uplift and inspiration. The proof of this lies scattered in every State of the South, and, above all, in the half-unwilling testimony of men disposed to decry our work.

Between the Negro college and industrial school there are the strongest grounds for co-operation and unity. It is not a matter of mere emphasis, for we would be glad to see ten industrial schools to every college. It is not a fact that there are to-day too few Negro colleges, but rather that there are too many institutions attempting to do college work. But the danger lies in the fact that the best of the Negro colleges are poorly equipped, and are to-day losing support and countenance, and that, unless the nation awakens to its duty, ten years will see the annihilation of higher Negro training in the South. We need a few strong, well-equipped Negro colleges, and we need them now, not tomorrow; unless we can have them and have them decently supported, Negro education in the South, both common-school and the industrial, is doomed to failure, and the forces of social regeneration will be fatally weakened, for the college to-day among Negroes is, just as truly as it was yesterday among whites, the beginning and not the end of human training, the foundation and not the capstone of popular education.

Strange, is it not, my brothers, how often in America those great watch-words of human energy—"Be strong!" "Know thyself!" "Hitch your wagon to a star!"—how often these die away into dim whispers when we face these seething millions of black men? And yet do they not belong to them? Are they not their heritage as well as yours? Can they bear burdens without strength, know without learning, and aspire without ideals? Are you afraid to let them try? Fear rather, in this our common fatherland, lest we live to lose those great watchwords of Liberty and Opportunity which yonder in the eternal hills their fathers fought with your fathers to preserve.

## Industrial Education—Will It Solve the Negro Problem (1904)

Fannie Barrier Williams

In this same series of articles, *The Colored American Magazine* published the position of only one woman, Fannie Barrier Williams. While clearly siding with Washington and the important of industrial education, she also emphasized the role of young women in the process.

Industrial Education is a much overworked term. Among the colored people, at least, it has caused no end of confusion of ideas and absurd conclusions as to what is the best kind of education for the masses of the people. Scarcely any subject, since emancipation, has been talked about and discussed as this one subject of Industrial Education. All sorts and conditions of people have their opinion as to the merits and demerits of this kind of education, and have been curiously eager to give expression to such opinions in the public press, in the pulpit and on the rostrum. Among these thinkers, writers, and speakers there are many who know absolutely nothing about the question, and there are others whose academic training has given them a fixed bias against any sort of mental training which does not include as a sine qua non the "humanities." On the other hand, there are those among the advocates of Industrial Education who insist that nothing else will solve the race problem. So the discussion goes on from one extreme to the other, with more or less earnestness and noise, truth and falsehood, sense and nonsense.

With the exception of occasional personalities and vindictive misrepresentations, this widespread discussion of the principles of Industrial Education has added enormously to the general interest in the subject of education for the colored and white people of the South. More than any other man, Dr. Booker T. Washington has made the subject of education in the South one of paramount interest to all the people. The helpful agencies that have been created and developed by this new propaganda of the training of the brawn as well as the brain of the people are quite beyond calculation. Industrial Education has long since ceased to be a theory. The discussion as to whether or not this kind of education is best for the Negro race may go on indefinitely; but, in the meantime, the industrial system of education has taken deep root in the needs of the people.

But what is this Industrial Education? The following are some of the answers given by persons who ought to know better: "To teach the Negro how to work hard"; "to teach the Negro how to be a good servant and forever hewers of wood and drawers of water"; "to teach the Negro how to undervalue his manhood rights."

Reprinted from: Fannie Barrier Williams, "Industrial Education—Will It Solve the Negro Problem," *The Colored American Magazine*, 7 July (1904): 491–495.

It is scarcely necessary to say that Industrial Education is immeasurably more than anything contained in these definitions. In the term Industrial Education, the emphasis is always upon education. Mathematics, drawing, chemistry, history, psychology, and sociology go along with the deft handling of the carpenter's and engineer's tools, with the knowledge of farming, dairying, printing, and the whole range of the mechanical arts. To the students in the industrial or manual training schools, their education means more than the mere names of the various trades imply. The carpenter has been given the foundation training by which he may well aspire to become an architect, the printer a publisher, the engineer a manufacturer, and the trained farmer a prosperous land owner. It can be readily seen that, by this kind of training, occupations that were once considered mere drudgery have become enlarged and ennobled by the amount of intelligence put into them. It was once thought that no one outside of the professions and other well deserved occupations needed to be educated. The tradesman or mechanic was not expected to know anything beyond the more or less skillful handling of his cash book or tools. An educated mechanic was the exception. Farming without the knowledge of forestry, dairying and the many other things that enter into the farmer's life, was regarded as drudgery.

What was true of masculine occupations was equally true of woman in the whole range of her special occupations and domestic concerns. It was thought that the only occupations for which women needed any sort of training were those which fitted her for the parlor and "society." Piano playing was an accomplishment; cooking and housekeeping, drudgery. A woman's apron was a badge of servility, and the kitchen a place not to be frequented by ladies. Poor woman! How narrow was her sphere! How wide the distance between the sphere of her every day home usefulness and the accomplishments of the "lady!" How different since the newer education has enlarged our sense of values. A new dignity has been added to the occupations that concern our health, our homes and our happiness. Through the influence of schools of domestic science, cooking has become a profession; the trained nurse divides honors with the physician, and the dressmaker and the milliner, by proper training, have become artists. In fact, Industrial training has dignified everything it has touched. It is not only banishing drudgery from the workshop and the home, but is widening the opportunities for talents of all kinds. There can be no such thing as caste in the everyday work of life, if that work is under the direction and control of trained intellects. Whether we do our share of the world's work with the pen or with the tool, in the office or in the shop, in the broad green acres on the hill slopes, or in the senate hall, the question is always the same—how much intelligence and character do you bring to your work? We believe that it is not too much to say that this is the spirit, the purpose and the result of Industrial Education.

Yet there are those who oppose this kind of education, as if it meant exactly the opposite of all this. It must be said that in a good deal of this opposition there is a curious blending of ignorance, envy and perversity. The best that can be said of those who think they are sincere is that they represent a belated conception of the higher and larger functions of education.

It should be stated in passing that nearly all of the most competent educators of the country, including presidents of the leading universities, believe in the Washington idea of Industrial Education, for white as well as colored people. That the idea has the encouragement and support of the best thought of the day is witnessed by the large number of industrial, polytechnical and agricultural schools that have been built and developed in the Northern states during the past ten or twelve years. These schools are always over crowded by white students. It is very difficult to keep a white boy in a high school long enough to enable him to graduate, but he will remain in a manual training school without persuasion. A leading professor in the Chicago University recently stated to his class that Booker T. Washington must always be regarded as the true leader of American education in its largest sense. The conception as to what is real and fundamental in education, has become so broadened that even the great universities are enlarging their curicula so as to include schools of technology. Such being the sphere and purpose and resulting possibilities of industrial education, can it be right or just to urge it as especially suited to the condition of the colored people?

It is claimed by the academician that the Negro is not essentially different from any other people, and, therefore, he should not be singled out for any special kind of education. We certainly all like to believe that the Negro is as good as any one else, but the important fact remains that the Negro is essentially different from any other race amongst us in the conditions that beset him. Just what these conditions are every intelligent Negro knows and feels. Among these conditions are illiteracy and restricted opportunities for the exercise of his talents and tastes. To multitudes of colored people illiteracy is a continuous night without a single ray of light. Inability to read and write is the least of his deficiencies; the ignorance of what to do to help himself and his kind is the pitiful thing. Any system of education that does not, in its helpful effect, reach from the school house back to the cabin is of small value in solving the race problem. The crying need of the multitude is, "Can you show me how to live, —how to raise more and better crops, —how to hold and use the benefits of my labor, —how to own and keep the land that I have earned over and over again by my labor, —how to appreciate the value of the earth's bounties and turn them into the currents of commerce? Any system of education that cannot give direct and helpful answer to this wail of despair, to this confession of incompetency and helplessness, falls far short of effectiveness. Industrial education aims to reach these conditions. It first aims to bring the benighted masses into conscious relationship with their own environments. It comes to teach these despairing people how to work out their own salvation by the tools and instrumentalities that are indigenous to their habitations. If agriculture must, for a long time to come, be the chief occupation of our people, then let their education for a long time to come be inclusive of all that which makes for thrift and intelligence in husbandry. If engineers, carpenters, plumbers, printers, wagon-makers, brick-makers, electricians, and other artisans are needed to build up and develope the rich resources of the communities in which they must live, is it not wise to train our own people to do all of this work so masterfully as to give them a monopoly against all others? It has been predicted already that the colored people will some day own the South, but this ownership can be realized only by the exercise of thrift, character and practical intelligence that can be gained in the best of the industrial schools.

It is not the contention of this article that Industrial Education must be the limit of education for colored people. We believe with Dr. Rankin, of Howard University, that "any system of education for the Negro that does not open to him the golden gate of the highest culture will fail on the ethical and spiritual side." At the same time the creators of wealth, —the great captains of industry, who are the real builders of communities, —have been those who wrought intelligently with their hands. The demand for colored artizans of all kinds is always in excess of the supply. The supply of lawyers, doctors and ministers and other professions, always exceeds the demand. The race is not only poor in the resources and means of wealth, but poor also in the practical intelligence that creates wealth.

It will prove an inmeasurable blunder if we shall now lack the foresight to provide for our young men and women the kind of training that will enable them to do everything in the line of industries that will equip them to become the real builders of the future greatness of the South. If by our neglect the master mechanics and skilled laborers of other races must be called into the South to do this work the Negro will be relegated to a position of hopeless servitude.

The advocates of industrial education are laying the foundation broad and deep for the future as well as providing for the present. They are wisely seeking to widen the Negro's sphere of usefulness. They realize the danger of equipping young colored men and women for occupations from which they are excluded by an unyielding prejudice. They are aiming to teach our aspiring young people that the positions and occupations from which they are now barred are not more honorable or more remunerative than those which they are permitted to enter, if they but carry the proper training and intelligence into those occupations. It teaches that the prizes of life lie along every pathway in which intelligence and character walk arm in arm. A professional man is not better than a mechanic unless he has more intelligence. An intelligent blacksmith is worth more to a community than an incompetent doctor, a hungry lawyer or an immoral minister.

The time is coming, aye, is now here, when a colored graduate from a school of domestic science will be more honored and better paid than are many white women who now hold the positions colored women cannot enter. The time is coming when there will be no excuse for a colored young woman to remain in soul-destroying idleness, because she cannot obtain a clerkship. She can be trained in an industrial school for positions that she can fill and still be socially eligible among those who make "society." An increasing respect is being shown to the young man or woman who is brave enough to learn a trade and follow it with pride and honor. The graduate from an industrial school finds a place awaiting him or her with a good salary. The graduates from Dr. Jones's Cooking School, in Richmond, Va., receive from $14 to $16 per week, while the untrained cook receives $5 per week. The graduates from Provident Hospital and Training School receive from $15 to $25 per week for their services; the untrained nurse not more than $6 per week. These instances are fair examples of how direct and immediate is the value of industrial training added to individual worth. These schools are every day creating new opportunities for honorable and well paid employment. The graduates of schools of this kind are seldom mendicants for employments. They have won their independence and their efficiency is a part of the good in every community in which they live and work.

The graduates of Hampton, Tuskegee and other industrial schools are the advance guard of efficiency and conquest. They touch more sides of the life of a community than any other class of our educated people. Rich and poor, black and white, prejudiced and unprejudiced, those who dread "Negro domination" and those who expect it, must all at one time or another ask for the service of the best trained artisan in the community. Along every pathway of material progress in that great undeveloped country south of the Ohio, we will soon begin to read all sorts of evidences of what industrial education has done for these black builders of a new empire of power.

The heroic efforts of Dr. Washington and others to furnish a system of education that shall be of the greatest good to the greatest number, should not and does not discourage what is called the higher education. In their tastes and aptitudes our young men and women are like those of other races. The doors of the universities are always open to the few who have the gifts and tastes for scholarship. The passion for higher education has not seemed to diminish as a consequence of the development of industrial schools. Every year witnesses a large number of Negro graduates from the best universities of the country. Many of these college graduates find their way down to Tuskegee, proving that Dr. Washington insists upon giving his Tuskegee students the advantage of studying under the best educated Negroes in the country.

The colored people are entitled to the best possible education that this country can afford, but this education should fit them for the life they must live. It should give as much encouragement to the would-be mechanic and agriculturist as to the would-be teacher, the lawyer or other professions. It should be the special aim to reach helpfully the lives of the thousands who live under conditions peculiarly their own and different from that of any other people.

These schools should educate their graduates toward and not away from the people. The evidence of this kind of education should make itself felt in every honorable relationship that the Negro bears to his community and to his government. Such an education will make the Negro efficient, self-respecting, proud, brave and proof against every prophecy of evil that would consign him to a destiny of "hopeless inferiority."

## *Working with the Hands* (1904)

Booker T. Washington

During the same year as his comments on the general importance of industrial education, Booker T. Washington wrote a long description, and defense, of his famous school, the Tuskegee Institute in Alabama. In this selection he describes the "industrial" education of women—an education that, while undeniably important and which employed tools, was hardly industrial as that term was commonly used. His description is, instead, of a two-year course in agriculture.

### Outdoor Work for Women

Seven years ago I became impressed with the idea that there was a wider range of industrial work for our girls. The idea grew upon me that it was unwise in a climate like ours in the South to narrow the work of our girls, and confine them to indoor occupations.

If one makes a close study of economic conditions in the South, he will soon be convinced that one of the weak points is the want of occupations for women. This lack of opportunity grows largely out of traditional prejudice and because of lack of skill. All through the period of slavery, the idea prevailed that women, not slaves, should do as little work as possible with their hands. There were notable exceptions, but this was the rule.

Most of the work inside the homes was done by the coloured women. Such a thing as cooking, sewing, and laundering, as part of a white woman's education, was not thought of in the days of slavery. Training in art, music, and general literature was emphasised. When the coloured girl became free, she naturally craved the same education in which she had seen the white woman specialising. I have already described our trials at the Tuskegee Institute, in attempting to get our girls to feel and see that they should secure the most thorough education in everything relating to the care of a home. When we were able to free them of the idea that it was degrading to study and practice those household duties which are connected with one's life every day in the year, I felt convinced that one other step was necessary.

New England and most of the Middle States are largely engaged in manufacturing. The factories, therefore, naturally give employment to a large number of women. The South is not yet in any large degree manufacturing territory, but is an agricultural section and will probably remain such for a long period. This fact confirmed

Excerpted from: Booker T. Washington, *Working with the Hands* (New York: Doubleday, Page & Co., 1904), 107–118; figure reprinted from p. 107.

**Figure 6.1**
A group of female students learning laundry work.

my belief that an industrial school should not only give training in household occupations to women, but should go further in meeting their needs and in providing education for them in out-of-door industries.

In making a study of this subject it became evident that the climate of every Southern State was peculiarly adapted to out-of-door work for women. A little later I had the opportunity of going to Europe and visiting the agricultural college for women at Swanley, England. There I found about forty women from some of the best families of Great Britain. Many of these women were graduates of high schools and colleges. In the morning I saw them in the laboratory and class room studying botany and chemistry and mathematics as applied to agriculture and horticulture. In the afternoon these same women were clad in suitable garments and at work in the field with the hoe or rake, planting vegetable seeds, pruning fruit trees or learning to raise poultry and bees and how to care for the dairy. After I had seen this work and had made a close study of it, I saw all the more clearly what should be done for the coloured girls of the South where there was so large an unemployed proportion of the population. I reasoned that if this kind of hand-training is necessary for a people who have back of them the centuries of English wealth and culture, it is tenfold more needful for a people who are in the condition of my race at the South.

I came home determined to begin the training of a portion of our women at Tuskegee in the outdoor industries. Mrs. Washington, who had made a careful study of the work in England, took charge of the outdoor work at Tuskegee. At first the girls were very timid. They felt ashamed to have any one see them at work in the garden or orchard. The young men and some of the women were inclined to ridicule those who were bold enough to lead off. Not a few became discouraged and stopped. There is nothing harder to overcome than an unreasonable prejudice against an occupation or a race. The more unreasonable it is, the harder it is to conquer. Mrs. Washington made a careful study of the girls and discovered the social leaders of a certain group. With this knowledge in hand she called the leaders together and had several conferences with them and explained in detail just what was desired and what the plans were. These leaders decided that they would be the pioneers in the outdoor work.

Beginning in a very modest way with a few girls, the outdoor work has grown from year to year, until it is now a recognised part of the work of the school, and the idea that this kind of labour is degrading has almost disappeared. In order to give, if possible, a more practical idea of just what is taught the girls, I give the entire course of study. In reading this it should be borne in mind that the theory is not only given, but in each case the girls have the training in actual work. Since the school year opens in the fall, the work naturally begins with the industries relating to the fall and winter. The course of study is:

First Year—Fall Term—Dairying—The home dairy is first taken up, and a detailed knowledge of the following facts taught: Kinds, use and care of utensils, gravity, creaming. A study of stone, wooden, and tin churns, ripening of cream, churning, working and salting butter, preparation and marketing of same. Feeding and care of dairy cows.

Poultry Raising—A working knowledge is required of the economic value of poultry on the farm, pure and mixed breeds, plain poultry-house construction, making of yards, nests, and runs.

Horticulture—Instruction is given as to the importance of an orchard and small fruits, varieties best suited, particular locality, selection and preparation of ground, setting, trimming, extermination of borers, lice, etc., special stress being laid upon the quality and quantity of peaches, pears, apples, plums, figs, grapes, and strawberries that should be planted in a home orchard.

Floriculture and Landscape Gardening—A study of our door-yards, how to utilise and beautify them. The kinds, care, and use of tools used in floriculture and landscape gardening. Trimming and shaping of beds and borders, and the general care of shrubbery and flowers. The gathering and saving of seed. Special treatment of rose bushes and shrubbery.

Market Gardening—Importance of proper management of the home garden, its value to the home, selection and preparation of ground; kinds, care and use of tools, planting, gardening and marketing of all vegetables. Gathering of seeds, drying of pumpkins, okra, and fruits.

Live Stock—Study is limited wholly to ordinary farm animals; the number and kind needed, how, when and what to feed; characteristics and utility of the various animals. . . .

I believe that all who will make a careful study of the subject will agree with me that there is a vast unexplored field for women in the open air. The South, with its mild climate and other advantages, is as well adapted to out-door labour for women as to that for men. There is not only an advantage in material welfare, but there is the advantage of a superior mental and moral growth. The average woman who works in a factory becomes little more than a machine. Her planning and thinking is done for her. Not so with the woman who depends upon raising poultry, for instance, for a living. She must plan this year for next, this month for the next. Naturally there is a growth of self-reliance, independence, and initiative.

Life out in the sweet, pure, bracing air is better from both a physical and a moral point of view than long days spent in the close atmosphere of a factory or store. There is almost no financial risk to be encountered, in the South, in following the occupations which I have enumerated. The immediate demands for the products of garden, dairy, poultry yard, apiary, orchard, etc., are pressing and ever present. The satisfaction and sense of independence that will come to a woman who is brave enough to follow any of these outdoor occupations infinitely surpass the results of such uncertain labor as that of peddling books or cheap jewelry, or similar employments, and I believe that a larger number of our schools in the near future will see the importance of outdoor handwork for women. . . .

Heretofore, one great drawback to farming, even in the North, has been the difficulty of keeping the farmers' sons on the farm. With trained and educated girls enthusiastically taking up the profession of farming, the country life will take on new charms, and the exodus of young men to cities will be materially lessened.

## How Electricity Is Taught at Tuskegee (1904)

Charles W. Pierce

The Electrical Engineering Department at Tuskegee was more clearly industrial and did not lead to a degree, but it exhibited the same mixture of theory and hands-on practice found in predominantly white engineering schools at the time. Note the use of demeaning "humor" directed at those who were ignorant of the nature and workings of electricity in a way identical to that shown by white electricians in a previous selection (see Using Jokes to Invent the Expert).

---

The course in Electrical Engineering at Tuskegee Institute was started in 1898, when an electric plant was installed in order to meet the need of the school in the lighting of the dormitories and other buildings, class rooms, and grounds.

There was great danger of fire when oil lamps were used, on account of the carelessness of students in handling the lamps, so that the Electrical Engineering Department was a necessity. After it was decided to have an electric plant, and the dynamo had arrived, a small number of students were permitted to work with the electrician in charge, in order to assist him and also learn Electrical Engineering. These students were first taught house wiring because none of the buildings had been wired. The simplest form of wiring—consisting of cleat and moulding wiring—was adopted, and the instructor employed these students to assist in running the plant at night.

The first dynamo used by the school was a 50-K. W. monocyclic alternator furnishing 43 amperes at 1,040 volts. This style of dynamo was chosen because the buildings to be lighted are scattered over a large area of ground, and this type of machine is adaptable to the use of motors.

A course of study covering three years was thought sufficient to teach the student the elements of Electrical Engineering, and to make first-class electricians out of them.

After the completion of this course of study a certificate is granted to the student, which states that he has completed the course in Electrical Engineering of the Tuskegee Normal and Industrial Institute. The school does not give any degree, but the certificate indicates that the bearer is a first-class electrician.

The electrical equipment of the Institute consists of a 50-K. W. monocyclic alternator, with exciter and switch-board; a 7-K. W. Brush arc-dynamo exciter; annunciator systems; watchman's time clock; electric storage battery; various primary batteries; galvanometers; voltmeters; ammeters; watt-meters; enclose arc lamps; and a telephone exchange.

---

Charles W. Pierce, "How Electricity Is Taught at Tuskegee," *The Colored American Magazine* 7 (November 1904): 666–673.

The Electrical Engineering Department has proved quite attractive to students from foreign countries as well as from the other states of the United States. In this department are enrolled students from Haiti, Jamaica, Porto Rico, Cuba, and from the states of Alabama, Georgia, Louisiana, Tennessee, Kansas, Texas, Indiana, Illnois, Montana, and the District of Columbia.

It will probably be interesting to know how the students first heard of Tuskegee; what they are able to do in their work; and what they intend doing after leaving the Institute. We quote from letters recently received, which cover these points more or less fully. The first letter is from a young man in Haiti, who writes as follows:

Electrical Engineering is one of the branches of scientific knowledge least in vogue in Haiti. There exist the telephone, the telegraph, and some small electric plants. While in school, I desired to come to America to study practical science and learn a trade. After finishing I heard that Tuskegee was a place which afforded excellent opportunities to the Negro for acquiring practical knowledge. I arrived at the Institute, September 1, 1903.

I can install electric bells and annunciators, trim arc lamps, wire buildings, and assist in running the dynamo. When I leave Tuskegee I shall attend some other school to pursue my studies still further, and shall then return to Port-au-Prince, Haiti, my home, for work at my trade.

This young man from the West Indian island could speak only French last September, but now he can speak English quite fluently.

Another young man writes:

I came to Tuskegee September 9, 1902. Hearing of the advantages which Tuskegee offered for one wishing to become an Electrical Engineer, I decided to go and avail myself of the opportunities. I can wire buildings, run direct and alternating current dynamos, trim arc lamps, and repair telephones. After leaving I intend to attend some other school, and then follow my trade either as a station manager or electrical contractor.

All students that enter the division are required to take studies in the Academic Department, exception being made in the case of post-graduates; and they must complete a certain requirement in academic studies before they can receive a certificate from the division. The Academic Department has two distinct schools—the day and night schools.

When a student enrolls in the day school, he spends one-half of his time in the Acemdemic Department, and the other half in the Industrial Department. When he enrolls in the night school he spends all day working at his trade, and his evenings in academic studies.

When a student enters the school and desires to learn a trade but is unable to pay his tuition, he may enter the night school. He receives small wages for the work that he does, and very soon establishes a credit in the School Treasury. When his credit is sufficient to pay his necessary tuition and expenses, he enters the day school, in which students advance much faster in their academic studies but not so fast in industrial work.

A great deal of stress is laid upon the quality of work the student does in the division, from the fact that the work is real work and must serve the purpose for which it is performed. This may be illustrated from the installation of an electrical plant. From the time the dynamo arrives on the ground, it is a part of the Electrical Division; and the installation, running, and all repairs must be done by the students working in the division.

When the present plant was installed, a pole line had to be constructed. The Instructor took his students to the woods and had them fell the trees. He then required them to prepare the poles for the line, make the cross-arms, and erect the electric line. The students were in this way taught line construction.

The transformer station was next installed, and the distributing lines laid out. The station was provided with all necessary appliances, such as primary cut-out boxes and fuses, lightning arresters, and secondary knife switches. The three-wire system from the transformers was used, on account of the distance of the buildings from the transformer station.

Special care and instruction are given to see that the students wire the buildings according to the latest rules of the National Board of Fire Underwriters.

New buildings are being constructed from time to time, and in these buildings the latest methods of wiring are adopted. The students have done cleat, moulding, brass-armored-conduit, flexible-metalic-conduit, and iron-armored-conduit wiring—a variety of forms of wiring being selected in order to familiarize the student with the different methods.

There are at present thirty-four buildings lighted by electricity with a total of 1,717 lights. The division supplies lights to several places off the school grounds, among these being the residence of the late Col. Charles Thompson, Congressman from this district, the Tuskegee railroad station, and a church. The residence of Colonel Thompson is situated two miles from the electric plant. There are 34,000 feet of primary line, 22,000 feet of secondary line, and 8,000 feet of street-light wiring used. Until recently the street lighting was done with 32-candle power incandescent lamps connected to the secondary lines wherever convenient; but, on account of the overload on the dynamo used for lighting the buildings, and also because the lights were lighted on the streets when not needed, these lamps were taken off the secondary line and connected to the Brush arc dynamo.

A course in Telephony was made possible on account of the school's need of quick communication from building to building and between the offices. A 25-subscriber exchange was therefore installed, and this branch of Electrical Engineering was added to the division, which gives the student experience in telephone work. The exchange was installed and managed by the school until recently, when the Southern Bell Company contracted with the school to furnish it long distance connection and with the company's switchboard.

The school had the first colored "hello girl" (to the writer's knowledge) in the world.

The laboratory methods are discussed and verified in the theory classes, and are finally made use of and pointed out in the work that the student does. While the student is at work he is encouraged to ask questions about the work; and he is questioned in turn, to make sure that he does understand all the operations that take place in the work. In this way the division is endeavoring to raise the workman above that of the plain, practical man by giving him a theoretical knowledge of his work.

The following will illustrate how this is being accomplished:

We are at present building a 100-light direct-current dynamo to furnish current at 110 volts. In the theory classes, the design of the machine, the drawings, and the calculations are made. The drawings then go to the patternmaker, who gets out the patterns; and these to the foundry, where the castings are made; the latter then go to the machine shop, where the necessary machine work is done; and finally, the parts are taken back to the Electrical Division, where the machine is assembled. During each operation the student follows the machine, and in the end sees a finished machine designed by himself and finally operated by himself.

The Electrical Division grows with the school. The present plant has become too small to light all the buildings; and at present another moncyclic alternator of 150-K. W. capacity is being installed. This new alternator is to work in parallel with the present one when needed, and will supply all the buildings with electric lights.

Although the alternating-current system is being used, there is a difficulty in teaching this system to the students, as many are in the lower academic classes. The students that enter the Electrical Division are not classified as they are in their academic work; and it often happens that a student in the lower academic classes takes theory class-work with the student of higher academic standing. The problem then arises as to the best methods of teaching these two groups of students in order that their advancement shall be equal.

The electric plant is run by the advanced students in the division, with the less experienced as helpers.

The duties assigned at the plant are those of chief electrician, first assistant electrician, second assistant electrician, and morning electrician. The hours of running are from sundown to 10:30 P. M., and from 4:30 A. M. to sunrise.

The plant has been operated very successfully, with no accident to anyone; neither has there been any time when a long stoppage was necessary. It must be remembered that the plant has been operated for six years, and that it is a single-unit plant. It requires very close inspection of the machinery and of the work of the students in order to keep the plant in running condition; and the students learn more from this system of inspection and criticism than would be possible by any other means.

The running expenses of the plant amount to about 4.5 cents per kilowatt-hour, although the price of coal is $4.00 per ton for a very poor grade of bituminous.

**Figure 6.2**
Students at work moving the 150-kilowatt alternator into the dynamo room. The instructor is at the right with a star on his coat.

Although there are 1,717 lights connected to the system, rarely ever do more than 850 lights burn at a time. This is made possible by the close schedule for use of the lights. There are three distinct consumers—the dormitories, the Academic Department, and the work-shops. No two of the light consumers use lights at the same time. The close supervision of the carrying out of the schedule falls to the Electrical Division, and the duties assigned to the young men of this division are to prevent a waste of electrical power.

The Academic Department is one of the largest consumers of electrical power, and when it is finally housed in its new building, it will use 600 lights. Flexible metalic conduit is now being installed in the building, and Frink reflectors will be employed in lighting it. There are three floors and a basement to this building, giving about 75 rooms for academic classes, laboratories, chapels, and a gymnasium. This is the largest building that the Electrical Division has attempted to wire.

The following antedotes will illustrate the popular conception of the Electric Engineering Department and the sentiment toward it:

It was during one of the Negro conferences which Principal Washington convokes every year at Tuskegee, that the following incidents happened:

We were at work in the dynamo room cleaning up the dynamo for an evening run, when one of the delegates was seen looking in at the door. I wished to make him feel welcome so I stepped to the door and asked him in. He said to me that he was afraid; and when I assured him that he would be perfectly safe, he went on to say that he had "no business fooling around the dynamite machine," because at his home three men had been "killed by the dynamite made from the dynamite machine."

Once, when we were installing a transformer in the grounds, a man wanted to know how the oil could get through the wire to the different buildings when the wire was solid.

Very often new students that enter the school come from rural districts where they never see any other light than pine knots and candles, and they are naturally quite amazed on seeing the electric lights.

A rule of the school requires all lights extinguished at 10 P. M. For several nights on one occasion a light had been noticed burning in one of the rooms; and the officer in charge of the building was ordered to investigate and see why the light was not put out on time. When he arrived in the building he found one of the new students almost out of breath and unable to answer any questions. The student finally recovered sufficient breath to answer the question of the officer, and explained to him the difficulty he experienced in blowing his lamp out, and said further that he had staid up half the night ever since his arrival blowing out his lamp. The real fact is that the dynamo stops at 10:30 P. M., and he thought that he had finally accomplished his task.

Finally, a word should be said as to what the graduates and students of the division do after leaving school. Young men of the division rarely stay to finish their course, on account of the demand for electricians such as the division turns out. Students who have not received certificates are working in Texas, Georgia, Indiana, Illinios, New York, Alabama, Kansas, and Missouri, working as electricians, linemen, dynamo-tenders, electrical repair men, and arc-light men, and in automobile stations. There is no doubt of a student receiving employment after completing the course, the only drawback being that the Institute cannot turn them out fast enough.

To sum up, we might say that at Tuskegee Institute students learn Electrical Engineering by doing Electrical Enginnecring work. The Engineering Division is therefore practicing the doctrine of the founder in teaching the dignity of labor and of learning to do things well.

## Racial Progress as Reported in *The Colored American Magazine* (1902)

African-American leaders attempted to keep track of the numbers of black workers in the various trades, and the way in which these figures went up or down. The annual Negro Conferences at Atlanta University at the turn of the century focused on this and other indicators of racial progress.

For the past six years Atlanta University has conducted through its annual Negro Conferences a series of studies into contain aspects of the Negro problems. The results of these conferences put into pamphlet form and distributed at a nominal price have been widely used and quoted. The first investigation in 1896 took up the "Mortality of Negroes in Cities." The following years the studies were:

1897—Social an Physical Condition of Negroes in Cities. 1898—Some Efforts of Negroes for Social Betterment. —1899—The Negro in Business. 1900—The College-bred Negro. —1901. —The Negro Common School.

Graduates of Atlanta, Fisk and Howard Universities, Hampton and Tuskegee Institutes and of many other schools have co-operated in this movement.

This year the Seventh Atlanta Negro Conference will meet May 27 at Atlanta University, and will take up the interesting subject of the Negro Artisan. There has been much discussion lately as to the Negro in mechanical industries, but few tangible facts. The census of 1890 gave 172,970 Negroes in the manufacturing and mechanical industries throughout the United States, but this includes many unskilled laborers and omits many artisans like miners and barbers. In detail there were the following skilled Negro laborers reported in 1890:

Negro Artisans in the United States, census of 1890: Carpenters, 22,318; barbers, 17,480; saw-mill operatives, 17,230; miners, 15,809; tobacco factory employees, 15,004; blacksmiths, 10,762; brick-makers, 10,521; masons, 9,647; engineers and firemen, 7,662; dressmakers, 7,479; iron and steel workers, 5,790; shoemakers, 5,065; mill and factory operatives, 5,050; painters, 4,396; plasterers, 4,006; quarrymen, 3,198; coopers, 2,648; butchers, 2,510; wood-workers, 1,375; tailors, 1,280; stone-cutters, 1,279; leather-curriers, 1,099.

The figures for 1900 are not yet available, but they will show a great increase in all kinds.

The investigation by the Atlanta Conference includes a personal canvass of some 2,000 Negro artisans, a study of general conditions in three hundred different cities and towns, a canvass of all the international trades unions and local assemblies,

Reprinted from: *The Colored American Magazine* 5 (June 1902): 133–134.

and a study of the opinions of employers, and tabulated returns from industrial schools.

Probably this will prove the most thorough investigation of the kind ever undertaken. Especially will light be thrown on the attitude of the trades unions. There are in the United States ninety-eight National Unions. In thirty-four of these there are Negro members; but in most cases very few. Only ten unions have any considerable number, viz., barbers, 800; brick-workers, 200; carpenters and joiners, 1,000; carriage builders, 500; coopers, 200; stationary firemen, 2,700; painters, 169.

The cigar makers, iron and steel workers, and miners also have considerable numbers. So that we have: Unions with no Negro members, 64; Unions with a few Negro members, 24; Unions with a considerable number of Negro members, 10.

Nearly all the unions with no Negro members refuse to receive Negroes; some by open discrimination, as in the case of the locomotive engineers, locomotive firemen, electrical workers, and boiler-makers, while others exclude them silently. In some cases, like the curtain operatives and jewelry workers, no Negro workmen have applied, so that question is unsettled. In nearly all cases any local union has a right to refuse an applicant, so that a single Negro workman would stand small chance of admission. On the other hand, the American Federation of Labor, with which most of these organizations are affiliated has taken strong ground for fair play toward Negroes and the union movement has greatly extended among them in the last ten years.

Among the speakers at the Seventh Atlanta Conference where this question will be thoroughly discussed will be Mr. Booker T. Washington of Tuskagee, President J. G. Merrill of Fisk University, Major R. R. Moton of Hampton Institute, Mr. William Benson of the Dixie Industrial Company, President Bumstead and Dr. W. E. B. DuBois of Atlanta, and a representative of the American Federation of Labor.

## The American Negro Artisan (1904)

Thomas J. Calloway

The statistics gathered at the Negro Conferences in Atlanta provided evidence for many that black workers could, and had, mastered a wide range of technological skills and that wherever allowed they were represented in the many trades that made up the working life of Americans.

---

Whether the artisans who built the pyramids of Egypt were Negroes or of Caucasian identity may not be well settled. Recent history is quite clear, however, that American Negroes, during slavery, developed marked mechanical skill, and in many cases, special inventive genius. Quite a large part of the mechanical work in the American Southern States before the Civil War was performed by slave artisans, these being valued much higher than ordinary farm hands and common laborers. This fact is so well known in the South that it has usually passed without comment; but Bruce, in his "Economic History of Virginia," makes the following reference to it:

> The county records of the seventeenth century reveal the presence of many Negro mechanics in the colony during that period, this being especially the case with carpenters and coopers. This was what might be expected. The slave was inferior in skill, but the ordinary mechanical needs of the plantation did not demand the highest aptitude. The fact that the African was a servant for life was an advantage covering many deficiencies; nevertheless, it is significant that the large slave-holders like Colonel Byrd and Colonel Fitzhugh should have gone to the inconvenience and expense of importing English handicraftsmen who were skilled in the very trades in which it is certain that several of the Negroes belonging to these planters had been specially trained. It shows the low estimate in which the planters held the knowledge of their slaves regarding the higher branches of mechanical work.

This historian might have added that the educational training that would have fitted the slaves for the higher grades of mechanical work would have unfitted them for their sphere of slavery. A somewhat different estimate of the Negro mechanic is held by ex-Governor Lowry, of Mississippi, who has said:

> Prior to the Civil War there were a large number of Negro mechanics in the Southern States; many of them were expert blacksmiths, wheelwrights, waggon-makers, brickmasons, carpenters, plasterers, painters, and shoemakers. They became masters of their respective trades by reason of sufficiently long service under the control and direction of expert white me-

---

Reprinted from: Thomas J. Calloway, "The American Negro Artisan," *The Colored American Magazine* 7 (May 1904): 319–332.

chanics. During the existence of slavery the contract for qualifying the Negro as a mechanic was made between the owner and the master workman.

The slavery system of training mechanics, which is sufficiently described in the two references, produced a type of Negro artisan which served the plantation needs fairly well, and to some extent supplied town carpenters and village blacksmiths. With the passing of slavery, however, that type of artisan is also passing, and the question now arises as to what extent post-bellum Negroes are finding their way into the trades. In the first place, it must be said that the slave descendants have not shown any special aptitude for the mechanic arts. This fact is all the more noteworthy because emancipation found the race in a practical monopoly in the South of all forms of labor, and it is to be regretted that some better form of leadership did not develop at that time so that the forty years past might have been more largely and beneficially utilized in holding the industrial foothold which the Negro has been so steadily losing. It may in the future appear that the political effort following enfranchisement was not in vain, but from the viewpoint of the present it looks very much as if the Negro had been repeating the experience of the Children of Israel who spent their forty years wandering in the wilderness.

Under conspicuous leadership, of which Booker T. Washington is the chief exponent, the colored people have had their attention directed to the urgent necessity of clinging to the industrial occupations. Then, too, the tremendous industrial growth of all sections of the United States has drawn into its service all forms of labor. While race prejudice has kept the Negro from some kinds of skilled labor, the greed for gold has forced him over the protest of prejudice into many of the trades.

There is an underlying trait of the African which enables the race to adjust itself to new conditions and environments, a trait that often leads him to become vicious when surrounded with vice just as easily as he becomes thrifty and progressive where the atmosphere in which he is thrown is that of thrift and progress. In fact, it is believed that no other race responds so quickly to external influences, be they climatic, moral, educational or industrial. The varied influences reaching out to the Negro are already producing results that cause the future to look brighter for the race.

In the seventh bulletin of the Atlanta University Conferences, edited by Dr. W. E. B. DuBois, the events following emancipation are summed up very admirably. The editor says:

After emancipation, in the midst of war and social upheaval, the first real economic question was the self-protection of freed workingmen. There were three classes of them: the agricultural laborers, chiefly in the country districts; the house servants in town and country; and the artisans who were rapidly migrating to town. The Freedmen's Bureau undertook the temporary guardianship of the first class; the second class easily passed from half-free service to half-service freedom; the third class, the artisans, however, met peculiar conditions. They had always been used to working under the guardianship of a master, and even though that guardianship in some cases was but nominal, yet it was of the greatest value for protection. This soon became clear, as the freed Negro artisan set up business for himself. If there were a creditor to be sued, he could no longer bring suit in the name of an influential white master; if there were a contract to be had there was no responsible white patron to answer for the good performance of the work.

Nevertheless, these differences were not strongly felt at first; the friendly patronage of the former master was often voluntarily given the freedman, and for some years following the war the Negro mechanic still held undisputed sway. Three occurrences, however, soon disturbed the situation: —the competition of white mechanics; the efforts of the Negro for self-protection; and the new industrial development of the South.

What the Negro mechanic needed, then, was social protection, —the protection of law and order, perfectly fair judicial processes, and that personal power which is in the hands of all modern laboring classes in civilized lands, viz., the right of suffrage. It has often been said that the freedman, throwing away his industrial opportunities after the war, gave his energies to politics and succeeded in alienating his friends and exasperating his enemies, and proving his inability to rule. It is, doubtless, true that the freedman laid too much stress upon the efficacy of political power in making a straight road to real freedom. And, undoubtedly too, a bad class of politicians, white and black, took advantage of this, and made the reconstruction Negro voter a hissing in the ears of the South.

Notwithstanding this, the Negro was fundamentally right. If the whole class of mechanics here, as in the Middle Ages, had been without the suffrage and half free, the Negro would have had an equal chance with the white mechanic, and could have afforded to wait. But he saw himself coming more and more into competition with men who had the right to vote; the prestige of race and blood; the advantage of intimate relations with those acquainted with the market and the demand.

The Negro saw clearly that his rise depended, to an important degree, upon his political power, and he, therefore, sought that power. In this seeking he failed primarily because of his own poor training, the uncompromising enmity and apprehensions of his white neighbors, and the selfishness and half-hearted measures of his emancipators. The result was that the black artisan entered the race heavily handicapped, —the member of a proscribed class, with restricted rights and privileges, without political and social power.

Any views of the Negro artisan as such would be incomplete did they not include a consideration of the influence of his educational training since emancipation. With the dawn of freedom came the spelling book and the opportunity to acquire the three R's. A tremendous demand at once ensued for a corps of teachers, and salaries unusually large for colored men and women up to that time soon drew the attention of the brightest-minded from all other pursuits into the profession of instructing the youth of their race. The demand was so great that schools of higher training than the public schools were established all over the South for the proper equipment of teachers. The entire thought was given at first to the simple work of instilling the elements of literary learning, or, in other words, the higher schools aimed to instruct the students in the very studies they were called upon to teach later on in the public schools.

As an additional impetus to the young people to educate themselves in this simple way there was an underlying conviction that all toil with the hands was a badge of slavery beyond which it was the duty of ambitious youth to aspire. It was an idea of slavery that physical labor was the duty of the bondman. Hence we find the emancipated parents, who were themselves fairly well skilled in trades, making heroic struggles and sacrifices to the end that their children might have book learning, with a dim conviction that a knowledge of books was a royal pathway to all the desires of life. So long as the abnormal demand for teachers continued, the conviction was not without an apparent proof of its truthfulness.

A thorough student of race development would have seen clearly that any permanent growth must have included the training for some special form of activity in life. In fact, a departure from the universal preparation for teaching had already shown itself in the way of schools of medicine, law, and theology, in which aspirants to become doctors, lawyers, and preachers were prepared.

That the educational trend following emancipation was the result of the intoxication of sudden manumission is made clear from the vain attempts by free colored people, as early as 1831, to establish trade schools where the free masses of their race might be equipped as artisans. Frederick Douglass, in 1853, led a formidable, but futile, effort to establish an industrial school. It was not, however, till the Hampton School, in Virginia, was established that industrial training for the Negro may be said to have had its beginning. General Armstrong opened the school at the close of hostilities between the North and the South, and struggled with it for many years before the marked success of the institution was finally reached.

The principles of the institute were so thoroughly in harmony with common sense that it eventually won the hearty confidence and the material aid of philanthropists. In addition to the large number of its own graduates it has created a sentiment that has led to the establishment of many similar schools, one of which, the Tuskegee Institute, in Alabama, has surpassed the parent school in point of numbers.

About one hundred such schools offer industrial courses. Only a modest proportion of these, however, may be said to be sufficiently equipped to do effective work in training graduates in any of the specialties set out. It would be difficult to select the most meritorious of these schools. A personal examination is pleasantly surprising in many instances in the case of schools of which little has been heard, and sadly disappointing in other cases where more was to have been expected. Without any attempt at an invidious comparison, it is believed that the large Hampton and Tuskegee schools are illustrating in their far-reaching spheres of influence what is being done in a lesser degree in many other places.

Industrial education, as the term has been generally employed in relation to Negro schools of the South, has meant a system of education serving not only to give the elements of academic learning, but likewise to train, in shop and farm, young people who may fill the demands of the industrial world for skilled labor, be that labor agricultural, domestic, or manufacturing. Exponents and friends of the system have never claimed that the young people which it graduates can possibly be of that high degree of skill which results only from long application to a specialty; but the claim has been made, and, it is believed, proven by the results, that in an elemental and primary way the better-equipped industrial schools are fitting a superior grade of artisans for the trades.

Each of the industrial schools gives instruction in all or a portion of the following trades: —agriculture, horticulture, carpentry, blacksmithing, wheel-wrighting, printing, painting, foundry and machine work, shoemaking, brick-laying, plastering, brick-making, saw-milling, tinning, harness-making, tailoring, plain sewing, dressmaking, millinery, cooking, laundering, nurse training, housekeeping, mechanical and architectural drawing, and perhaps others. The actual number of graduates from these schools who have entered into the trades is comparatively small, due to the greater demand for teachers and for persons to take other occupations for which a knowledge of the trades is a material aid. It has been estimated that the industrial schools have sent out about a thousand actual artisans, —not enough to man a single large manufacturing plant. As representing an awakening process, these thousand graduates must be regarded not so much for the actual proportion they represent to the millions of their race, but their true estimate is that of the compressed yeast cake which is leavening the infinitely greater race problems.

Having considered as fully as the plan here permits, the formative influences of slavery, politics and education upon the Negro artisan, it is proper now to consider the actual status of the Negro in the skilled industries. According to the United States Census of 1890, the distribution of bread-winning American Negroes was in the following occupations:

| | | Percent |
|---|---|---|
| Agriculture, fishing and mining | 1,757,403 or | 57 |
| Domestic and personal service | 963,080 | 31 |
| Manufacturing and mechanical industries | 172,970 | 6 |
| Trade and transportation | 145,717 | 5 |
| Professional service | 33,994 | 1 |

Taking the six percent of wage-earners found in manufacturing and mechanical industries, we find them, according to the same census, distributed as follows, omitting those occupations containing less than one thousand:

| | |
|---|---|
| Carpenters | 22,318 |
| Barbers | 17,480 |
| Saw-mill hands | 17,230 |
| Miners | 15,809 |
| Tobacco factory employees | 15,004 |
| Blacksmiths | 10,762 |
| Brick-makers | 10,521 |
| Masons | 9,647 |
| Engineers and firemen | 7,662 |
| Dressmakers | 7,479 |

| | |
|---|---|
| Iron and steel workers | 5,790 |
| Shoemakers | 5,065 |
| Mill and factory operatives | 5,050 |
| Painters | 4,396 |
| Plasterers | 4,006 |
| Quarrymen | 3,198 |
| Coopers | 2,648 |
| Butchers | 2,510 |
| Wood-workers | 1,375 |
| Tailors | 1,280 |
| Stone-cutters | 1,279 |
| Leather-curriers | 1,099 |

The temptation is strong to pursue a further statistical view of the subject, but it is deemed more in harmony with the present discussion to make use of a special investigation carried out in connection with the Negro exhibit at the Paris Exposition of 1900. That exhibit was planned with a view to furnishing to Europeans a concise and carefully studied representation of life among American Negroes.

It was felt by the writer of this article, who was the agent in charge of the preparation and superintendence of that exhibit, that no phase of the Negro's life in America would so claim the attention of Europeans, as was afterwards proven, as his industrial status in the United States. Among other features, therefore, the effort was seriously made to present a narrative exhibit of the Negro artisan. To this end an expert made a tour of inspection. The cities and towns visited were Atlanta and Augusta, in Georgia; Anniston, Hobson City, Pratt City and Birmingham, in Alabama; Chattanooga and Knoxville in Tennessee; Asheville, Charlotte, Concord, Greensboro and Durham in North Carolina; Columbia and Charleston in South Carolina; and Richmond, in Virginia. Fifty-seven manufacturing plants were visited, employing 7,244 colored males and 1,620 colored females.

At Atlanta colored men were found doing skilled labor in the manufacture of fertilizers and chemicals, the making of bricks, as coopers, waggon and buggy makers, shoemakers, and as size mixers and cotton classers in the cotton mills. They were generally employed in all branches of the building trades. At Anniston and Birmingham, Ala., and vicinity many thousands were employed in the mines and in the production of coke, pig-iron and steel. Colored men were observed to be holding the responsible positions of foremen, cupola men, furnace keepers, and iron graders. At Knoxville, Chattanooga, Richmond, and indeed at all the places visited, Negroes were employed as blacksmiths, black-smiths' helpers, machinists' helpers, moulders, puddlers, rollers, roughers, catchers, furnace men, and boiler-makers. At Knoxville, Tenn., hundreds of Negroes were employed in cutting, shaping and polishing "Tennessee marble," —a work requiring close attention and considerable skill. In the various places visited in

**Figure 6.3**
In the machine shop of the Newport News Shipbuilding and Dry Dock Co., Newport News, Virginia.

North Carolina and Virginia colored men were engaged in tanneries and tobacco factories; and throughout the South they are almost entirely used in the cotton-seed oil mills.

Interesting and valuable statistics were gathered during the tour of inspection, but it is believed that the following signed letters from some of the factories visited will prove of special interest:

As general manager of the Evans Marble Company I have been employing colored men for the past twelve years, and have found them steady, reliable, and always to be depended upon. My experience with them as laborers has been that, under proper treatment, they give the best satisfaction, are willing, contented, and rarely strike or join labor unions, and that with proper instruction they show great capability in marble finishing and the different processes thereof, such as bed rubbing, polishing, both manual and by machine, and in manipulating machinery for cutting and preparing marble to be polished, all of which requires close attention and considerable skill and experience. —J. E. Willard, Supt., Evans Marble Co., Knoxville, Tenn.

In answer to your letter of the 12th inst. we take pleasure in saying that for nearly fifty years we have employed Negroes at these works in many occupations requiring skill and intelligence, with results eminently satisfactory. Two of the best blacksmiths we ever had were Negroes. Another Negro was for many years in charge of the air furnace in which we melted pig iron for heavy guns. In the rolling mills we have had, and still have, excellent rollers, charged with responsibility for trains of rolls and their product, heaters' helpers, roughers, catchers, and straighteners of this race. Negroes are also employed in punching fish bars and regulating our water wheels. In the foundry all our cupolas are manned by Negroes. We may add that our relations with our Negro workmen have always been pleasant. —Archer Anderson, President, The Tredegar Iron Co., Richmond, Va.

Our company have worked Negroes in about all our departments since the year 1881. We found considerable trouble in teaching them to do professional work, and it cost us a very large amount of money to instruct them. We find, however, that since they have been properly taught they give equally as good service as the white labor.

We have had them employed as puddlers, heaters, rollers, roughers, and, in fact, have tried them in all departments of our works. We manufacture bar iron in its various forms, and, as stated before, we find that a colored man gives equally good service and is really more adapted to the climate than the white labor of the North. —T. I. Stephenson, V. P. and G. M., Knoxville Iron Co., Tennessee.

The Richmond stemmery of the American Tobacco Company employs 1,000 Negroes for stemming tobacco, whose wages are $4.50 per week. For the class of work for which we employ them there is no other help in the world as good. —T. J. Walker, Richmond, Va.

We employ almost exclusively colored laborers, who take up the work from the stage that requires very little skill up to the finishing point, where it does require some degree of skill and experience. Our experience with this labor has been very satisfactory indeed. —R. S. Bosher, Treas., T. C. Williams Tobacco Co., Richmond, Va.

Noteworthy examples are not wanting of individuals and enterprises that fully illustrate the upward strivings of the Negro in mechanics. The Coleman Manufacturing Company is a capitalized corporation composed of colored stockholders, and is managed by a board of directors of the same race, of whom Warren C. Coleman, an ex-slave, is the president and general manager, as well as the largest stockholder. The company has built, equipped and is conducting a cotton mill at Concord, N. C. As race pride and race improvement were the moving spirits in the enterprise, a special effort was made to employ Negro mechanics in all features of the construction and operation. The architect, the brick-makers and masons, the carpenters and the factory employees now engaged in turning out a fair grade of cotton goods, sheetings, etc., are all colored.

"Reformers' Hall," a business building, containing a theatre, office rooms, an armory, etc., has just been completed in the heart of Washington, D. C., at a cost of a hundred thousand dollars, by an architect, J. A. Lankford, contractors and mechanics, all of the Negro race. The building is finely appointed and thoroughly modern.

R. R. Taylor, instructor in mechanical and architectural drawing at the Tuskegee Institute, in Alabama, and the architect of the numerous buildings of that institution, is probably the best equipped in training and experience of any Negro architect in the United States. He graduated from the Massachusetts Institute of Technology several years ago, and has designed a score or more of the buildings for the Tuskegee school, ranging in value from $50,000 downward. A chapel seating 2,000 persons and a Carnegie library are his best designs.

The United States Patent Office was able to identify in 1900, through correspondence with patent attorneys, 357 patents issued to Negroes. Probably as many more existed which could not be identified, inasmuch as the color or race of the patentee does not enter into the application. This appears the more probable for the reason that 126 of the patents, or more than a third of the total number discovered, were issued in the five years immediately preceding the year 1900. The inventions are of a great variety of subjects, and while the number is not large, they are sufficient to show that the Negro artisan, in common with other American mechanics, is contributing a portion to the remarkable growth of the United States.

Allowing for the recent emancipation of his body and the dense thraldom of ignorance in which chattel slavery left him, from which he is even now but half free, it is believed that the thoughtful "captains of industry" will see in the Negro artisan a mine of valuable and reliable labor that is well worth their efforts to exploit.

## Results of Some Hard Experiences: A Plain Talk to Young Men (1902)

William H. Dorkins

William H. Dorkins, apparently a sailor in the engine room of the U.S.S. *Amphitrite*, encouraged young black men to become financially successful as a tool for breaking down racial barriers. Here he gives the example of a young man (himself?) who joined the U.S. Navy to gain mechanical experience and then rose to a position of responsibility.

I find in my travels a great many of our young men have trades but are not putting them to use. I know good mechanics who are farming, waiting in hotels, and occupied in various ways that are foreign to their talents.

It is a good thing for us to have industrial schools and other institutions, that tend toward the elevation of the race, but do we fully realize the fact that after a boy has finished his education, it is only his starting point in life? That is the time he is most in need of advice, and should be in close contact with someone who has a thorough knowledge of the world. Someone who is able to point out the best place for him to make his start.

Take a green boy from the country. He may finish his school career as one of the finest tradesmen, yet, as a rule, his knowledge of the outside world is not of that practical character that would enable him to settle in the spot where his trade would be of the most benefit.

The general idea prevails among us that a boy, when his education is completed, must go home, regardless of the trade he has learned. But a machinist or mechanical draughtsman is of no benefit in a log cabin twenty or more miles away from the city or place where he could use his talents.

We learn our trades that we may make the forces of nature work for us, and to be successful we must connect ourselves with those forces. In other words, a mechanic wants to get as far away from "mule power" as possible.

We need money, our people need it, and the more we get of it the better it will be for us and the race. To be more successful in acquiring it, let us be found where it is most plentiful, especially if it is in connection with our particular branch of business.

Some of us become disheartened because we can't get a first-class position immediately after leaving school. That's a wrong idea. If you are a machinist, get a job in a machine shop, regardless of position or pay. Start in sweeping the floor if you can't get anything better (you know we don't take on sight), and I will guarantee if you have the proper mettle you will win.

Reprinted from: William H. Dorkins, "Results of Some Hard Experiences: A Plain Talk to Young Men," *The Colored American Magazine* 5 (August 1902): 271–272.

The object is to connect ourselves with that business for which time and money has been spent to fit us. Unless we do, the purpose toward which our race leaders are working will become a farce. Those men are working for results.

The results are what you and I make them.

I know a young man who served an enlistment in the Engineer's Department of the U. S. Navy. He entered as a coal passer, and by hard work and study he advanced as high as it was possible for a man without a trade to do. Seeing better opportunities, he went to Hampton (after the expiration of his enlistment), to learn the machinist's trade. After staying two years he left to enter the service as a machinist, but they claim he failed in his examination—on account of color (there being no colored machinists in the Navy). The idea was to discourage him from enlisting at all, but being determined he went in as a fireman, and at the expiration of one year he was promoted to second class machinist, six months later to first class with pay at $55.00 per month and expenses, and a first class chance for further promotion. This brings me to say once more that there is nothing like connecting ourselves with the business that interests us the most. A good blacksmith wouldn't make a successful barber, for the simple reason that he wouldn't be interested enough to make it pay.

If a boy aspired to become a railroad president he should be satisfied to clean cuspadores in the ticket office, if nothing better presented itself in that line. He would then be on the bottom round of the ladder whose top he expected to reach, and he would always be on the scene to study the duties of the man on the next round above. First let us find our ladder. Second, lie around it as "Grant lay around Richmond," until we get on it. Third, stay on it. Fourth, strive for the top round, and if we never reach it our efforts to do so will be crowned with success. If we don't find our own ladder we will be "hanging around" one belonging to some one else, all our lives, and be classed with that fraternity whose banner reads "Misery Loves Company."

The U. S. Navy furnishes the best opportunities for a man to educate himself along mechanical lines of any institution in the country. If a man is physically sound he can enter the Engineer's Department without any knowledge of the business at all. He starts as a coal passer at $22.00 per month, and can, with proper push, advance himself to chief machinist at $70.00 per month and expenses. Should he be so unfortunate as to serve a full enlistment as a coal passer and is a keen observer, he will be in possession of knowledge that will enable him to get a position in civil life as a fireman or engineer. This is brought about by his being surrounded by pumps of all descriptions, engines of various designs, besides dynamos and motors. The running and repairing of these is constantly going on with the assistance of all in the department regardless of position.

I explain these things not only to impress the reader with the fact that we must branch out, but that we might have one more place to add to our list of opportunities.

It is a fine thing for us to know how to save a dollar; but you tell me where and how I can make it, and I assure you that I will be an interested listener when you come to tell me how to save it.

**Figure 6.4**
William H. Dorkins at the throttle of the Amphrite's powerful engine.

Making money and opposition are two opposing forces. My experience has taught me that where the most money is to be made there is where the greatest opposition is found. Not only is this true of the black man, but any man regardless of color is opposed when he tries to rise. It is human nature to try to crush the ambitious. To be successful means the calling forth and putting into practice those qualities that go toward the making of men, being patient and putting behind us those things that tend toward strife.

When two men fight aboard ship, both are punished. One or the other may be right, but we mustn't lose sight of the fact that "it takes two to make a quarrel." By learning to talk more and cultivating those qualities within us that go toward the making of friends, we let down a barrier that has hindered us for quite a while.

We can't afford to lose time with such things as the so-called color question, especially when there is a dollar in sight. My experience has taught me that the only time my neighbors bothered me about my color was when I became "broke." My color was the most interesting thing I had, but when I got a dollar they dropped the color question, and joined the "Ways and Means Committee" to try and figure out how they could get that dollar.

The dollar is what they all want, and the man who has it, be he "any old color," will go through the world with the feeling "They all take their hats off to me."

## Manufacturing Household Articles (1904)

Samuel R. Scottron

This firsthand account describes how a self-styled "merchant and trader" became the inventor and manufacturer of a range of household articles.

Certainly the most difficult of any of my undertakings in the way of a public address is now before me. Difficulty overtakes me in the beginning, and I presume that before long difficulty will be still mine when I should stop and, in common parlance, give you a rest. Nevertheless, regarding the honor which our worthy president has done me in inviting me to address the League, I shall do my utmost to say something interesting and instructive.

I regard this as a very different assemblage from that which meets to hear a political address, or one of a religious nature, both of which must be delivered with force and eloquence, which in the serious matters of manufacturing any particular article we are all prepared to eliminate the oratorical feature and subordinate it to the just and true presentation of ascertained facts.

Thirty-six years of experience in any particular line ought certainly present some features worthy of talking about, even if it be only to say I have manufactured such and such an article for that length of time, for it will be taken for granted that one either possesses an unusual degree of persistency or that manufacturing has been reasonably profitable.

I did not begin business life as a manufacturer, but as a merchant and trader, being associated with others as sutler of a regiment toward the end of the civil war, at and about the same time as our honored friend who has resided in this city of Indianapolis for so many years, William H. Furniss. We were each of us regimental sutlers in the Department of the South. We quit it probably at the same time, when mustered out in 1865. He to enter upon a professional line in which he became an instructor in mathematics, while I finally and at once entered upon my course as a manufacturer.

There was just one thing that I learned beyond all doubt while engaged as a merchant trader and sutler, and that was that in whatever line, professional or business, a man enters upon, he should have such education as befits that line, if he wishes to succeed. Consequently when inventive genius gave me possession of my first patent upon adjustable mirrors, and wishing to manufacture the same, I set about acquiring the education necessary to a mechanic, entered the shop of a pattern maker, and studied at night for quite a period, under a master mechanic, mechanical drawing.

Reprinted from: Samuel R. Scottron, "Manufacturing Household Articles," *The Colored American Magazine* 7 (October 1904): 621–624.

This I continued fully seven years at night school until the thing had such a hold upon me that I couldn't give it up.

There is possibly no shop where one can serve and get a broader knowledge of applied mechanics than a well patronized pattern-making shop, bringing one as it does into a consideration of the various elements, substances, etc., used in manufacture; their nature and possibilities. There you will soon learn the difference between cast and rolled metals. In both brass and iron, not to speak of other more valuable, or, as I should say, more costly metals. The pattern-maker must absolutely know the qualities of both, and which is necessary when certain desired results are to be obtained. He must be equally well informed as to the nature of and possibilities in working glass, for that too may be either moulded or blown, either pressed into shape, or blown with or without a seam, or rolled into peats, or drawn into tubes, polished and bent into shape. He must know something of the nature of woods, of their varieties, structural and workable differences. The difference to be found in every kind of hardwood, such as mahogany, lignum-vitae, ebony, oak, birch or walnut, and the softer woods, such as pine, white-wood, spruce, red-wood, bass or cedar, rendering either of greatest advantages for a certain work, of their peculiar qualities necessary to take into consideration in modeling a certain machine. The sciences generally, and in particular, chemistry, civil engineering and mathematics, must be somewhat familiar to the pattern maker. All of this I became well aware of as time went on, and for seven years never missed a possible lesson from an instructor in one or more of these studies. It grounded me and gave me confidence in myself, and an actual knowledge of possibilities, which prevented many costly ventures and foolish mistakes, such as the patenting of things absolutely useless. Spending ones life on an ignus fatnus, pursuing the ever retreating and never overtakable, or the useless if overtaken things, that spoil the lives of so many, both men and women.

Well, my first patent, an adjustable mirror, —mirrors so arranged opposite each other as to give the view of every side at once, —was new and useful, and so simple too, that it was impossible to accomplish the same result in a simpler or cheaper manner. A thing very necessary in patent articles. A patent which can be simplified by another is worth nothing. A knowledge of mechanics will show you how to use the simplest methods in obtaining certain desired results, mechanically. How not to use three motions where two will do the work. Contact with the market will show you whether what you wish to accomplish will be worth anything in the market even if you do succeed in making the thing. Years are spent by some in trying to do something by way of a patented article, that even if made, wouldn't be wanted by a sufficient number of persons to make it desirable.

My mirrors had the good fortune to be quite a salable article, and I obtained afterwards, other patents on mirrors which were serviceable. All these led me finally into the general business of manufacturing looking glasses, pier and mantel mirrors, mouldings, etc. Wood working had ever a fascination for me, a moulding or planing machine buzzing away, moulding the face of a board, or a saw buzzing away, always had a sweeter sound for my ear than the finest organ. The organ I could grow tired of, but the thirty years of buzzing saws are still the most delightful music to my ear. I never see one of the building or furniture wood trees in the forest but what the most wonderful possibilities in it as a piece of furniture pass through my mind. If oak, I think of the beautiful table, sideboard, sofa, desk, room-casing or moulding it will make. The same if it be walnut or mahogany. If it be pine I think of the fine unknotted part near the ground, from which the finest shelving will come, or above, where floor boards will come from, or still higher up, common boxing lumber, or piers for small turnings, or, lastly, firewood.

The looking glass business made necessary a knowledge of preparing mouldings by the overlaying of white coating, putty work gilding, bronzing, gold gilding, silver gilding, metal gilding, burnishing. Staining, varnishing and polishing of woods, and in particular, the silvering of peats, both by the old method of tinfoil and quicksilver, and the newer method, that of precipitating a solution of the nitrate of silver.

Very soon, however, a new patent having great possibilites was granted me, for an extension cornice. I received several patents for these and abandoned the mirror, putting these out on a royalty, and entering upon the manufacture of extension cornices, which coined thousands while it lasted, an excellent thing in every way; but it came to grief through one of those causes that will sometime lay out, stiff dead, the best thing in the market, viz.: the capriciousness of fashion. Curtain poles came into fashion and killed the cornice business entirely, in less than six months of activity in opposition. The American woman, whose desire for change makes many fortunes, also helps many to lose them. Extension cornices were helping me to thousands of dollars, until curtain poles came as suddenly into fashion as a meteor strikes the earth. Forced to abandon cornices I lived upon my royalties, and entered upon the curtain pole business as manager and salesman in a new firm, in which line I continued nearly fifteen years, covering in each year, the territory both in Canada and the States between the two great oceans, and as far south as the Ohio river.

The curtain pole manufacture might seem extremely simple to the superficial observer, but it is very intricate, and only a certain knowledge of mechanics could warrant one in entering upon it, requiring as it does a full knowledge of the working properties of wood and brass. These poles were first made in moulding machines and hand-turning lathes, both of which were driven out completely by the central cutter, automatic feeding machine; these being still further improved by the automatic sander for sandpapering the stick as it emerged from the cutter, and these further improved by passing the pole, still automatically, into another drum-like machine, which colored and polished it, and, in the course of twelve years, reducing the cost of an inch and a half wood pole from six cents per foot to one cent per foot. Wood rings which we started making by hand at four cents to six cents each, one man turning out one hundred furnished rings per day, were made by machine finally, at the rate of five thousand furnished rings per day by a single attendant.

But, as wonderful as is the story of the rapid changes in the wood-making part of curtain poles, still more wonderful is the story of the brass pole and brass pole trimmings. The brass pole we made of solid metal in the beginning, metal of heavy guage, drawn and brazed at the centre joining line, of English manufacture, was principally used, costing a dollar per foot. The rings of brass were also English, largely from Birmingham. Made of drawn brass, spiralled about a round hard stick cut through and then eyed and soldered, costing a dollar and a half to three dollars per dozen, six to ten dollars for a pair of complete brass poles, even twenty dollars per pair. American inventive genius under the fostering care of a favoring tariff reduced the cost of these steadily, month by month, not year after year, but monthly reduced the price, so that it was always unsafe to carry a large stock, until finally a complete brass pole and trimming was sold for less than fifty cents, that a few years earlier cost six dollars. To follow such a business through all its changes for years I think you will appreciate the difficulties to be overcome.

A chance discovery, however, twelve years ago, set me off in still another direction, that of manufacturing what is known to the trade as "procelain onyx," which when sure it was a marketable article I taught my daughters to make, and at which we have now continued twelve years with greater or less profit, but, as a rule, forging ahead somewhat. These procelain onyx cylinders have been used by lamp and candlestick manufacturers to an extent that I doubt not that each one of you have seen the article, even if you have not realized the fact that the article was an imitation of the real stone. There is probably no part of the world where these articles have not gone along with other articles of American manufacture.

The combination of this vitreous or glassy substance with wood, to make a pedestal, is the very latest application of the procelain onyx, and involves more difficulties than would seem probable to the inexperienced. To put together in permanent form articles or substances so different in their nature as wood and glass, or as wood and earthenware presents many difficulties which could not be overcome without such knowledge as comes from contact with varying mechanical appliances.

These things seem to be growing in public favor, for while we made hundreds last year we are making thousands this year to meet the demand.

We shall not stop at pedestals and tables, but in a short time, hope to have the procelain onyx tubes used inside architectural decorations, such as are made for church ornamentation, bar room and barber shop mirrors, mantle mirrors, pier mirror fronts and many ways too numerous to mention.

## The Career of Mechanic John G. Howard (1902)

Craft training and white prejudice shaped the career of the mechanic, John G. Howard.

John G. Howard, was born at Raleigh, N. C., December 25, 1883. His father died when he was nine months old, and his early childhood was spent in Port Royal, Ala. At the age of ten years he returned to Raleigh, N. C., and spent four years at the Washington Graded School. From there he entered the Agricultural and Mechanical College at Greensboro, N. C., and spent three years in the mechanical department. From Greensboro, Mr. Howard went to Philadelphia, and served an apprenticeship in plumbing and gas fitting, under Jos. Smith, a Contractor and Builder. To thoroughly learn the scientific part of the trade, Mr. Howard entered the New York Trade School, on December 9, 1901, and graduated with the class of 1902, in practical and scientific sanitary plumbing, gas fitting and lead burning. In a class of one hundred and thirty-seven students from all parts of the United States and Canada, he was the only colored student.

Mr. Howard then passed a satisfactory examination before the Board of Health of New York City. From New York he went to Johnstown, Pa., to work for the Johnstown Supply House as a sanitary plumber, but the white mechanics would not work with him, and on the second day of his stay there, twenty-five plumbers pulled off their overalls, dropped their tools and posively refused to work with a negro. This action caused Mr. Howard to leave Johnstown and he went to Camden, N. J., and engaged to work for the Camden Gas, Electric & Traction Co., at gas and pipe fitting. Mr. Howard did not remain long here, but soon started in business for himself, in the city of Philadelphia, where he is at the present time doing a good business.

Reprinted from: *The Colored American Magazine* 5 (August 1902): 298–299.

**Figure 6.5**
John G. Howard at work in Philadelphia, Pennsylvania, setting up two circulation boilers.

# 7 Inventors

*The Colored Inventor: A Record of Fifty Years* **(1913)**
Henry E. Baker

**Clara Frye, A Woman Inventor (1907)**

**U.S. Patent to Garret A. Morgan for a "Breathing Device" (1914)**

## *The Colored Inventor: A Record of Fifty Years* (1913)

Henry E. Baker

In a nation that prided itself on technical innovation as well as ingenuity, the role of the inventor continued to be privileged. In an age when Thomas A. Edison raised that profession to a new height of prominence and surrounded it with an aura of romantic mystery, it is little wonder that the identification and listing of black inventors was a project that attracted wide and urgent attention.

---

The year 1913 marks the close of the first fifty years since Abraham Lincoln issued that famous edict known as the emancipation proclamation, by which physical freedom was vouchsafed to the slaves and the descendants of slaves in this country. And it would seem entirely fit and proper that those who were either directly or indirectly benefited by that proclamation should pause long enough at this period in their national life to review the past, recount the progress made, and see, if possible, what of the future is disclosed in the past.

That the colored people in the United States have made substantial progress in the general spread of intelligence among them, and in elevating the tone of their moral life; in the acquisition of property; in the development and support of business enterprises, and in the professional activities, is a matter of quite common assent by those who have been at all observant on the subject. This fact is amply shown to be true by the many universities, colleges and schools organized, supported and manned by the race, by their attractive homes and cultured home life, found now in all parts of our country; by the increasing numbers of those of the race who are successfully engaging in professional life, and by the gradual advance the race is making toward business efficiency in many varied lines of business activity.

It is not so apparent, however, to the general public that along the line of inventions also the colored race has made surprising and substantial progress; and that it has followed, even if "afar off," the footsteps of the more favored race. And it is highly important, therefore, that we should make note of what the race has achieved along this line to the end that proper credit may be accorded it as having made some contribution to our national progress.

Standing foremost in the list of things that have actually done most to promote our national progress in all material ways is the item of inventions....

All these great achievements have come to us from the hand of the inventor. He it is who has enabled us to inhabit the air above us, to tunnel the earth beneath,

---

Excerpted from: Henry E. Baker, *The Colored Inventor: A Record of Fifty Years* (New York: The Crisis Publishing Co., 1913), 1–12.

explore the mysteries of the sea, and in a thousand ways, unknown to our forefathers, multiply human comforts and minimize human misery. Indeed it is difficult to recall a single feature of our national progress along material lines that has not been vitalized by the touch of the inventor's genius.

Into this vast yet specific field of scientific industry the colored man has, contrary to the belief of many, made his entry, and has brought to his work in it that same degree of patient inquisitiveness, plodding industry and painstaking experiment that has so richly rewarded others in the same line of endeavor, namely, the endeavor both to create new things and to effect such new combinations of old things as will adapt them to new uses. We know that the colored man has accomplished something—indeed, a very great deal—in the field of invention, but it would be of the first importance to us now to know exactly what he has done, and the commercial value of his productions. Unfortunately for us, however, this can never be known in all its completeness.

A very recent experiment in the matter of collecting information on this subject has disclosed some remarkably striking facts, not the least interesting of which is the very widespread belief among those who ought to know better that the colored man has done absolutely nothing of value in the line of invention. This is but a reflex of the opinions variously expressed by others at different times on the subject of the capacity of the colored man for mental work of a high order. Thomas Jefferson's remark that no colored man could probably be found who was capable of taking in and comprehending Euclid, and that none had made any contribution to the civilization of the world through his art, would perhaps appear somewhat excusable when viewed in the light of the prevailing conditions in his day, and on which, of course, his judgment was based; but even at that time Jefferson knew something of the superior quality of Benjamin Banneker's mental equipment, for it is on record that they exchanged letters on that subject....

Less than ten years ago, in a hotly contested campaign in the State of Maryland, a popular candidate for Congress remarked, in one of his speeches, that the colored race should be denied the right to vote because "none of them had ever evinced sufficient capacity to justify such a privilege," and that "no one of the race had ever yet reached the dignity of an inventor." Yet, at that very moment, there was in the Library of Congress in Washington a book of nearly 500 pages containing a list of nearly 400 patents representing the inventions of colored people.

Only a few years later a leading newspaper in the city of Richmond, Va., made the bold statement that of the many thousands of patents annually granted by our government to the inventors of our country, "not a single patent had ever been granted to a colored man." Of course this statement was untrue, but what of that? It told its tale, and made its impression—far and wide; and it is incumbent upon our race now to out-run that story, to correct that impression, and to let the world know the truth.

In a recent correspondence that has reached nearly two-thirds of the more than 12,000 registered patent attorneys in this country, who are licensed to prosecute applications for patents before the Patent Office at Washington, it is astonishing to have nearly 2,500 of them reply that they never heard of a colored inventor, and not a few of them add that they never expect to hear of one. One practising attorney, writing from a small town in Tennessee, said that he not only has never heard of a colored man inventing anything, but that he and the other lawyers to whom he passed the inquiry in that locality were "inclined to regard the whole subject as a joke." And this, remember, comes from practising lawyers, presumably men of affairs, and of judgment, and who keep somewhat ahead of the average citizen in their close observation of the trend of things.

Now there ought not to be anything strange or unbelievable in the fact that in any given group of more than 10,000,000 human beings, of whatever race, living in our age, in our country, and developing under our laws, one can find multiplied examples of every mental bent, of every stage of mental development, and of every evidence of mental perception that could be found in any other similar group of human beings of any other race; and yet, so set has become the traditional attitude of one class in our country toward the other class that the one class continually holds up before its eyes an imaginary boundary line in all things mental, beyond which it seems unwilling to admit that it is possible for the other class to go.

Under this condition of the general class thought in our country it has become the fixed conviction that no colored man has any well-defined power of initiative, that the colored man has no originality of thought, that in his mental operations he is everlastingly content to pursue the beaten paths of imitation, that therefore he has made no contribution to the inventive genius of our country, and so has gained no place for himself in the ranks of those who have made this nation the foremost nation of the world in the number and character of its inventions.

That this conclusion with reference to the colored man's inventive faculty is wholly untrue I will endeavor now to show.

In the world of invention the colored man has pursued the same line of activity that other men have followed; he has been spurred by the same necessity that has confronted other men, namely, the need for some device by which to minimize the exactions of his daily toil, to save his time, conserve his strength and multiply the results of his labor. Like other men, the colored man sought first to invent the thing that was related to his earlier occupations, and as his industrial pursuits became more varied his inventive genius widened correspondingly. Thus we find that the first recorded instances of patents having been granted to a colored man—and the only ones specifically so designated—are the two patents on corn harvesters which were granted in 1834 and 1836 to one Henry Blair, of Maryland, presumably a "free person of color," as the law was so construed at that time as to bar the issuance of a patent to a slave.

With the exception of these two instances the public records of the Patent Office give absolutely no hint as to whether any one of the more than 1,000,000 patents granted by this government to meritorious inventors from all parts of the world has been granted to a colored inventor. . . .

There have been two systematic efforts made by the Patent Office itself to get this information, one of them being in operation at the present time. The effort is made through a circular letter addressed to the thousands of patent attorneys throughout the country, who come in contact often with inventors as their clients, to popular and influential newspapers, to conspicuous citizens of both races, and to the owners of large manufacturing industries where skilled mechanics of both races are employed, all of whom are asked to report what they happen to know on the subject under inquiry.

The answers to this inquiry cover a wide range of guesswork, many mere rumors and a large number of definite facts. These are all put through the test of comparison with the official records of the Patent Office, and this sifting process has evolved such facts as form the basis of the showing presented here.

There is just one other source of information which, though its yield of facts is small, yet makes up in reliability what it lacks in numerousness; and that is where the inventor himself comes to the Patent Office to look after his invention. This does not often happen, but it rarely leaves anything to the imagination when it does happen.

Sometimes it has been difficult to get this information by correspondence even from colored inventors themselves. Many of them refuse to acknowledge that their inventions are in any way identified with the colored race, on the ground, presumably, that the publication of that fact might adversely affect the commercial value of their invention; and in view of the prevailing sentiment in many sections of our country, it cannot be denied that much reason lies at the bottom of such conclusion.

Notwithstanding the difficulties above mentioned as standing in the way of getting at the whole truth, something over 1,200 instances have been gathered as representing patents granted to colored inventors, but so far only about 800 of these have been verified as definitely belonging to that class.

These 800 patents tell a wonderful story of the progress of the race in the mastery of the science of mechanics. They cover inventions of more or less importance in all the branches of mechanics, in chemical compounds, in surgical instruments, in electrical utilities, and in the fine arts as well.

From the numerous statements made by various attorneys to the effect that they have had several colored clients whose names they could not recall, and whose inventions they could not identify on their books, it is practically certain that the nearly 800 verified patents do not represent more than one-half of those that have been actually granted to colored inventors, and that the credit for these must perhaps forever lie hidden in the unbreakable silence of official records. . . .

We come now to consider the list of more modern inventions, those inventions from which the element of uncertainty is wholly eliminated, and which are represented in the patent records of our government.

In this verified list of nearly 800 patents granted by our government to the inventors of our race we find that they have applied their inventive talent to the whole range of inventive subjects; that in agricultural implements, in wood and metalworking machines, in land conveyances on road and track, in seagoing vessels, in chemical compounds, in electricity through all its wide range of uses, in aeronautics, in new designs of house furniture and bric-à-brac, in mechanical toys and amusement devices, the colored inventor has achieved such success as should present to the race a distinctly hope-inspiring spectacle.

Of course it is not possible, in this particular presentation of the subject, to dwell much at length upon the merits of any considerable number of individual cases. This feature will be brought out more fully in the larger publication on this subject which the writer now has in course of preparation. But there are several conspicuous examples of success in this line of endeavor that should be fully emphasized in any treatment of this subject. . . .

Another very interesting instance of an inventor whose genius for creating new things is constantly active, producing results that express themselves in terms of dollars for himself and others, is that of Mr. Joseph Hunter Dickinson, of New Jersey. Mr. Dickinson's specialty is in the line of musical instruments, particularly the piano. He began more than fifteen years ago to invent devices for automatically playing the piano, and is at present in the employ of a large piano factory, where his various inventions in piano-player mechanism are eagerly adopted in the construction of some of the finest player pianos on the market. He has more than a dozen patents to his credit already, and is still devoting his energies to that line of invention.

The company with which he is identified is one of the very largest corporations of its kind in the world, and it is no little distinction to have one of our race occupy so significant a relation to it, and to hold it by the sheer force of a trained and active intellect.

Mr. Frank J. Ferrell, of New York, has obtained about a dozen patents for his inventions, the larger portion of them being for improvements in valves for steam engines.

Mr. Benjamin F. Jackson, of Massachusetts, is the inventor of a dozen different improvements in heating and lighting devices, including a controller for a trolley wheel.

Mr. Charles V. Richey, of Washington, has obtained about a dozen patents on his inventions, the last of which was a most ingenious device for registering the calls on a telephone and detecting the unauthorized use of that instrument. This particular patent was only recently taken out by Mr. Richey, and he has organized a company for placing the invention on the market, with fine prospects of success.

Hon. George W. Murray, of South Carolina, former member of Congress from that State, has received eight patents for his inventions in agricultural implements, including mostly such different attachments as readily adapt a single implement to a variety of uses.

Henry Creamer, of New York, has made seven different inventions in steam traps, covered by as many patents, and Andrew J. Beard, of Alabama, has about the same number to his credit for inventions in car-coupling devices.

Mr. William Douglass, of Kansas, was granted about a half dozen patents for various inventions in harvesting machines. One of his patents, that one numbered 789,010, and dated May 2, 1905, for a self-binding harvester, is conspicuous in the records of the Patent Office for the complicated and intricate character of the machine, for the extensive drawings required to illustrate it and the lengthy specifications required to explain it—there being thirty-seven large sheets of mechanical drawings and thirty-two printed pages of descriptive matter, including the 166 claims drawn to cover the novel points presented. This particular patent is, in these respects, quite unique in the class here considered.

Mr. James Doyle, of Pittsburgh, has obtained several patents for his inventions, one of them being for an automatic serving system. This latter device is a scheme for dispensing with the use of waiters in dining rooms, restaurants and at railroad lunch counters. It was recently exhibited with the Pennsylvania Exposition Society's exhibits at Pittsburgh, where it attracted widespread attention from the press and the public. The model used on that occasion is said to have cost nearly $2,000.

In the civil service at Washington there are several colored men who have made inventions of more or less importance which were suggested by the mechanical problems arising in their daily occupations.

Mr. Shelby J. Davidson, of Kentucky, a clerk in the office of the Auditor for the Post Office Department, operated a machine for tabulating and totalizing the quarterly accounts which were regularly submitted by the postmasters of the country. Mr. Davidson's attention was first directed to the loss in time through the necessity for periodically stopping to manually dispose of the paper coming from the machine. He invented a rewind device which served as an attachment for automatically taking up the paper as it issued from the machine, and adapted it for use again on the reverse side, thus effecting a very considerable economy of time and material. His main invention, however, was a novel attachment for adding machines which was designed to automatically include the government fee, as well as the amount sent, when totalizing the money orders in the reports submitted by postmasters. This was a distinct improvement in the efficiency and value of the machine he was operating and the government granted him patents on both inventions. His talents were recognized not only by the office in which he was employed by promotion in rank and pay, but also in a very significant way by the large factory which turned out the adding machines the government was using. Mr. Davidson has since resigned his position and is now engaged in the practice of the law in Washington, D. C.

Mr. Robert Pelham, of Detroit, is similarly employed in the Census Bureau, where his duties include the compilation of groups of statistics on sheets from data sent into the office from the thousands of manufacturers of the country. Unlike most of the other men in the departmental service, Mr. Pelham seemed anxious to get through with his job quickly, for he devised a machine used as an adjunct in tabulating the statistics from the manufacturers' schedules in a way that displaced a dozen men in a given quantity of work, doing the work economically, speedily and with faultless precision, when operated under Mr. Pelham's skilful direction. Mr. Pelham has also been granted a patent for his invention, and the proved efficiency of his devices induced the United States government to lease them from him, paying him a royalty for their use, in addition to his salary for operating them.

Mr. Pelham's mechanical genius is evidently "running in the family," for his oldest son, now a high-school youth, has distinguished himself by his experiments in wireless telegraphy, and is one of the very few colored boys in Washington holding a regular license for operating the wireless.

Mr. W. A. Lavalette, of the Government Printing Office, the largest printing establishment in the world, began his career as a printer there years before the development of that art called into use the wonderful machines employed in it to-day; and one of his first efforts was to devise a printing machine superior to the pioneer type used at that time. This was in 1879, and he succeeded that year in inventing and patenting a printing machine that was a notable novelty in its day, though it has, of course, long ago been superseded by others....

In the credit here accorded our race for its achievements in the field of invention our women as well as our men are entitled to share. With an industrial field necessarily more circumscribed than that occupied by our men, and therefore with fewer opportunities and fewer reasons, as well, for exercising the inventive faculty, they have, nevertheless, made a remarkably creditable showing. The record shows that more than twenty colored women have been granted patents for their inventions, and that these inventions cover also a wide range of subjects—artistic, utilitarian, fanciful.

The foregoing facts are here presented as a part only of the record made by the race in the field of invention for the first half century of our national life. We can never know the whole story. But we know enough to feel sure that if others knew the story even as we ourselves know it, it would present us in a somewhat different light to the judgment of our fellow men, and, perhaps, make for us a position of new importance in the industrial activities of our country. This great consummation, devoutly to be wished, may form the story of the next fifty years of our progress along these specific lines, so that some one in the distant future, looking down the rugged pathway of the years, may see this race of ours coming up, step by step, into the fullest possession of our industrial, economic and intellectual emancipation.

## Clara Frye, A Woman Inventor (1907)

As the brief account of Clara Frye makes clear, inventors usually have made improvements in those areas of activity with which they are most familiar. It is not surprising therefore that Frye, like many female inventors, contributed to the improvement of activities that were traditionally gendered as appropriate for women.

---

That the woman of this hustling and progressive age is a safe competitor with man in every avocation of life is demonstrated daily. In art, science, law, medicine and in business we see woman applying her energy to success. It was the boasted opinion of man, until the dawn of this advanced era, that the hand of woman was good only to rock the cradle and attend household affairs but now he concedes that she can go hand in hand with him in the attainment of fame and wealth in all the pursuits of life.

The Negro race has subscribed to all of these avocations, now another woman inventor has come forth, Mrs. Clara Frye of Tampa, Florida, who has invented a "Combination Bed, Air and Bed Pan" Mrs. Frye was born in Albany, New York, where her father had come to live, in 1872. Soon after her birth her father decided to come South. They located in Montgomery, Alabama, and here Mrs. Frye grew to womanhood. No girl was better loved than she; her kindly disposition and lovable ways winning friends at once. The father died within a few years, leaving four children; the care of all falling upon the mother and elder child. Because of this, Mrs. Frye did not have the advantages of a high education, only attending the city schools for a short time.

In 1888, she married S. H. Frye, of Atlanta, Georgia, who had located at Montgomery. A few years later, Mrs. Frye decided to adopt the profession of trained nurse, and went to Chicago to pursue the course. After finishing she worked in some of the leading hospitals of the city and with some of the leading physicians. In 1900, the Fryes located in Tampa, Florida.

Mrs. Frye's constant work in the sick room made her see the need of a device to aid the physician and nurse and for comfort of the patient; hence, this invention. At this account, she is daily expecting successful news from her invention, which has been before the Patent Officials since last June. This invention has been endorsed by leading physicians of Tampa and Hot Springs, Arkansas, to which place Mrs. Frye went in care of a patient. There seems to be no record of anything similar to this and it bids fair to rank with inventions of service to the medical profession along this line. The Patent Office has reported favorably upon the application and the patent granted.

---

Reprinted from: "A Woman Inventor," *The Colored American Magazine* 12 (April 1907): 292–293.

**Figure 7.1**
Portrait of Mrs. Clara Frye.

## U.S. Patent to Garret A. Morgan for a "Breathing Device" (1914)

One of the best-known African-American inventors of the twentieth century was Garrett Morgan. Using this early gas mask, Morgan made a daring rescue of trapped workers caught in a salt-mine explosion deep under Cleveland in 1916.

**United States Patent Office**

**Garrett A. Morgan, of Cleveland, Ohio. Assignor to the National Safety Device Company, of Oberlin, Ohio, a Corporation of Ohio**
Breathing Device

1,113,675. Patented Oct. 13, 1914.
Specification of Letters Patent. Application filed August 19, 1912. Serial No. 715,697.

To all whom it may concern:

Be it known that I, Garrett A. Morgan, a citizen of the United States, and resident of Cleveland, in the county of Cuyahoga and State of Ohio, have invented certain new and useful Improvements in Breathing Devices, of which I hereby declare the following to be a full, clear, and exact description, such as will enable others skilled in the art to which it appertains to make and use the same.

The objects of the invention are to provide a portable attachment which will enable a fireman to enter a house filled with thick suffocating gases and smoke and to breathe freely for some time therein, and thereby enable him to perform his duties of saving life and valuables without danger to himself from suffocation.

The device is also efficient and useful for protection to engineers, chemists, and working men who are obliged to breathe noxious fumes or dust derived from the materials in which they are obliged to work.

The invention has for its further objects to provide a device which can be quickly and easily attached and carried upon the person without the delay caused by buckling straps or the use of fastening devices of any kind and thus will be serviceable for immediate use in emergencies, since a little delay will often endanger life beyond recovery.

The invention comprises a hood to be placed over the head of the user, from which depends a tube provided with an inlet opening for air and the tube is long

Reprinted from: U.S. Patent No. 1,113,675, dated October 13, 1914, granted to G. A. Morgan for a "Breathing Device."

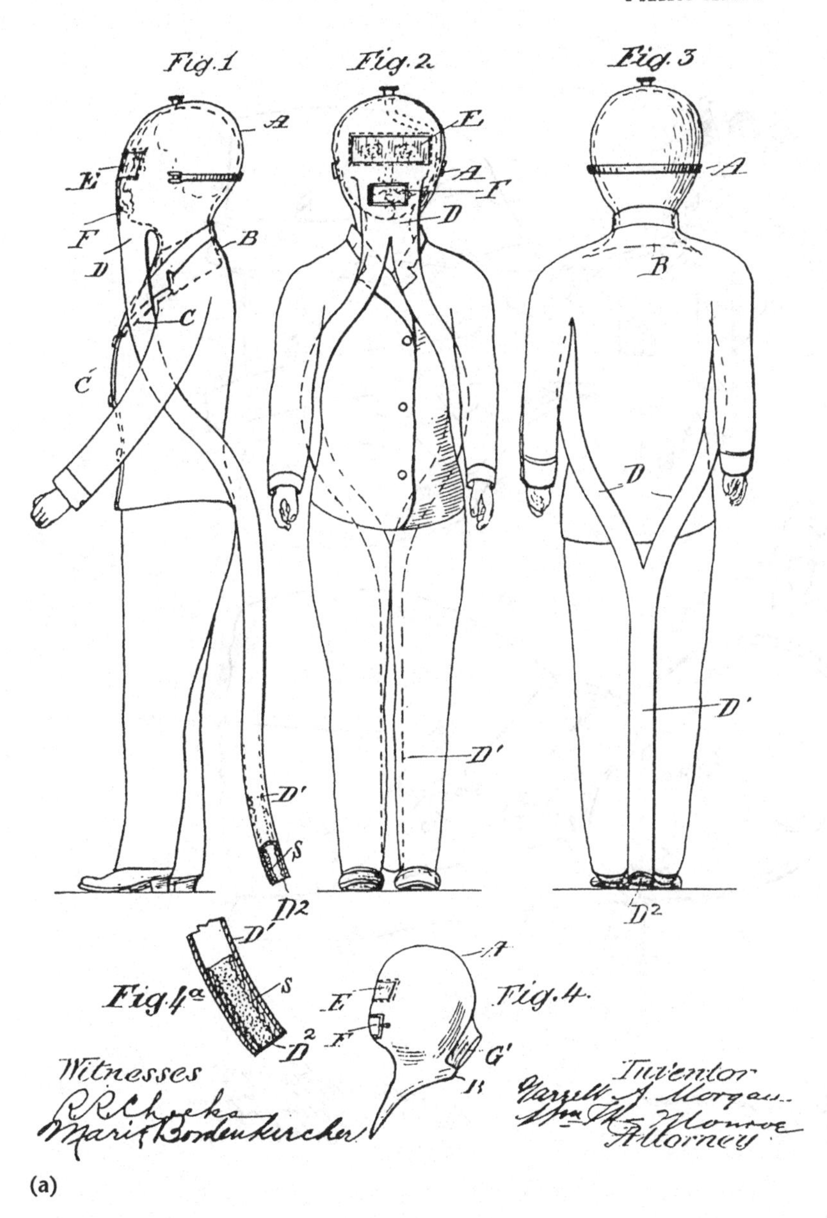


(a)

**Figure 7.2a–b**
Patent drawings for Garrett A. Morgan's Breathing Device.

G. A. MORGAN.
BREATHING DEVICE.
APPLICATION FILED AUG. 19, 1912.

1,113,675. Patented Oct. 13, 1914.
2 SHEETS—SHEET 2.

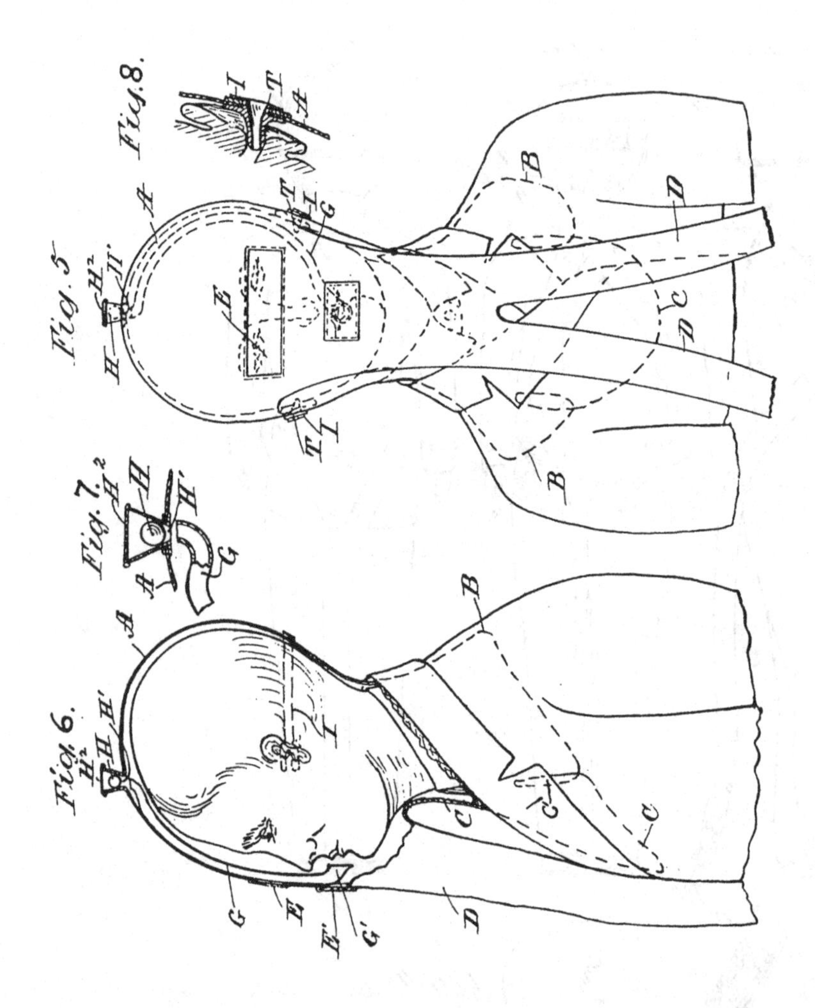

Witnesses
Mary Bodenkircher
Geo. S. Cole

Inventor
Garrett A. Morgan
by Wm. M. Monroe
Attorney

(b)

**Figure 7.2a–b**
(continued)

enough to enter a layer of air underneath the dense smoke within the hall or apartment entered by the fireman, and which can be placed beyond the reach of gaseous fumes or dust, and through which pure or much purer air can be furnished to the user. The hood is also provided with an appliance enabling the operator to hear clearly what is going on about him.

The invention further consists in the combination and arrangement of parts and manner of construction of the various details hereinafter described, shown in the accompanying drawings and specifically pointed out in the claims. . . .

In testimony whereof, I hereunto set my hand this 14th day of August 1912.

Garrett A. Morgan

In presence of—

Wm. M. Monroe

C. L. Case

# V BETWEEN THE WORLD WARS

The two decades between World War I and World War II are usually neatly divided into the Roaring Twenties and the Great Depression of the 1930s. The prosperity of the twenties was, however, spotty. Agriculture in general, after a large wartime expansion, struggled with a loss of markets and went into an economic depression a decade before the rest of the nation. But the country had come out of the war a creditor nation, and the great cornucopia of consumer durable goods (those expected to last more than just a few years) encouraged a buying binge (also encouraged by the relatively new industries of advertising and consumer credit) that brought the economy to a fever pitch. Automobiles were perhaps the most important of these durable goods, but a wide range of electrical appliances for the home—radios, toasters, vacuum cleaners, refrigerators, and so forth—contributed also. For those who had the money, and access to the technological infrastructures found in most cities, it was a time of a rising standard of living based on new tools and machines.

The Depression showed the limits of this boom and plunged the nation into the worst economic hardship it had ever experienced. Hundreds of thousands lost their jobs, and the government was slow to provide a safety net for the newly unemployed. The New Deal undertook large projects of technological improvement like those of the Tennessee Valley Authority (TVA), the building of larger dams in the West including the Boulder Dam on the Colorado River, and a widespread effort to bring electricity to rural areas through the Rural Electrification Administration (REA). A massive low-tech program was also undertaken: armies of the unemployed were given picks and shovels and set to work improving park lands and building trails and barbecue pits.

During the twenties, African-Americans who could afford to bought automobiles in an effort to escape the handicap of Jim-Crow public transportation. Those who lived in cities aspired to the same household technologies that white America was buying, but dwellings not properly wired, or not wired at all, had no way to support such amenities. And, again not surprisingly, African-Americans, being the last hired, were the first fired when the Depression struck. Some found work in segregated work brigades of the Civilian Conservation Corps (CCC), but it was not until the beginnings of the defense buildup at the end of the decade that industrial jobs opened up again for black or white Americans.

Those who had not been a part of the Great Migration continued to be a critical part of southern agriculture, but New Deal policies crafted to combat the depression in agriculture worked against African-American farmers even as it improved the lot of better-off whites. Government checks to limit crops struck at tenant farmers and sharecroppers in two ways: fields not planted did not need to be tended, and federal subsidies often were used to buy tractors and other labor-saving (job-destroying) technologies for use on the farm. The cotton-picking machine was only the most obvious and devastating of these "improved" machines.

# 8 The Rural South

*You May Plow Here: The Narrative of Sara Brooks*
Sara Brooks

**Cotton Picking Machines and Southern Agriculture**

***We* (1927)**
Charles Lindbergh

## *You May Plow Here: The Narrative of Sara Brooks*

Sara Brooks

Sara Brooks was born in rural Alabama in 1911. Her family owned its own farm and one of the first cars in the area. Nevertheless, as with perhaps a majority of southern African-Americans, their farm life was still heavily dependent on traditional tools and techniques. Many years later, when Brooks was working as a domestic in Shaker Heights, Ohio, she told her life story to the daughter of one of her employers.

### The Things He Knew How to Do, He Did Em

My father would get up early in the mornin, and after he ate his breakfast, he might have one strap fastened on his overalls, and one maybe be hangin down cause we laughed at him so many times—he'd be in a hurry and he'd be gone! He would go off to work like that *early* in the mornin. Sometime he would say he didn't want no breakfast and just get his snuff and go, and then he'd be in the field when we'd get there. See, he'd go first unlessen he's gonna drive the wagon. If he's gonna drive the wagon, everybody had to be ready when he got ready. You see, he probably wanna take his plow-stocks and other things because Rhoda would plow, and he would plow. He wouldn't drag the plow all the way to field hitched onto the mule—he would put it in the wagon, hitch the mules to the wagon. When he get there, he take the plow off and go to work.

So he'd plow all day and dip that snuff. And he never seemed as if he was tired—that's one thing he never did. I don't know why, but my father never did seem as if he was tired. And oh, he has been a *good* man cause he always looked out for us. And sometime if there wasn't enough food on down in the year, he didn't want nothin. He'd say, "Well, I don't want nothin this mornin." That's in order that we'd have it. See what I mean? We didn't understand it like that *then*, but we've learned since that if anybody went, likely it would be him. He always wanted us to have, and we never had nothin fine, but we had.

Some mornings my father'd get up *real* early and make him up a big fire outside the palin fence—the yard had a palin fence around it—and he'd sharpen all his plows by putting em in the fire till they get hot. And he had some tongs with a long handle, and when that plow get red hot, then he'd reach in that fire and pull it out and he'd put it on the anvil—he had an anvil sittin out there. Then he would beat

Reprinted from: Sara Brooks, "The Things He Knew How to Do, He Did Em," in *You May Plow Here: The Narrative of Sara Brooks*, ed. Thordis Simonsen (New York: W. W. Norton & Co., 1986), 43–52. Reprinted by permission of W. W. Norton & Company, Inc.

that plow with a great big sledgehammer and dip it in the water what he had in this old hooped barrel—after he done sharpen it, he'd dip it in the water to harden it. That's the way he'd sharpen his plow. *Bam* bam *bam* bam *bam.* He'd beat it. You could hear it. In the country, wide open country, it was quiet, you know, and you could hear so *far!* We could hear other peoples when they'd be sharpenin their plows, too, like we say, "Mr. Harrison sharpenin his plows this mornin," or Mr. Jackson or Mr. Strong.

Oh, he was busy—he was a busy man. He really was. The things he knew how to do, he did em. He did all his wagon makin—he'd make all the bodies for his wagons. And sometime he had to replace spokes in his wagon wheels. And I've seen him fix collars for his mules and sew straps on his saddle and fix his plowstock. He'd fix things—he used to keep things goin. He'd do what he could.

At night when everybody's in and everybody ate and everybody sittin around, well, my father'd be done got this white oak—some long tall pieces of white oak. He'd shave em down into strips, and he'd bottom our chairs with that. And he used to make baskets, too—great huge cotton baskets outa that oak. He would do that in the wintertime cause he'd be makin em for the next year. He'd make em for the next year cause nothin else to do in the wintertime. So he'd make baskets, and we'd be in our room, and he'd be in the other room, and we'd be runnin to and fro—the kids—we'd be runnin to and fro.

And I know when I was smaller I remember he used to have a fish basket he made outa white oak. You make it so the fish can't come out when they get in. My father used to set his basket at night. He'd set it in the branch—it wasn't too far from our house—and he'd go the next mornin and pull that basket, and oh, he'd have some fish! Really! He'd have a *basket* of fish! And he bring em home and we clean em up, and boy, we'd have more fried fish! Oh, I wish those days would come back.

But he'd make baskets, do chair bottoms, make brooms—he used to make palmetto brooms. Do you ever know what palmetto is? Kinda got little sticky things on it? Well, my father used to go in the big swamp and he cut these leaves off, bring em home, and let em dry. He'd always cut em in the fall of the year and he let em season in the smokehouse. Then when they get dry, he would get an old hoe handle or he would take a piece of a limb off a tree, and then he would tack these leaves on, one on top of the other one, all the way round till he make a broom. It'd be a big broom, and that's what we used to sweep our floors with.

We only had bare floors cause we didn't have no rugs. It's just boards, board floors. We scrubbed the floors often with a big old scrubbin mop. My daddy made it outa board. He had put holes in it, and then you take shucks from the corn and you wet the shucks, and then you twist em and you push em through those holes, and when you get it all filled—full of shucks—then you could take it, put a handle in it, and then you scrub with it. We would use sometimes sand and lye soap, put the scrubbin mop in that water and scrub the floors—me or Rhoda or Molly—and they'd be just as clean as they could be. But that's the way we would scrub our floors.

And so what else did my father do? He used to make axe handles and he used to make hoe handles out of white oak by shavin it, shavin it, shavin it until he get it the shape he want. And he made his own palins. What he would do, he would cut down small oak trees and that would be the posts he'd put in the ground. And this palin came from pine wood. He cut it in blocks as you want your palins so high, and then he would split these blocks, and then he would go with his chisel to what size he wanted em. Just like now peoples go up to the lumberyard, but all our posts and palins came through by his hand-makin.

So my father did all that. And he used to go muddyin for fish with a hoe and a gig. You make a board and you nail nails down through it—sharpen the nails and nail down through and put a handle on it. That's a gig. Then you take your hoe and your gig and you go to a branch where the water's almost dried up when it's real dry in the summer. You go to the deepest part of the branch, and there where most of the bigger fish, they'll be in that particular spot. You take your hoe and you get in the water and you start stirrin up the water until you make it real muddy. When you make it real muddy, then the fishes'll go to jumpin up, and when they start jumpin up you take this gig—*slap*, just *slap* with the gig because it's got the sharp nails. You know, every time you slap at a fish, he'd stick on the gig. You'd pull it off and put it in the bag—you have a croker sack bag hangin on your shoulder—or sometime they'd have an old big tub or bucket where you can throw it on the bank. So you keep on like that until you get all the bigger fish you want. Then my father'd come out and he'd get in the bushes and he'd pull his wet pair of overalls off and put on a dry pair, and he pulled off his muddyin shoes and dry off and put his other pair on. So that's the way he used to catch fish. That was muddyin for fish with a gig and a hoe.

And he used to get his lantern and go in the woods at night and cut down a tree—he had observed this tree already seein bees flyin around it—and he would get bee honey. We call it "bee tree" because bees'll be in it. They'd have a hollow in it and that hollow is where the bees would be done made the honey. So he had to cut the tree down, and then he'd make a smoke so the bees wouldn't bother him too bad. He'd have his match and he had lightwood to make a fire and rags to make the smoke. That smoke from the rags would run the bees away while he'd get the honey. And my father would put old screen over his head to keep from gettin his face stung. Then he would rob that bee tree—we'd call it "robbed it"—and bring that honey home. Sometime he'd get about a lard can full of the honeycomb with the honey in it, and when we get up the next mornin, oh, boy! We'd have a good time eatin honey and chewin that honeycomb. We'd eat biscuit and honey—hot biscuits and honey was *real good.* We'd eat that honey and we'd chew that honeycomb, and then we'd have us a chewing gum for all day long. And we'd chew sweet gum from the sweet gum tree. Very seldom we had store-bought gum, but if we got a piece of sweet gum, that'd last us a long time cause we'd always stick it somewhere and go back and get it. Now it's true!

But my father used to get bee honey, and he used to go possum huntin, too, at night. My father'd go huntin around where a lotta simmon trees would be because they always would go up the simmon tree because they love simmons and they eat simmons. But you don't catch possums in the summer cause in the summer that's breedin time for em and they have young ones, so they hunt possum in the wintertime.

My father would catch that possum and they'd break that possum's neck—put a hoe across that neck and *pull* the tail. And there always was a fire going, so my mother'd pull out some coals and take that possum by the tail and roll that possum in them hot coals until it burn the hair off it, and then she'd take a knife and scrape that hair off the skin. Then she'd gut him, you know, and then when she get through guttin him, she'd cut off the feet, but she left the head on. Then she would wash him real good and put some salt on it, and that would be to preserve this possum until she's ready to cook it cause we didn't have no freezer or refrigerator neither. So she'd let it go over night soakin in the salt water. Then she would boil this possum, and when she get through boilin it she would roast this possum in the coals—hot coals—and she'd have boiled potatoes all around it, and the head would be sittin up there and the possum would be grinnin at you. That's what I couldn't stand—I didn't want no possum till the head was took off cause the possum would always be grinnin. But when she took the head off, that possum would be *so* good. Boy, I'd be so anxious to get home so we could eat up this good possum!

And my father used to kill rabbits, too. That's another thing we used to have in the evenin. That was a delicious dish! My father'd go huntin for rabbits in the fall, and sometime he'd come back with lotta rabbits in his sack. If they're young, you fry em. If they'se old, you par boil em. And sometime my father'd catch a squirrel and my mother'd make a squirrel pie. Oh, boy! That's good! Yup, we had that, too.

It wasn't fine, but we had food to eat. And we wasn't well off—you can *never* believe that! But we did come up with *some* things that some didn't come up with. On Sundays we always had a nice dinner. We never lived in a rented place—we had our own home. And we had horses and we had mules. I reckon it's because my father—well, a lot of people didn't work, just like they don't work now. But I bet *we* worked! We'd be workin, and peoples be downtown meetin the train. It's a little town, Orchard. Used to be a train come through, and people would be down there at the depot sittin around downtown. My father never let us go down there like that. We didn't go places, but we'd have plenty at home.

My father always wanted to have his own and he always told us, "Always work for what you get." Cause I remember somebody went and broke in my father's smokehouse once and stole a whole can of lard. He knew who it was, and he used to steal from us. But my father would say, "Well, that's all right. If they need it that bad, let em have it." So he'd say that and go on.

I don't remember my daddy ever mentioning nobody else stealin nothin from us, but peoples would borrow from us—they'd borrow anything they didn't have that they needed. They even borrowed so much they was borrowin fire coals. You ever heard of that? We livin so far from town, if matches give out—you look in the box and see you don't have but two or three matches to make a fire? Well, we always used oak wood on the fire and oak coals'll stay hot a long time. So while we didn't have no matches, we'd dig a hole in the hearth and put a big coal in there and cover it over, and then in the morning you can take lightwood and hold it to this coal and blow till you get a blaze. So I know all about that, too! It's just like I said—this is what we did.

So they keep these coals wrapped up in the ashes, and if they should happen to go out and they didn't have any matches, they'd send over to borrow some fire. We used to borrow some fire every once in awhile, too. "Go over Mrs. Harrison, get a coal of fire." The Harrisons lived over cross the field. We'd borrow a fire coal and carry it in a cup, a salmon cup. We'd hold it in a rag so it didn't get too hot and bring it home. Isn't that funny? We run outa matches and the fire went out in the house. We'd borrow some fire coals every once in awhile.

Or if my mother ran out of something she would send us down to Aunt Georgia's for it. But we didn't go around borrowing like a lotta peoples did. Peoples run out and they borrow cause maybe they didn't have the money at that time and town was a long ways to go. But they'd always wanta borrow somethin—flour, meal, lard, bakin powder. Name it! And my mother was always a good-hearted person, so if she had, she'd lend. Salt. You know it give you bad luck to pay back salt if you borrow it—it's bad luck. But it ain't too bad a luck to come and get it! Milk. We always had milk because we've had as many as four or five milk cows. Some cows go out, some be coming in. So we'd milk twice a day and we'd churn twice a day. We'd milk the cows in the morning, put that milk in the churn, it'll be churned this evenin. You gonna milk this evenin and put *that* in the churn, and it can be churned in the mornin. So long as it's not churned, it's sweet milk. We'd put this sweet milk into our churn, and then you put some buttermilk that you already have in the sweet milk to make it sour. We'd call it "turn"—it would clabber. Now in the winter the churn had to sit by the fireplace in order that the milk would clabber, but in the summertime we could leave the churn in the kitchen because it would be warm enough in the summer that it would clabber anywhere. *Then* you churn it, and that makes that sweet milk buttermilk which we'd drink at night for supper, and butter which we'd wash and put in a mold, and we had the butter for breakfast with hot biscuits and syrup. But you got to churn the milk to make the butter come.

My grandmother would go out on the porch and churn the milk because it would be hot in the summertime, and when she would get ready to churn, she would tell us to break a branch off the chinaberry tree and come fan the flies. So we'd get a branch off the chinaberry tree, or either we'd get a peach tree switch, and we'd stand up over the churn and we'd just fan the flies to keep the flies from botherin the churn. So my gramma would be churnin—boom de boom de boom de boom. And Mrs. Harrison would send Winifred over or either Aunt Georgia would send Lydia and Irene over for some milk. Everybody had their own cows, like Mrs. Harrison and them, but sometime the cows go dry so they used to send over and my grandmother used to give milk. She never sold it. Then if our cows go dry, I never remember my mother sending us to nobody's house to get no milk. But they'd come after the milk about the time we'd churn and they be standin at the gate—they be standin at the gate waitin. So my grandmother made a song she used to sing to the churn:

Come, butter, come, come, come.
The calves are bleating,
The cows are lowing.
Little boy standin at the gate
Waitin for the butter,
And the butter won't come, come, come.

She was just churning the milk and she'd be singing it. She made this song up because somebody always standin around, comin with a bucket to get milk. Sometimes somebody be there with a four-pound bucket which is a lard bucket, and then sometime it was a syrup bucket which we'd call a can bucket. So somebody's always waitin for the milk cause we always had a lotta cows. And my father had a lotta peoples to feed, but he was blessed—I think he was blessed to feed all of us.

## Cotton Picking Machines and Southern Agriculture (1937–1938)

The traditional farming methods Sara Brooks's family practised on their own land were not typical of the thousands of black farm families who cultivated crops, often cotton, on land owned by large (usually white) planters. The storied "forty acres and a mule" that could support a family of tenant farmers was based on a long-standing method of cotton cultivation that was largely done by hand—especially the laborious picking of the crop. Attempts to mechanize the picking of cotton threatened the livelihoods of tens of thousands of poor, rural, southern black families. After World War II, the Rust cotton-picking machine described here proved to be successful and triggered a large outflow of former tenant farmers and sharecroppers from the South.

---

During 1935 and 1936, there was a great deal of discussion concerning a cotton picking machine invented by John D. Rust and Mack D. Rust of Memphis, Tennessee. Predictions were made that this machine would revolutionize cotton growing in the South, since it is said to be able to do the work of from fifty to one hundred persons. It was stated that tenant farmers, both white and Negro, would be directly affected.

In a demonstration in September, 1936, of the machine on a Mississippi Delta cotton field lasting one hour, it picked four hundred pounds; as much as one average hand picker could gather in four days. Opinions were divided as to the success or failure of the machine.

One opinion was, for the picker to be successful, cotton stalks of standard height and size must be developed with bolls at the top, middle and bottom opening together. Another opinion advanced was that farmers would have to let their cotton stay in the field until nearly every boll on each stalk was opened, whereas, the hand pickers now go over the same field two or three times. The cotton picked by the machine is also more trashy than that picked by hand.

From an economic standpoint, the picker must be cheaply produced. More important, it must be simple to operate and replacement parts easily available. A farmer cannot leave his cotton standing in a Louisiana field and wait a week for new parts to be brought from Atlanta or New Orleans.

Only the big farms will be able to afford a picker—the only way small farmers can use them is on a share system, and southern farmers are not familiar with the practice of small grain farmers who give the thresher owner part of the crop for his services. A

---

Excerpted from: "Cotton Picking Machines and Southern Agriculture," *Negro Yearbook: An Annual Encyclopedia of the Negro, 1937–1938* (Tuskegee: Negro Year Book Publishing Co., n.d.), 41.

collectively-owned picker would not work because the joint owners could not agree on who would use it first.

Two other cotton picking machines were being demonstrated in Mississippi in the fall of 1936. L. C. Stukenborg, of Memphis, has an invention which he calls the "poor man's mechanical picker," that picks the cotton from the boll by a hand operated brushing process. The picker is operated by an individual walking down a row, and shifting the picker from one open boll on a stalk to another until all the open bolls are picked from the stalk. The outfit appears to be inexpensive and either horse drawn or tractor drawn. A defect of this machine appears to be that it picks the cotton from only one boll at a time.

C. R. Berry, of Greenville, Mississippi, and L. E. Worth, of Pittsburgh, Pennsylvania, had a machine with a nipple mechanism that stripped the cotton from the barbed revolving spindles that plucked the cotton from the plant. The outfit is mounted on an auto tractor.

Account should be taken of the fact that efforts have been made for more than a hundred years to invent a practical cotton picking machine. More than seven hundred unsuccessful cotton picking devices have been patented at Washington since the Civil War. In 1912, it was stated that the success of the cotton picking machine seems assured. At an exhibit of the Campbell Cotton Picking Machine in Boston, April 12, several speakers referred to the probable effect it would have on Negro labor. Two prominent white men of Texas, Mr. Mike H. Thomas, of Dallas, and E. A. Calvin, of Houston, were of the opinion that the cotton picking machine, instead of injuring Negro labor, would ultimately help it just as any other invention had eventually helped labor. Dr. Booker T. Washington, speaking on this occasion, said that the cotton picking machine when extensively employed, would give the Negroes, and in fact the whole South, more time to raise other things. At present over three-fourths of the whole year is spent in planting, cultivating and gathering the cotton crop.

## *We* (1927)

Charles Lindbergh

Like electricity before it, the airplane was a marker of modern efficiency and rationality, but at the same time it was also a magical technology. Once again, those who claimed the appropriate knowledge and skills demonstrated their superiority by contrasting it with the inferiority of others who lacked knowledge and skill. African-Americans, both men and women, made convenient foils for the smug "humor" of those white men who made claim to being modern and could prove it by their knowledge of modern machines.

---

That afternoon a group of whites chipped in fifty cents apiece to give one of the negroes a hop, provided, as they put it, I would do a few "flip flops" with him. The negro decided upon was perfectly willing and confident up to the time when he was instructed to get in; even then he gamely climbed into the cockpit, assuring all of his clan that he would wave his red bandanna handkerchief over the side of the cockpit during the entire flight in order to show them that he was still unafraid.

After reaching the corner of the field I instructed him, as I had the previous passengers, to hold the throttle back while I was lifting the tail around. When I climbed back in my cockpit I told him to let go and opened the throttle to take off. We had gone about fifty yards when it suddenly occurred to him that the ship was moving and that the handle he was to hold on to was not where it should be. He had apparently forgotten everything but that throttle, and with a death grip he hauled it back to the closed position. We had not gone far enough to prevent stopping before reaching the other end of the field and the only loss was the time required to taxi back over the rough ground to our starting point. Before taking off the next time, however, I gave very implicit instructions regarding that throttle.

I had promised to give this negro a stunt ride yet I had never had any instruction in aerobatics. I had, however, been in a plane with Bahl during two loops and one tailspin. I had also been carefully instructed in the art of looping by Reese who, forgetting that I was not flying a Hisso standard with twice the power of my Jenny, advised me that it was not necessary to dive excessively before a loop but rather to fly along with the motor full on until the plane gathered speed, then to start the loop from a level flying position.

I climbed up to three thousand feet and started in to fulfill my agreement by doing a few airsplashes, steep spirals and dives. With the first deviation from straight

---

Excerpted from: Charles Lindbergh, *"We"* (New York: G. P. Putnam's Sons, 1927), 58–60. 

flight my passenger had his head down on the floor of the cockpit but continued to wave the red handkerchief with one hand while he was holding on to everything available with the other, although he was held in securely with the safety belt.

Finally, remembering my ground instructions, I leveled the plane off and with wide open motor waited a few moments to pick up maximum speed, then, slowly pulling back on the stick I began to loop. When I had gotten one-fourth of the way around, the ship was trembling in a nearly stalled position; still, the Curtiss motor was doing its best and it was not until the nose was pointing directly skyward at a ninety degree angle that the final inertia was lost and for an instant we hung motionless in the perfect position for a whipstall. I kicked full right rudder immediately to throw the plane over on its side but it was too late, the controls had no effect.

The Negro meanwhile decided that the "flip flops" were over and poked his head over the side of the cockpit looking for mother earth. At that instant we whipped. The ship gathered speed as it slid backwards towards the ground, the air caught the tail surfaces, jerked them around past the heavier nose and we were in a vertical dive; again in full control, but with no red handkerchief waving over the cockpit. I tried another loop in the same manner but just before reaching the stalling point in the next one I kicked the ship over on one wing and evaded a whipstall. After the second failure I decided that there must be something wrong with my method of looping and gave up any further attempt for that afternoon. But it was not until we were almost touching the ground that the bandanna again appeared above the cowling.

I remained in Maben for two weeks carrying over sixty passengers in all or about three hundred dollars worth. People flocked in from all over the surrounding country, some travelling for fifteen miles in oxcarts just to see the plane fly.

One old negro woman came up and asked, —

"Boss! How much you all charge foah take me up to Heaben and leave me dah?"

# 9 Industrial Employment

## Tuskegee Ideals in Industrial Education (1926)

Joseph L. Whiting

The Great Migration of southern African-Americans to the North to find work during World War I gave new meaning to the long struggle to establish industrial training as a bridge to the large manufacturing centers of the country. After the war, Tuskegee Institute continued to train young people for industrial jobs, which were by then often in the North.

---

All work has in it two elements; manipulative skill and job or technical knowledge. The proportion varies according to the difficulty of the processes involved and the quality of the product to be turned out.

A Tuskegee Institute graduate printer, arriving in Cleveland, Ohio, answered the following advertisement:

Wanted: A No. 1 Cylinder Pressman.

As he entered the operating plant he was ushered into the presence of the foreman of the pressroom. The foreman, after a casual observation, directed the young printer to perform a certain job spread upon the cylinder press—a daily report sheet for one of the electrical plants. He was told to "make ready" and submit a "proof" on this job. When the proof was taken it was dispatched upstairs to the proofroom, where it was inspected and shortly returned with the proofreader's approval. The foreman, also, scrutinized the copy and gave it his O. K. The young man was assigned a press and at once began running off copy. Although the young apprentice had come to answer this advertisement before eating breakfast, not expecting to be so speedily and fortunately drafted into the industry, he completed the first half-day's work and was permanently employed at $29.50 per week, the wages to reach $32.00 after three weeks employment.

This young man was required to produce a standard product. Here was an opportunity to be grasped at once, and a probable opening into an important industry demanding as preliminary qualifications such elements of economic prepardness as may be required in many of our industrial and technical schools. The applicant must possess personality. He must also have technical knowledge and manipulative skill. Experience has shown that manipulative skill can best be trained for, produced, and developed by skillful instruction on the job. Usually, related information and auxiliary knowledge can in part be given on the job, and in part, away from the job.

---

Reprinted from: Joseph L. Whiting, "Tuskegee Ideals in Industrial Education," *Opportunity*, 4 (February 1926), 58–60.

The value of buildings and grounds occupied by Tuskegee is $1,892,303.59; the value of scientific apparatus, furniture and other equipment used alone for educational purposes is $252,987.95; and the estimated value of the apparatus and shop equipment used exclusively for giving industrial and technical training is $138,163.00. With ample facilities, excellent equipment, and a trained, devoted, and loyal teaching staff, Tuskegee has steadily aimed to contribute the largest possible service to the hundreds of young men and women who have sought and found it possible to enter into its academic life and industrial activities. Its dual educational system, and its varied vocational offerings afford an attractive and wide choice well adapted to answer many of the complex educational problems that confront the equitable administration and conduct of training large groups of young men and women possessing a wide degree of differences in aptitude and capacities, yet, each capable in some particular field of usefulness of successfully contributing to the general happiness and economic life of the community in which he or she should play a worthwhile part.

Tuskegee Institute's educational and industrial training program furnishes a cross-section of a number of the most important activities to be found in industry. In the illustration is seen a group of young men in auto trimming. During the process of application the instructor may be seen watching the learner while he is working at the job; he notes where the student has not fully grasped some part of the operation, and gives him further instruction under such favorable conditions to learning. The carrying out of this stage of the learning process effectively requires care and skill on the part of the instructor to determine just when to assist the student and just how to assist him; but in no case does the instructor do the work for the student.

Effective instructional conditions are maintained through the inter-relation of the various industry divisions of the vocational training plant. One instance, the important position of the blacksmith shop will suffice. The smith, among all mechanics, it is said, enjoys the distinction of producing his own tools. "By the hammer and hands all the arts do stand," is a trite proverb and one quite frequently quoted in modern mechanical literature.

The institute blacksmith shop during a single month was called upon to furnish to the agricultural division several dozen pairs of horse shoes; to reset *tires*; for forge bolts; to repair wagon and buggy wheels; to shrink tires and repair wagon front. To the carpentry division, to forge yokes, chisels anchors, nail pullers, vise collars and braces; to the landscape architectural division, to dress picks, mattocks, to repair plows, forge wedges, repair stretchers and repair one double tree plow, and to perform characteristic or similar operations for the plumber, the electrician, the machinist, the power plant operator, the sheet metal worker, the brickmason and the auto mechanic.

A few years ago, a group of young men at Tuskegee completed the course in carpentry instruction. They were also graduated from the literary courses in the academic department. Out of this group was one who had shown above the average ability, both in mechanical aptitude and literary subjects. This young man was encouraged to continue his technical studies in a higher institution and now occupies an important position as an assistant superintendent of industries in a large trade school. His fellow graduates from the vocational carpentry class are, also, successful, skilled mechanics, pursuing an important calling and contented. They are economically fit. Thus, Tuskegee Institute, attempts to serve not only as very reliable clearing house in which many young men and women may obtain an opportunity to test their vocational bent but, also, enables them during the period of instruction to ascertain a high degree of assurance whether their choice promises success consistent with their capacities.

In a recent survey made by a member of the Institute teaching staff it was found that 70% of the carpenters, 80% of the tailors, 95% of the shoe repairers, 70% of the sheet metal workers, 55% of the bricklayers, 69% of the plumbers and 67% of the printers who completed their respective trade courses at Tuskegee were actively engaged in the industry. The fact that these efficient recruits to industry entered so readily into the industries for which they had been skillfully trained and prepared is very clearly due to the practical method of their instruction and training, to contact with production on a commercial basis, and to the systematic performance of purposeful acts in a natural setting. The transition from the school and instruction shop to industry was normal, and the experience during the brief period of adjustment was progressive and continuous.

An intelligent observer and distinguished educator remarked, after an inspection of the educational methods in practice at Tuskegee Institute, that Tuskegee Institute had helped to teach the friends of industrial and agricultural education in this country the fundamental principle that effective vocational education of any kind requires practice or experience; requires the doing of things combined with the study about that thing. That Tuskegee Institute seems to have recognized from the first the necessity that a vocational school should be a finishing school, and not a preparatory school. That Tuskegee seems to have recognized from the start that the pupil in his practical training should participate in productive work. That Tuskegee Institute seems also to have recognized that there are two fundamental principles in teaching which must be followed in a vocational school. These principles seem to be equally applicable to liberal and cultural education. These two principles are that the boy or girl must be taught on the basis of what each knows, what each thinks, what each is doing, and on the basis of the things in which he or she is interested.

Tuskegee Institute has been fortunate through the wisdom of its founder, Booker T. Washington, in laying the foundation of an industrial training and educational plant in the very beginning to embrace the ideals that are now so universally accepted as fundamental in any scheme of radical awakening and economic ability. The modified social conditions, the curtailment of the total influx of alien-born populations to the reservoirs of industry, the national cleavage of large groups of the Negro population toward urban centers, the enforced demand since the World War for more employees skillfully and technically trained are tremendous economic factors to be encountered and resolved by welfare, industrial and educational institutions. At Tuskegee Institute it has been necessary only to intensity the instructional processes; to make more thorough the methods of educational procedure, and to heighten the degree of cultural and technical attainment required in a more concentrated social environment and a more exacting economic situation.

In a very interesting book on hotel management, by Mr. L. M. Boomer, president of the Boomer-Dupont properties, is the following statement:

> It is safe to predict that many of the department heads of the future will be secured as they have in the past, by promoting minor executives to positions of great responsibility. They will be men and women with broader general education and specialized education because the opportunities to secure both will be more accessible. . . . It is clear, in view of the decrease in foreign-trained employees, that department heads must give more time and attention to organized and systematic training of subordinates. They must develop the instruction and supervision phases of their work. . . . These skilled employees can learn the fundamentals of their work more rapidly under well organized school training than by any system of modified apprenticeship and practice likely to be adopted in this country.

It has been comparatively impossible, or not without great difficulty, for our young men to be admitted into opportunity schools to obtain apprenticeship training under the auspices of the large industrial organizations. This has been partly due, perhaps, to the very small number of applicants to seek admission at one time. However, since a shift of the Negro population has set in and more significant groups are penetrating the industrial centers, taken together with the falling off of both skilled and unskilled immigrant employees, Mr. Boomer's prediction can be equally applied to a large number of important industries than the hotel business. One instance will suffice.

Under the agricultural extension service, promoted by the Federal Government, the several States, and the agricultural institutions in their respective localities, an increasingly large group of competently trained Negroes are employed as agricultural experts in scientific farming and related home economic pursuits. These young men and women exert an educational influence and the economic point of view over an extensive area of the agricultural South, conducting demonstrations and rendering practical advice and conveying scientific information from the agricultural colleges to the people in the remote rural sections in order that agriculture may be made not only more profitable but that the environment and the home more attractive and habitable. There is an instance, recently, where one of these agricultural experts, a Tuskegee graduate, had become so successful in one section that his fame spread to another and created a keen competition for his services. The result was an increase of 66 2/3% in his salary and a general determination to multiply the number of such experts throughout the agricultural section. Similar examples can be cited of a group of young women, cooperating in home economics, creating more beautiful homes, and bringing cheerfulness and greater contentment to rural life.

## *Women at Work: A Century of Industrial Change* (1934)

During the depths of the Depression in the 1930s, the Women's Bureau of the U.S. Department of Labor surveyed women's work in the country and found, not surprisingly, that beyond the gender bias experienced by all women, the double burden of gender and racial discrimination black women suffered made their working lives particularly difficult and unrewarding.

While women workers in general have been restricted by lack of opportunities for employment, by long hours, low wages, and harmful working conditions, there are groups—the latest comers into industry—upon whom these hardships have fallen with doubled severity. As the members of a new and inexperienced race arrive at the doors of industry the jobs that open up to them ordinarily are those vacated by an earlier stratum of workers who move on to more highly paid occupations. Negro women constitute such a new and inexperienced group among women workers.

Added to the fact that they came late into the job market; they have borne the handicap of race discrimination. Slavery placed a stigma on their capabilities and they were considered unfit for factory or skilled work. White men and women, partly because of this and partly because they resented the competition of cheap Negro labor, were unwilling to be engaged on the same work processes with them. To the Negro woman have fallen the more menial, the lower paid, the heavier and more hazardous jobs. Her story has been one of meeting, enduring, and in part overcoming these difficulties.

Previous to the Civil War few Negroes were employed in manufacturing and mechanical pursuits. As slaves in the South, where more than nine tenths of the Negro population of the United States was to be found, they had worked on plantations—raising cotton, tobacco, sugar, rice, and hemp; or had done the household service of maids, cooks, washer-women, and seamstresses. Some Negroes had gained industrial experience as slave labor in cotton, tobacco, and bagging factories, in iron furnaces and charcoal plants, but their numbers were small, as the industrial development of the South was almost negligible at that time.

With the close of the Civil War and the freeing of the slaves, the majority settled down as farmers or share-croppers. Others turned to domestic and personal service. Both these types of work they had done formerly as slaves.

White men and women were entering industry in increasing numbers, and because of their priority and because of race consciousness, factory opportunities

Excerpted from: Women's Bureau, U.S. Department of Labor, *Women at Work: A Century of Industrial Change*, Bulletin No. 115 (Washington: GPO, 1934), 33–37.

were restricted to the whites. Thus, manufacturing was closed to Negro women, whose employment was almost entirely limited to farm work and domestic and personal service—a condition that continued down through the years. As late as 1910, 95 percent of all Negro women workers were in these occupations. Up to the time of the World War the only manufacturing industry to employ any large number of Negro women was the making of cigars and cigarettes.

With the shortage of labor created by the World War, the opportunity came for Negro women to join the growing army of American women in industry. They entered in large numbers those occupations that white women were leaving as new opportunities opened. In other cases Negro women filled the places of men who had gone to the front. The greatest gains were made in textile and clothing factories, the food industries, tobacco factories, and wood-products manufacture. The war industries, too, recruited Negro women in the making of shells, gas masks, and parts of airplanes. The census of 1920, taken immediately after the war period, showed that Negro women in the manufacturing and mechanical industries had increased by over one half. In the professions (as teachers), in office work, and as salesgirls, Negro women also found new work opportunities during the war.

With the return of men from the front and the end of the labor shortage, many of these gains were lost. According to the census of 1930, however, Negro women have increased their war gains in trade, professional service, and clerical occupations. While small numerically, these large proportional increases represent real achievement in the occupational progress of Negro women. That they are finding a place in the growing laundry business is shown by the fact that about 30 percent of the women laundry operatives are Negroes.

The wages of Negro women workers have been on even lower levels than those of white women. A study of Negro women in 15 States, published by the Women's Bureau in 1929, shows that in only 2 of 11 States was the median of the week's earnings—that is, one half of the women receiving more and one half receiving less—as high as $9. In 4 of these States the median of the earnings was below the pitifully small sum of $6.

Scattered wage figures of a more recent date are found in Women's Bureau studies of women in slaughtering and meat packing and in the cigar and cigarette industries. In the first, the wages of Negro women compare favorably with those of white women, but in the second the median earnings of Negro women, most of whom stripped the leaf, were $10.10 in cigars and $8 in cigarettes. For white women, most of whom were makers and packers, the corresponding medians were $16.30 and $17.05.

In the fight to improve their working conditions through organization, Negro women workers have met with even greater failure than have women workers as a whole. To an even greater extent than all women workers, they are concentrated in the unorganizable and unskilled occupations, and few unions have made any attempt to include them as organized workers.

In the garment trades the influx of women workers began in Chicago in 1917 when Negro girls were brought in as strike breakers, and some 500 remained in the trade when the strike was finally broken. Negro women were also used as strike breakers in New York and in Philadelphia, as well as in the less important garment centers. On entering the needle trades the overwhelming majority of the Negro women worked at the unskilled jobs. But in spite of this, both the International Ladies Garment Workers and the Amalgamated Clothing Workers have made every effort to include them in the unions. However, none of the unions that are open to Negroes have made much progress in organizing Negro women workers.

Today almost 2,000,000 Negro women are wage earners. Their employment is so general that 39 are at work in every 100 who are as much as 10 years old. This is practically double the percentage of white women. Nine tenths of the employed Negro women are in agriculture or domestic and personal service. The majority of the others are in the manufacturing and mechanical industries. The industrial depression that has devastated the lives of millions of workers has fallen with particular severity on the Negro workers. Although not giving figures by sex, a recent study made by the National Urban League has shown that the proportion of Negroes is much greater among the unemployed than among the employed. One result of the depression, according to reports from various cities, is that Negro waitresses and other domestics are being displaced by white workers. Not only has the Negro worker taken the ragged edges of employment in times of prosperity, but in times of depression her unemployment is the most acute.

# The Typewriter (1926)

Dorothy West

The following prize-winning short story appeared in the mid-1920s in the pages of *Opportunity*, the journal of the National Urban League. At the time, business was growing and with it came openings for young, unmarried women as clerks, secretaries, and other kinds of office workers. Few of these jobs were open to black women, but hope for some measure of economic and social independence had a powerful attraction nonetheless. This story underscores the gendered nature of both the workplace and of technology. The typewriter belonged in an office, not in a home, but it promised to give the daughter access to a commercial world of which her father could only dream.

---

It occurred to him, as he eased past the bulging knees of an Irish wash lady and forced an apologetic passage down the aisle of the crowded car, that more than anything in all the world he wanted not to go home. He began to wish passionately that he had never been born, that he had never been married, that he had never been the means of life's coming into the world. He knew quite suddenly that he hated his flat and his family and his friends. And most of all the incessant thing that would "clatter clatter" until every nerve screamed aloud, and the words of the evening paper danced crazily before him, and the insane desire to crush and kill set his fingers twitching.

He shuffled down the street, an abject little man of fifty-odd years, in an ageless overcoat that flapped in the wind. He was cold, and he hated the North, and particularly Boston, and saw suddenly a barefoot pickaninny sitting on a fence in the hot, Southern sun with a piece of steaming corn bread and a piece of fried salt pork in either grimy hand.

He was tired, and he wanted his supper, but he didn't want the beans, and frankfurters, and light bread that Net would undoubtedly have. That Net had had every Monday night since that regrettable moment fifteen years before when he had told her—innocently—that such a supper tasted "right nice. Kinda change from what we always has."

He mounted the four brick steps leading to his door and pulled at the bell; but there was no answering ring. It was broken again, and in a mental flash he saw himself with a multitude of tools and a box of matches shivering in the vestibule after supper. He began to pound lustily on the door and wondered vaguely if his hand would bleed if he smashed the glass. He hated the sight of blood. It sickened him.

---

Reprinted from: Dorothy West, "The Typewriter," *Opportunity*, 4 (July 1926), 220–222, 233–234. 

Some one was running down the stairs. Daisy probably. Millie would be at that infernal thing, pounding, pounding.... He entered. The chill of the house swept him. His child was wrapped in a coat. She whispered solemnly, "Poppa, Miz Hicks an' Miz Berry's orful mad. They gointa move if they can't get more heat. The furnace's bin out all day. Mama couldn't fix it." He said hurriedly, "I'll go right down. I'll go right down." He hoped Mrs. Hicks wouldn't pull open her door and glare at him. She was large and domineering, and her husband was a bully. If her husband ever struck him it would kill him. He hated life, but he didn't want to die. He was afraid of God, and in his wildest flights of fancy couldn't imagine himself an angel. He went softly down the stairs.

He began to shake the furnace fiercely. And he shook into it every wrong, mumbling softly under his breath. He began to think back over his uneventful years, and it came to him as rather a shock that he had never sworn in all his life. He wondered uneasily if he dared say "damn." It was taken for granted that a man swore when he tended a stubborn furnace. And his strongest interjection was "Great balls of fire!"

The cellar began to warm, and he took off his inadequate overcoat that was streaked with dirt. Well, Net would have to clean that. He'd be damned—! It frightened him and thrilled him. He wanted suddenly to rush upstairs and tell Mrs. Hicks if she didn't like the way he was running things, she could get out. But he heaped another shovelful of coal on the fire and sighed. He would never be able to get away from himself and the routine of years.

He thought of that eager Negro lad of seventeen who had come North to seek his fortune. He had walked jauntily down Boylston Street, and even his own kind had laughed at the incongruity of him. But he had thrown up his head and promised himself: "You'll have an office here some day. With plate-glass windows and a real mahogany desk." But, though he didn't know it then, he was not the progressive type. And he became successively, in the years, bell boy, porter, waiter, cook, and finally janitor in a down town office building.

He had married Net when he was thirty-three and a waiter. He had married her partly because—though he might not have admitted it—there was no one to eat the expensive delicacies the generous cook gave him every night to bring home. And partly because he dared hope there might be a son to fulfil his dreams. But Millie had come, and after her twin girls who had died within two weeks, then Daisy, and it was tacitly understood that Net was done with child-bearing.

Life, though flowing monotonously, had flowed peacefully enough until that sucker of sanity became a sitting-room fixture. Intuitively at the very first he had felt its undesirability. He had suggested hesitatingly that they couldn't afford it. Three dollars the eighth of every month. Three dollars: food and fuel. Times were hard, and the twenty dollars apiece the respective husbands of Miz Hicks and Miz Berry irregularly paid was only five dollars more than the thirty-five a month he paid his own Hebraic landlord. And the Lord knew his salary was little enough. At which point Net spoke her piece, her voice rising shrill. "God knows I never complain 'bout nothin'. Ain't no other woman got less than me. I bin wearin' this same dress here five years, an' I'll wear it another five. But I don't want nothin'. I ain't never wanted nothin'. An' when I does as', it's only for my children. You're a poor sort of father if you can't give that child jes' three dollars a month to rent that typewriter. Ain't 'nother girl in school ain't got one. An' mos' of 'ems bought an' paid for. You know yourself how Millie is. She wouldn't as' me for it till she had to. An' I ain't going to disappoint her. She's goin' to get that typewriter Saturday, mark my words."

On a Monday then it had been installed. And in the months that followed, night after night he listened to the murderous "tack, tack, tack" that was like a vampire slowly drinking his blood. If only he could escape. Bar a door against the sound of it. But tied hand and foot by the economic fact that "Lord knows we can't afford to have fires burnin' an' lights lit all over the flat. You'all gotta set in one room. An' when y'get tired settin' y'c'n go to bed. Gas bill was somep'n scandalous last' month."

He heaped a final shovelful of coal on the fire and watched the first blue flames. Then, his overcoat under his arm, he mounted the cellar stairs. Mrs. Hicks was standing in her kitchen door, arms akimbo. "It's warmin'," she volunteered.

"Yeh," he was conscious of his grime-streaked face and hands, "it's warmin'. I'm sorry 'bout all day."

She folded her arms across her ample bosom. "Tending a furnace ain't a woman's work. I don't blame your wife none 'tall."

Unsuspecting he was grateful. "Yeh, it's pretty hard for a woman. I always look after it 'fore I goes to work, but some days it jes' ac's up."

"Y'oughta have a janitor, that's what y'ought," she flung at him. "The same cullud man that tends them apartments would be willin'. Mr. Taylor has him. It takes a man to run a furnace, and when the man's away all day—"

"I know," he interrupted, embarrassed and hurt, "I know. Tha's right, Miz. Hicks tha's right. But I ain't in a position to make no improvements. Times is hard."

She surveyed him critically. "Your wife called down 'bout three times while you was in the cellar. I reckon she wants you for supper."

"Thanks," he mumbled and escaped up the back stairs.

He hung up his overcoat in the closet, telling himself, a little lamely, that it wouldn't take him more'n a minute to clean it up himself after supper. After all Net was tired and prob'bly worried what with Miz Hicks and all. And he hated men who made slaves of their women folk. Good old Net.

He tidied up in the bathroom, washing his face and hands carefully and cleanly so as to leave no—or very little—stain on the roller towel. It was hard enough for Net, God knew.

He entered the kitchen. The last spirals of steam were rising from his supper. One thing about Net she served a full plate. He smiled appreciatively at her unresponsive back, bent over the kitchen sink. There was no one could bake beans just like Net's. And no one who could find a market with frankfurters quite so fat.

He sank down at his place. "Evenin', hon."

He saw her back stiffen. "If your supper's cold, 'tain't my fault. I called and called."

He said hastily, "It's fine, Net, fine. Piping."

She was the usual tired housewife. "Y'oughta et your supper 'fore you fooled with that furnace. I ain't bothered 'bout them niggers. I got all my dishes washed 'cept yours. An' I hate to mess up my kitchen after I once get it straightened up."

He was humble. "I'll give that old furnace an extra lookin' after in the mornin'. It'll las' all day to-morrow, hon."

"An' on top of that," she continued, unheeding him and giving a final wrench to her dish towel, "that confounded bell don't ring. An'—"

"I'll fix it after supper," he interposed hastily.

She hung up her dish towel and came to stand before him looming large and yellow. "An' that old Miz Berry, she claim she was expectin' comp'ny. An' she knows they must 'a' come an' gone while she was in her kitchen an' couldn't be at her winder to watch for 'em. Old liar," she brushed back a lock of naturally straight hair. "She wasn't expectin' nobody."

"Well, you know how some folks are—"

"Fools! Half the world," was her vehement answer. "I'm goin' in the front room an' set down a spell. I bin on my feet all day. Leave them dishes on the table. God knows I'm tired, but I'll come back an' wash 'em." But they both knew, of course, that he, very clumsily, would.

At precisely quarter past nine when he, strained at last to the breaking point, uttering an inhuman, strangled cry, flung down his paper, clutched at his throat and sprang to his feet, Millie's surprised young voice, shocking him to normalcy, heralded the first of that series of great moments that every humble little middle-class man eventually experiences.

"What's the matter, poppa? You sick? I wanted you to help me."

He drew out his handkerchief and wiped his hot hands. "I declare I must 'a' fallen asleep an' had a nightmare. No, I ain't sick. What you want, hon?"

"Dictate me a letter, poppa. I c'n do sixty words a minute. —You know, like a business letter. You know, like those men in your building dictate to their stenographers. Don't you hear 'em sometimes?"

"Oh, sure, I know, hon. Poppa'll help you. Sure. I hear that Mr. Browning—Sure."

Net rose. "Guess I'll put this child to bed. Come on now, Daisy, without no fuss. —Then I'll run up to pa's. He ain't bin well all week."

When the door closed behind them, he crossed to his daughter, conjured the image of Mr. Browning in the process of dictating, so arranged himself, and coughed importantly.

"Well, Millie—"

"Oh, poppa, is that what you'd call your stenographer?" she teased. "And anyway pretend I'm really one—and you're really my boss, and this letter's real important."

A light crept into his dull eyes. Vigor through his thin blood. In a brief moment the weight of years fell from him like a cloak. Tired, bent, little old man that he was, he smiled, straightened, tapped impressively against his teeth with a toil-stained finger, and became that enviable emblem of American life: a business man.

"You be Miz Hicks, huh, honey? Course we can't both use the same name. I'll be J. Lucius Jones. J. Lucius. All them real big doin' men use their middle names. Jus' kinda looks big doin', doncha think, hon? Looks like money, huh? J. Lucius." He uttered a sound that was like the proud cluck of a strutting hen. "J. Lucius." It rolled like oil from his tongue.

His daughter twisted impatiently. "Now, poppa—I mean Mr. Jones, sir—please begin. I am ready for dictation, sir."

He was in that office on Boylston Street, looking with visioning eyes through its plate-glass windows, tapping with impatient fingers on its real mahogany desk.

"Ah—Beaker Brothers, Park Square Building, Boston, Mass. Ah—Gentlemen: In reply to yours of the seventh instant would state—"

Every night thereafter in the weeks that followed, with Daisy packed off to bed, and Net "gone up to pa's" or nodding inobtrusively in her corner, there was the chamelion change of a Court Street janitor to J. Lucius Jones, dealer in stocks and bonds. He would stand, posturing, importantly flicking imaginary dust from his coat lapel, or, his hands locked behind his back, he would stride up and down, earnestly and seriously debating the advisability of buying copper with the market in such a fluctuating state. Once a week, too, he stopped in at Jerry's, and after a preliminary purchase of cheap cigars, bought the latest trade papers, mumbling an embarrassed explanation: "I got a little money. Think I'll invest it in reliable stock."

The letters Millie typed and subsequently discarded, he rummaged for later, and under cover of writing to his brother in the South, laboriously, with a great many fancy flourishes, signed each neatly typed sheet with the exalted J. Lucius Jones.

Later, when he mustered the courage, he suggested tentatively to Millie that it might be fun—just fun, of course! —to answer his letters. One night—he laughed a good deal louder and longer than necessary—he'd be J. Lucius Jones, and the next night—here he swallowed hard and looked a litte frightened—Rockefeller or Vanderbilt or Morgan—just for fun, y'understand! To which Millie gave consent. It mattered little to her one way or the other. It was practise, and that was what she needed. Very soon now she'd be in the hundred class. Then maybe she could get a job!

He was growing very careful of his English. Occasionally—and it must be admitted, ashamedly—he made surreptitious ventures into the dictionary. He had to, of course. J. Lucius Jones would never say "Y'got to" when he meant "It is expedient." And, old brain though he was, he learned quickly and easily, juggling words with amazing facility.

Eventually he bought stamps and envelopes—long, important-looking envelopes—and stammered apologetically to Millie, "Honey, poppa thought it'd help you if you learned to type envelopes, too. Reckon you'll have to do that, too, when y'get a job. Poor old man," he swallowed painfully, "came round selling these envelopes. You know how 'tis. So I had to buy 'em." Which was satisfactory to Millie. If she saw through her father, she gave no sign. After all, it was practise, and Mr. Hennessey had promised the smartest girl in the class a position in the very near future. And she, of course, was smart as a steel trap. Even Mr. Hennessey had said that—though not in just those words.

He had got in the habit of carrying those self-addressed envelopes in his inner pocket where they bulged impressively. And occasionally he would take them out—on the car usually—and smile upon them. This one might be from J. P. Morgan. This one from Henry Ford. And a million-dollar deal involved in each. That narrow, little spinster, who, upon his sitting down, had drawn herself away from his contact, was shunning J. Lucius Jones!

Once, led by some sudden, strange impulse, as an outgoing car rumbled up out of the subway, he got out a letter, darted a quick, shamed glance about him, dropped it in an adjacent box, and swung aboard the car, feeling, dazedly, as if he had committed a crime. And the next night he sat in the sitting-room quite on edge until Net said suddenly, "Look here, a real important letter come to-day for you, pa. Here 'tis. What you s'pose it says," and he reached out a hand that trembled. He made brief explanation. "Advertisement, hon. Thassal."

They came quite frequently after that, and despite the fact that he knew them by heart, he read them slowly and carefully, rustling the sheet, and making inaudible, intelligent comments. He was, in these moments, pathetically earnest.

Monday, as he went about his janitor's duties, he composed in his mind the final letter from J. P. Morgan that would consummate a big business deal. For days now letters had passed between them. J. P. had been at first quite frankly uninterested. He had written tersely and briefly. Which was meat to J. Lucius. The compositions of his brain were really the work of an artist. He wrote glowingly of the advantages of a pact between them. Daringly he argued in terms of billions. And at last J. P. had written his next letter would be decisive. Which next letter, this Monday, as he trailed about the office building, was writing itself on his brain.

That night Millie opened the door for him. Her plain face was transformed. "Poppa—poppa, I got a job! Twelve dollars a week to start with! Isn't that *swell!*"

He was genuinely pleased. "Honey, I'm glad. Right glad," and went up the stairs, unsuspecting.

He ate his supper hastily, went down into the cellar to see about his fire, returned and carefully tidied up, informing his reflection in the bathroom mirror, "Well, J. Lucius, you c'n expect that final letter any day now."

He entered the sitting-room. The phonograph was playing. Daisy was singing lustily. Strange. Net was talking animatedly to—Millie, busy with needle and thread over a neat, little frock. His wild glance darted to the table. The pretty, little centerpiece, the bowl and wax flowers all neatly arranged: the typewriter gone from its accustomed place. It seemed an hour before he could speak. He felt himself trembling. Went hot and cold.

"Millie—your typewriter's—gone!"

She made a deft little in and out movement with her needle. "It's the eighth, you know. When the man came to-day for the money, I sent it back. I won't need it no more—now! —The money's on the mantle-piece, poppa."

"Yeh," he muttered. "All right."

He sank down in his chair, fumbled for the paper, found it.

Net said, "Your poppa wants to read. Stop your noise, Daisy."

She obediently stopped both her noise and the phonograph, took up her book, and became absorbed. Millie went on with her sewing in placid anticipation of the morrow. Net immediately began to nod, gave a curious snort, slept.

Silence. That crowded in on him, engulfed him. That blurred his vision, dulled his brain. Vast, white, impenetrable.... His ears strained for the old, familiar sound. And silence beat upon them.... The words of the evening paper jumbled together. He read: J. P. Morgan goes—

It burst upon him. Blinded him. His hands groped for the bulge beneath his coat. Why this—this was the end! The end of those great moments—the end of everything! Bewildering pain tore through him. He clutched at his heart and felt, almost, the jagged edges drive into his hand. A lethargy swept down upon him. He could not move, nor utter sound. He could not pray, nor curse.

Against the wall of that silence J. Lucius Jones crashed and died.

# 10 The Automobile

## Automobiles and the Jim-Crow Regulations (1924)

Like all technologies, automobiles were not used by all for the same reasons and did not always have the same effects. We know, for example, that African-Americans in Atlanta, were more likely than white people of the same level of wealth to buy automobiles. One clear explanation is that Jim Crow streetcars were perfectly acceptable to white riders but a constant humiliation and inconvenience to blacks. The following news story from *Negro World* was reprinted from the *Pittsburgh American.*

### From The Pittsburgh American

The automobile has revolutionized industry. This is no less true in the railway passenger service, where the leading railroad companies in the country have resorted to car advertisement as a means of counterbalancing the financial loss attributed to the automobile industry. The motor car is regarded as a necessity rather than a luxury and this is the rule even among small families with moderate means. The Negro has long since had the same standard of living of other well regulated American citizens, and consequently, it is common experience to see a Negro or a member of his family driving a high-powered automobile.

In the South the Negro must ride on Jim Crow cars. He has rightly regarded it as a violation of his constitutional rights. The automobile enables him to enjoy personal liberty and to go from one city or state to another unrestrained. He is making use of the opportunity afforded. During the Masons' and Elks' conventions in Pittsburgh, Negroes drove their Cadillacs, Pierce-Arrows, MacRarlans and Buicks from every State east of the Rocky Mountains. In the South, this mode of travel is becoming general. The result is a wholesale loss of patronage to the Southern railroad companies.

At one time, the Negro lost—today the railroad companies are losing. Within the near future, we expect the railroads of the South to make concessions to regain their lost trade, even if it should mean open support to anti–Jim Crow legislation. Republican and Democratic administrations have ignored the Negro's demand for abolition of the Jim Crow laws. There is absolutely no excuse for the present administration to condone this un-American Southern policy. This is especially true in interstate travel. Here is an opportunity for the American Government and the Southern railroad capitalists to retrieve one of their unlawful and prejudicial practices.

Reprinted from: "Automobiles and the Jim Crow Regulations," *Negro World* (October 11, 1924) 6.

# Through the Windshield (1933)

Alfred Edgar Smith

Again, like all technologies, the automobile (itself a bundle of many separate technologies) has always existed in an intricate web of social, political, economic, cultural and technological contexts. They were, for example, of little use without roads, gasoline stations, mechanics' garages, insurance, and traffic laws and courts; long-distance travel also demanded tourist amenities such as diners and motels. Escaping Jim Crow streetcars was only one part of the equation.

---

Good roads beckon to you and me, daily we grow more motor-wise. The nomad in the poorest and the mightiest of us, sends us behind the wheel, north, south, east, and west, in answer to the call of the road. And in addition to invitation, there is propulsion. The ever-growing national scope of modern business commands; pleasure suggests; and (in down-right selfish frankness) it's mighty good to be the skipper for a change, and pilot our craft whither and when we will. We feel like Vikings. What if our craft is blunt of nose and limited of power and our sea is macadamized; it's good for the spirit to just give the old railroad Jim Crow the laugh.

Nevertheless, with transcontinental and intersectional highways, with local roads growing nationally good, with good cars growing cheaper, and with good tires within the reach of everyone; there is still a small cloud that stands between us and complete motor-travel freedom. On the trail, this cloud rarely troubles us in the mornings, but as the afternoon wears on it casts a shadow of apprehension on our hearts and sours us a little. "Where," it asks us, "will you stay tonight?" An innocent enough question; to our Nordic friends, of no consequence. But to you and me, what a peace-destroying world of potentiality. It is making motor-"interurbanists" out of us. We have a friend in Atlanta and another in Jacksonville, so what? So we must be off like a rocket at break of dawn from Atlanta and drive like fiends and fanatics to reach Jacksonville before midnight. Or maybe it's Cincinnati and St. Louis, or Dallas and El Paso. We must not tarry, cannot, to see the wonders that thrust themselves at us 'round each bend; and why? Listen to the echoes in your own experiences as I tell you why.

The Bugbear is the great uncertainty, the extreme difficulty of finding a lodging for the night, —a suitable lodging, a semi-suitable lodging, an unsuitable lodging, any lodging at all, not to mention an eatable meal. In a large city (where you have no friends) it's hard enough, in a smaller city it's harder, in a village or small town it's a gigantic task, and in anything smaller it's a matter of sheer luck. And, in spite of

---

Reprinted from: Alfred Edgar Smith, "Through the Windshield," *Opportunity* 11 (May 1933), 142–144.

unfounded beliefs to the contrary, conditions are practically identical in the Mid-West, the South, the so-called North-east, and the South-southwest.

The typical confronting-condition and procedure is as follows: After a day of happy carefree meandering along good roads to the tune of from four to sixteen cylinders, a pleasant physical sense of tiredness makes itself felt and it is decided to stop at the next village, hamlet, or junction to seek a place to dine and to lay the weary head. Lights presently come to view, and assuming a look of confidence for the benefit of the wife, we search anxiously and somewhat furtively for a dark-hued face. Presently we spy one and make inquiry in a low voice as to the possibility of securing a night's lodging. This first individual invariably answers you with a blank stare, and you suddenly remember certain mannerisms of your long-lost boyhood in the South (those of us who are from a month to a generation removed from residence in the heart of this pleasant much lyricized section) and we repeat the question slowly and in a way you know as well as I do. This elicits the information that there is no hotel for "us," there is no rooming or boarding house, but Mrs. X has a place where "folks stay sometimes." So away to Mrs. X's across the railroad tracks.

Mrs. X when aroused from dinner or from a rocker, regards our proposals with an expressionless stare, which we, being of kindred blood understand quite well. In our desperate need our early training again comes to the rescue and we disarm our hostess after a time and put her completely at her ease, so that it is guessed that we can be provided with a bed, and later under the influene of our overworked charm we are provided with a meal. It doesn't seem so hard in the telling; but remember we were lucky, any one of a number of things could have gone wrong forcing a continuation of the journey for twenty or thirty miles and a repetition of the procedure. And also, Mrs. X in the above case probably provides for us to the best of her ability, and her provisions, meals and otherwise, are without a doubt, the best to be had. But "best" is a relative thing.

Sometimes the procedure varies a little. Maybe we are fortunate and there is a hotel after a fashion. Or maybe we spy a druggist or some individual who directs us to a really nice home. Or maybe one of the members of the professions takes us in out of the kindness of his heart and the memories of his own experiences. Maybe, and maybe not. How many homes where "folks stay sometimes," there are, with sewage but no indoor toilet or bath, water pipes but no running water, feather beds to trap the unsuspecting, minute bed inhabitants, no screens, and so on without end. And imagine our embarassment when we have had the good fortune to find a hotel after a sort, and then notice the proprietor unashamedly passing out halfpint bottles of some colorless fluid to an evidently well-established clientele; or the time when we found a surprisingly clean room through the good offices of a young lady drug clerk only to be kept awake by a certain type of music downstairs and the trooping in of young ladies and their escorts to occupy the other rooms during the hectic night; or pick one out of your own experiences. Remember the time they wanted to move the old man (who had been ailing with tuberculosis) from his bed to make a place for you, or the time you had to sit up all night in a chair with a light on to ward off the organized attacks of the vermin. Remember? Of course you do. And what to do about it?

In the light of personal experience and of questioning friends, it seems that the real motorists' nightmare is the uncertainty of finding a lodging, rather than the type of lodging itself. The more so, as I am led to believe that a traveller must inure himself to a certain amount of discomfort and hardship, be he in any section of the United States, Canada, Europe, or wherever he may be. So the first and major problem is how and where to be sure of a place to stay. Obviously the answer lies in the compilation of an authentic list of hotels, rooming houses, private homes catering to the occasional traveller, tourist camps, and every type of lodging whatsoever, including those run by members of other races and open to Negroes; and the availability of such a list to our growing army of motor-travellers. Such a list would if complete, be invaluable (and I can hear your fervent amens) for I am convinced that within the area of every fifty square miles of the more frequently travelled sections there are lodgings to be found at all times. If we just knew exactly where they were, what a world of new confidence would be ours.

Up to date there have been, to my knowledge, two efforts to supply this need. The preparation and sale of one such list, which included hotels, rooming houses and the like, resulted in a bankruptcy. The other list includes only hotels, Y. M. C. A.'s and Y. W. C. A.'s, but it is most complete and authentic. It was compiled by Mr. James A. Jackson who is now the director of the Small Business Section of the Marketing Service Division of the United States Department of Commerce, with offices in the new Commerce Building in Washington. This list contains some 120 hotels, 31 Y. M. C. A.'s and 14 Y. W. C. A.'s in 35 States, the District of Columbia and Canada. I heartily recommend this list to you. Also I live in hope that some individual, organization, or publication with unlimited publicity at his or its command will attempt a more inclusive and complete list. I am sure it would not lack contributors. I think with what great pleasure I could recommend a private home in a small town in Ohio, one in Missouri, one in Tennessee, and a hotel in Virginia.

The summer past, we turned the blunt nose of our Viking craft westward on the concrete-asphalt seas, and sailed away for new conquests and adventures. Starting from Washington, D. C., the first half of our journey was punctuated with the usual very short or very long runs between Pittsburgh, Chicago, St. Louis, and Ft. Riley, to take advantage of the presence of known friends and accommodations. And pleasant are the memories of our stay with real friends at these objectives. But somehow it takes the joy out of gypsying about, when you have to be at a certain place by a certain time, and there are one or two unpleasant memories of way stops between stations. Westward from St. Louis, we were advised to turn southward and wend our way through Texas and thenceward. But we refused and pointed our nose toward northern Colorado and Utah, if for no other reason than we knew that there were stretches of mountainous and desert country that boasted no Negro inhabitants, and we wanted the feeling of stopping when and where we wished, since accommodations must come from the other group, and one place was probably as good as another. We tried to run away from the Bugbear, but it merely changed its form and stuck close.

As we neared the Colorado State line we began to grow Tourist Camp-Conscious. (The West has reduced this business to a science). We found that cabins were cabins in name only, and ran the scale from one room cottages with a wood stove, to a modern detached apartment with sunken bath and locked garage with a connecting door. So following flivvers and Rolls Royces, we patronized the Tourist Camps. And here old Bugbear appeared in his new form.

It would seem that our sensibilities would be somewhat dulled by the continuous hurts they receive in this land of ours; but not so. Every time that a camp manager announced his camp full for the night, or showed us to a cabin that was second rate in appearance, the thought persisted, "Is he doing this because I am a Negro?" Probably we imagined slights where none were intended. Certainly we have the pleasantest of memories of these camps, of the Mormons, of the Utah Indian (named Smith) who welcomed us royally and wanted to talk far into the night about Senator Smoot and things politic, of the Colorado hostess who insisted we would be more comfortable in her hotel than in a cabin, of cordial Californians, and others. Of course, occasionally we struck a Texan who had emigrated, and our sensibilities received new hurts when he would advise us that no accommodations were to be had for "us" within a radius of fifty miles. But it was balm to our wounds, when invariably we would go next door to a better camp (usually run by a chain syndicate) and be welcomed right royally.

And speaking of chain syndicates, a last word on this matter. As a group we have no reason to love these organizations as colored labor within their ranks is usually taboo. But having switched to a well known brand of gasoline because it could be found in any and all localities the length and breadth of this country and others, in all fairness let it be known that the employees were uniformly courteous, attentive, anxious to serve, and never failed to welcome us with a cherry good morning or evening, and to speed us on our way with well wishes and an invitation to revisit. And this regardless of locale (we returned through the South). Only the attendants at privately-owned stations were discourteous or apathetic. And yet, only the privately-owned stations employed an occasional Negro. A paradox, no? A typical example of a changing spirit in the cotton belt of the South, was the deft covering by an attendant at a chain station, of a sign on the drinking water fountain reading "For White Only," when the colored tourist drew up for gas and oil. This was three Summers ago, last summer the sign had disappeared.

## Running the Red Light (Tilman C. Cothran)

Since the days of slavery, humor has been a powerful weapon among African-Americans, not least as a way of defying and mocking white "supremacy" and the pervasive racism of American society. Like any new technology, the automobile had the potential to redistribute power within society or to reassert existing power structures. Stories persist that at least some southern states contemplated "separate but equal" highway systems not only to separate the races but to prevent white drivers from having to wait at red lights while black drivers had the right-of-way. This joke, collected and published after World War II, probably dates back to the early days of the automobile in the South.

---

The colored man was arrested in a town in Mississippi for crossing a red light. He explained, "I saw all the white folks going on the green light; I thought the red light was for us colored folks." The judge let him off.

(Frequently educated Negroes will pretend to be ignorant to fool the white man.)

---

Reprinted from: "Running the Red Light (Tilman C. Cothran)," *American Negro Folktales*, collected by Richard M. Dorson (Greenfield, Conn.: Fawcett Publications, 1967), 309. 

## U.S. Patent to G. A. Morgan for a "Traffic Signal" (1923)

One of the ironies of the preceding joke is that African-American inventor Garrett Morgan by 1922 already had made a significant improvement in the design of traffic signals, such as the one through which the Mississippi driver "ran." Although Morgan's signal was hand operated, he was able to sell it to the General Electric company.

**United States Patent Office**

**Garrett A. Morgan of Cleveland Ohio**
Traffic Signal

Patented Nov. 20, 1923
Application Filed February 27, 1922. Serial No 539,403

To whom it may concern:

Be it known that I, Garrett A. Morgan, a citizen of the United States, residing at Cleveland, in the county of Cnyahoga and State of Ohio, have invented a certain new and useful Improvement in a Traffic Signal, of which the following is a full, clear, and exact description, reference being had to the accompanying drawings.

This invention relates to traffic signals, and particularly to those which are adapted to be positioned adjacent the intersection of two or more streets and are manually operable for directing the flow of traffic.

One of the objects of my invention is the provision of a visible indicator which is useful in stopping traffic in all directions before the signal to proceed in any one direction is given. This is advantageous in that vehicles which are partly across the intersecting streets are given time to pass the vehicles which are waiting to travel in a transverse direction: thus avoiding accidents which frequently occur by reason of the over-anxiety of the waiting drivers, to start as soon as the signal to proceed is given.

Another object is the provision of a semaphore signal which is useful by night as well as by day and which is arranged to be easily and automatically operable by the traffic director. In addition, my invention contemplates the provision of a signal which may be readily and cheaply manufactured.

To this end, I provide a signal wherein the direction indicating arms are pivotally supported and adapted to be moved vertically for stopping the flow of traffic and

Reprinted from: U.S. Patent No. 1,475,024, dated November 20, 1923 to G. A. Morgan, for a "Traffic Signal."

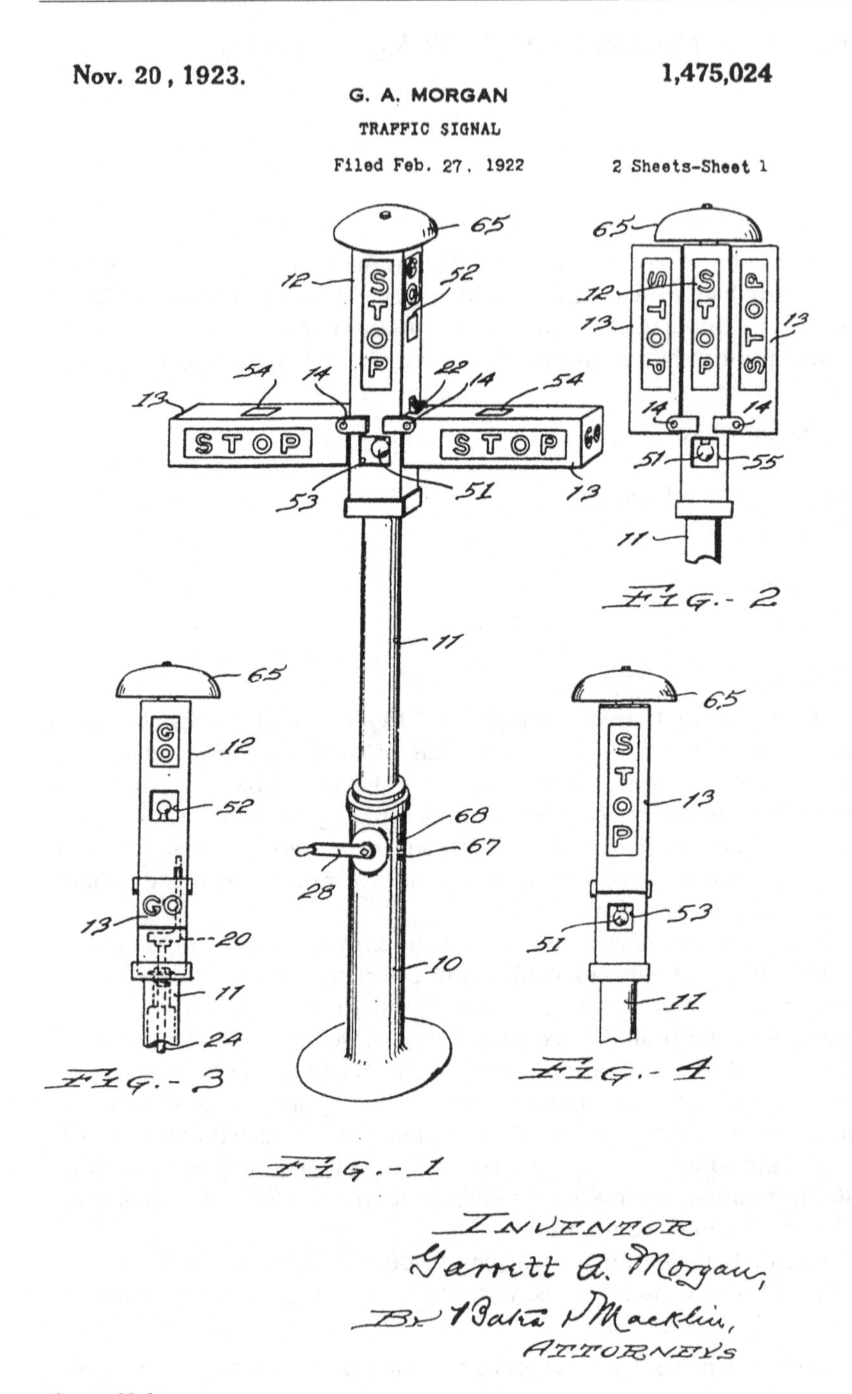


**Figure 10.1**
Patent drawing for Garrett A. Morgan's Traffic Signal.

then to be revolved and dropped to indicate a right of way to vehicles moving in another direction. The raising and revolving movements of the arms are adapted to be accomplished by the turning of a crank, and suitable mechanism actuated by the crank is provided for automatically indexing the arms to the required position, and for permitting their proper functioning in accordance with the wishes of the traffic director.

The means for accomplishing the above objects will be fully set forth in the following description which relates to the drawings, and the essential characteristics of my invention will be summarized in the claims....

To adapt a signal constructed according to my invention for use at night, I have shown two electric lamps 50 and 51 which are mounted within the vertical indicator. These lamps may receive electrical energy either from a battery mounted within the standard 10 (not shown) or from any other suitable source of supply, such as through leads depending from an overhead line. The lamp 50 is positioned adjacent the openings 52 above the point of pivotal connection, while the lamp 51 is adjacent the openings 53 below the point of pivotal connection....

In testimony whereof, I hereunto affix my signature.

Garrett A. Morgan

# VI World War II and the Cold War

Historians commonly acknowledge that the mobilization for World War II finally put an end to the Great Depression and that spending for the arms race of the Cold War fueled the sustained prosperity of the 1950s and 1960s. With increased federal spending for armaments, and the acceptance of more and more men into the armed services, manufacturing jobs opened up and, as usual in such instances, women and African-Americans were gradually recruited to fill many of the new or recently vacated positions. Some African-Americans and women, of course, wound up in the forces themselves, though in both cases in carefully segregated units.

In the factories and shipyards, too, old patterns persisted among new conditions. Trades unions were reluctant to admit members of color, whole jobs categories (usually the higher paying ones) were reserved for white men, and when entirely new facilities were built, like the super-secret Manhattan District city of Oak Ridge, Tennessee, all of the worst manifestations of racist planning were preserved. But, also as usual, muddy waters made good fishing, and thousands of technical jobs were opened up to workers who previously had been barred from employment.

Pushed by wartime demands, the technologies themselves evolved, in some cases with great speed. Electronics in the form of radar and proximity fuses were only the most obvious instances of the creation of a whole new "high-tech" battlefield environment. At one end of the spectrum, scientists performed the research and development that made these new weapons possible. At the other end, thousands of young soldiers and sailors were given intensive courses in how to operate, maintain, and repair them. Too late for the war itself, but pregnant with possibilities for the second half of the century, the first computers were being assembled at the University of Pennsylvania and other campuses around the country. If World War I had been a chemists' war, resulting in a large postwar growth of the chemical industry in the country, World War II was thought of as a physicists' war, and the electronic future would prove equally expansive.

Long-established industries, such as the automobile and steel, simply increased production, though in the first case of newly designed products. It was mass production on an unprecedented scale, and some of the most obvious techniques of mass production, like the use of subassemblies, were applied now to building ships and airplanes on an equally large scale. Appeals to "Rosie the Riveter" brought African-American women out of the white homes where they served as domestics and turned them into welders at union wages. Both jobs, of course, used tools, but the rewards (social and emotional as well as economic) of the latter were several steps up the hierarchy which society assigned to technologies.

The window of opportunity for greater participation in the technological life of the nation that wartime opened for African-Americans lasted a scant half-decade but could not be closed again quite a tightly as before. The war years were a period not only of state-sponsored opportunity but one of increasing civil-rights demands from black Americans themselves. New technologies both created and destroyed jobs and cultural opportunities; their applications are always contested and their results always contingent.

After the war, again the last hired were the first fired. Women who had been urged out of their homes to take war jobs were now just as insistently told to return to husbands and babies. African-Americans also were moved aside so that "the boys" who had fought overseas could have "their" jobs back. The great fear that having millions of demobilized workers suddenly dumped on the market, and the closing of war industries as their products were no longer needed, would lead quickly straight back to depression turned out not to be soundly based. The pent-up demand for consumer technologies—cars, refrigerators, single-family houses—that had gone unfulfilled during the war now created a seemingly bottomless market for civilian goods and services. Such federal initiatives as the G. I. Bill to send ex-military personnel to college, the program patterned on the Rural Electrification Administration to bring telephones to rural America, the project to lace the country together with interstate highways, and a host of others pumped money into the economy, creating infrastructure and jobs together.

The threat of technological unemployment on the farm, epitomized by the Rust Brothers' cotton-picking machine of the 1930s, was fulfilled on a staggering scale after the war. Cotton quickly went from a labor-intensive crop that sustained thousands of (admittedly poor, and predominantly black) families to one that was highly capitalized and almost totally mechanized. Tomato-picking machines had a similar effect on Latino field workers in California. Field hands from the South had little choice but to move to cities in the North or the South to look for new industrial jobs.

The aerospace industry of southern California became a model of the new economy. Substantially reconfigured during the war, the old airplane manufacturing sector, under the twin stimuli of wartime technological developments and a Cold War arms race, metamorphosed into a high-tech aerospace industry producing aircraft and now also missiles for the military as well as space probes and satellites for the National Aeronautics and Space Administration (NASA). Centered originally in California and Seattle, Washington, it spread by the mid-1960s to the South, forming, in those years, a new "fertile crescent" stretching from Atlanta, through Florida and Texas to Southern California, the San Francisco Bay Area, and on to Seattle. The combination of new jobs and federal rules barring discrimination made the industry one where women and African-Americans had, again, a better than usual chance to join the beneficiaries of this new technology.

# 11 Learning to Fly

*Soaring above Setbacks: The Autobiography of Janet Harmon Bragg, African American Aviator*
Janet Harmon Bragg

**Coffey School of Aeronautics (1944) and 99th Pursuit Squadron, Air Unit (1942)**

**Lt. Clarence "Lucky" Lester**

## *Soaring above Setbacks: The Autobiography of Janet Harmon Bragg, African American Aviator*

Janet Harmon Bragg

**Figure 11.1**
The romance of flying was contagious in the years following the Wright brothers' first flight, and African-American pilots were among the first to experience the thrill and emancipation of this new technology. In fact, one of the Challenger Aero Club's founders flew for the Ethiopian government in its 1936 war against Italian invaders. Janet Harmon Bragg was one of the first female pilots in the club. In this photograph, some club members pose at their hanger in Robbins, Illinois; Bragg is standing in the center, next to the airplane. Reprinted from Janet Harmon Bragg, *Soaring above Setbacks: The Autobiography of Janet Harmon Bragg, African American Aviator* (Washington: Smithsonian Institution Press, 1996).

## Coffey School of Aeronautics (1944) and the 99th Pursuit Squadron, Air Unit (1942)

As with other technologies, the airplane proved not to be immune from the social constructions of a racist and segregated society. At the time of World War II the separate flying facilities established with the black community were swept up into a (still segregated) push for national defense. The most famous African-American pilots of World War II were the airmen from Tuskegee.

### Coffey School of Aeronautics

The Coffey School of Aeronautics, owned and operated by Negroes, was founded in 1934 at Oaklawn, Ill., a suburb of Chicago, by Cornelius R. Coffey and Miss Willa Brown, aviators, as a flying service for sightseeing trips. It later took on the function of a private training school for commercial pilots. In its early days the majority of its students were white. The number of colored students gradually increased, however, after Miss Brown circled the country by plane in 1939 in order to interest Negroes in flying.

The training staff is composed of both colored and white instructors; but the school is operated and controlled by the two colored founders. Mr. Coffey is president and Miss Brown (who is his wife) is business manager and co-ordinator of the war training program, under the Civil Aeronautics Administration.

In January, 1940, before the United States entered the war, the school signed its first contract with the government to train non-college students in civil flying. With the aid of the National Airmen's Association (colored), organized in 1939, the school negotiated another contract with the CAA, in December, 1941, to train selected colored students "as an experiment" to determine whether the government would use them in the army air forces.

Under this program, twenty students were entered in the school, under government contract. When the first colored Army air unit was established at Tuskegee Institute in 1942, the graduates were sent there for advanced training. After that time, the school functioned as a feeder base for the air unit at Tuskegee.

At the same time, the Coffey School was engaged in giving refresher courses to flight instructors, white and colored, who were entering the Army. Up to July, 1943, the school had trained 250 government pilots.

Reprinted from: "Coffey School of Aeronautics," *The Negro Handbook, 1944*, ed. Florence Murray (New York: Current Reference Publications, 1944), 122; and "99th Pursuit Squadron (Air Unit)," *The Negro Handbook*, ed. Florence Murray (New York: Wendell Malliet and Co., 1942), 78.

When the school became a part of the Army's program, Miss Brown succeeded in having the CAA authorities authorize the formation of a complete squadron of flyers under the regulations of the Army. Following the necessary written and practical tests, she was commissioned second lieutenant in the Army Air Forces, the only colored woman to receive this commission. (There are about six white women.) Mr. Coffey was commissioned first lieutenant.

In 1943, a squadron of the Civil Air Patrol was organized at the school. This unit (the CAP), now government-sponsored, was originated by private pilots who volunteered to organize it. It now consists of some 75,000 members, who for various reasons do not meet the requirements of the armed forces.

The duties of the CAP consist primarily of patrolling the home coast to watch for enemy submarines, enemy planes, and other hostile activity. Other activities are towing targets for the army air forces to shoot at in practice maneuvers, watching for forest fires and aiding in fighting them, guarding junk piles, protecting defense areas, performing courier service, co-operating with the Office of Civilian Defense in mock air raids and in performing other tasks that relieve army flyers for war duties.

## 99th Pursuit Squadron, Air Unit

Continuing their efforts to open up equal opportunities to serve in the armed services of the country, during the latter part of 1940 Negroes began to concentrate their efforts on breaking down color barriers in the Air Corps branch of the Army. In January, 1941, under the direction of the National Association for the Advancement of Colored People a Howard University student, Yancey Williams, filed suit against the War Department to compel his admission to an air-training center. Almost immediately following the filing of the suit, the War Department announced that it would establish an air unit near Tuskegee Institute, Ala.; in cooperation with Tuskegee Institute, for the training of colored pilots for the Army, to be called the 99th Pursuit Squadron.

Negroes were divided in their opinions on the fairness and efficacy of a racially separate pilot unit. The National Airmen's Association, composed of colored licensed civil pilots, took the lead in opposing this "segregated" air squadron.

Nevertheless, plans went forward for the training of the men and construction of the air base near Tuskegee Institute, with the War Department, in April, appropriating $1,813,942 for its buildings and facilities. Land was purchased about six miles from the school and Tuskegee was given a contract of $80,000 for the first year to house and feed the cadets and ground crew.

The first 75 colored men recruited for training as airplane mechanics were sent to Chanute Field in Illinois for technical training. The air base, completed in December, consisted of hangars, repair shops, classrooms, facilities for administrative personnel, facilities for flight surgeons, an infirmary, dining hall, fire house and dormitories. The entire personnel, except a few Army flight instructors, are colored.

Capt. Benjamin O. Davis was appointed military instructor and flight commander and was given flight training for the job. The architect and contractor for the air base were also colored.

The air training center was dedicated at Tuskegee Institute on July 19, 1941, and the first ten pilots were enrolled for preliminary military and ground training.

## Lt. Clarence "Lucky" Lester

**Figure 11.2**
Lieutenant. Clarence "Lucky" Lester, left, with his crew chief, on the wing of his P-51 Mustang fighter. Reprinted from Stanley Sandler, *Segregated Skies: All-Black Combat Squadrons of WWII* (Washington, D.C.: Smithsonian Institution Press, 1992).

# 12 War Work

Negro Workers and the National Defense Program (1942)

A Black Woman at Work in a Wartime Airplane Assembly Plant

Graduate Technicians and National Technical Association (1944)

Defense and Wartime Employment (1946–1947)

Bay Area Council against Discrimination, San Francisco (1942)

## Negro Workers and the National Defense Program (1942)

With the increase in defense work after 1939, the employment situation began to look remarkably like that of World War I. The combination of increased orders for goods for the American military as well as its allies and the draft of so many white men into the armed services created labor shortages in critical industries that could be filled only by women and people of color. Mexican nationals were imported to work the agricultural fields of the West, and the new shipyards and airplane factories of California began, however grudgingly, to recruit women and African-Americans. Early in the war, it still was not clear how far this would go and what would be the results.

---

In spite of the increasing difficulty of finding experienced skilled and semiskilled workers, many employers continue to refuse to hire available Negroes for production work. Specific instances that have already come to the attention of the employment service indicate that the failure to use available skilled and semiskilled Negroes has accentuated the shortage of experienced workers in certain occupations and slowed up production.

Failure to use locally available Negroes has not only delayed production in certain instances, but it is compelling employers to recruit labor from distant areas, with two serious consequences: it is intensifying existing housing shortages, and it is increasing the amount of labor turnover. Moreover, the failure to use Negro labor exaggerates the shortage of labor and results in competition for white labor and, what is perhaps more important, results in the under-employment of vocational training facilities in some communitites where there is a shortage of white applicants for training.

As the defense program moves from its present preliminary phase to full production, enormous demands are going to be made on the unemployed reserve. The very sharp reduction during the past year in the number of unemployed job seekers gives some indication of the effects of the armament program in its early stages, before the full effects have been felt.

Negroes constitute a far more important proportion of the unemployed labor reserve than is commonly recognized; this is true both in the South, where they form from one-third to one-half of the unemployed labor reserve, and in many Northern urban-industrial areas, where they form from one-fourth to one-third of the most employable reserve. Consequently, the failure to use Negro labor presents an important

---

Excerpted from: "Negro Workers and the National Defense Program," *The Negro Handbook*, ed. Florence Murray (New York: Wendell Malliet and Co., 1942), 86–93.

problem in certain areas, and the continued refusal to hire or train them for defense production is likely to generate increasingly serious labor market problems.

Responsibility for the failure to use Negro labor in defense or other industries cannot be laid exclusively on employers. The traditional prejudice of white workers, both organized and unorganized who are not accustomed to working with Negroes, frequently influences employer policy. There are innumerable instances, however, in which the opposition of white workers to the employment of Negroes has been overcome when employers have consciously sought to enlist the cooperation of the white workers in an intelligent and enlightened way.

### Importance of Negro Workers in the Unemployed Labor Reserve

It is the official labor market policy of the government for the present emergency to use the local labor reserve insofar as possible to meet the local demand before widening the areas of recruitment beyond the commuting area to the State, the region or the Nation. The underlying reason for the emphasis on this order of recruitment is the desire to minimize some of the serious difficulties which developed during 1917–18 when the failure to follow a systematic plan of recruitment resulted in unnecessary competition for labor, unnecessary migration, labor turnover, and housing shortages, which seriously disrupted the Nation's labor market and interfered with production.

The success of such a policy depends on the willingness of employers to make full use of the available labor supply in each community, which in certain areas would mean extensive use of Negro labor. The importance of Negroes in the total labor force is sometimes overlooked, and their disproportionate importance among the unemployed, largely due to the displacement of Negroes by white workers during the depression years, has made them a factor of the first magnitude in the labor reserve in many areas.

Furthermore, the residual unemployed labor reserve rapidly becomes even more heavily weighted with Negro workers as a result of preferential hiring of white workers during a period of re-employment, partly on racial grounds and partly because white workers, by and large, are more likely to have acquired training and experience in the skilled and semiskilled occupations that are in demand. This is true not only in the South but in the industrial areas of the North as well.

### Job Opportunities for Negroes in the Defense Program

The experience of Negroes during the rapid expansion of job opportunities, which began in the spring of 1940, can be described in the following terms:

1. In Construction Throughout the country there was a heavy demand for unskilled labor needed for the construction of airports, military cantonments, barracks, arsenals, manufacturing plants, et cetera. Negro unskilled labor shared substantially in these em-

ployment opportunities and benefited by the wage rates paid on these projects, which were generally higher than wages usually earned in unskilled work.

The extraordinary demand for skilled construction workers, particularly carpenters (one of the building trades in which Negroes have established a substantial foothold) resulted in many job opportunities, particularly for union members. There is evidence, however, that in some localities, in spite of the acute shortage of carpenters, Negroes were not employed. In skilled building trades occupations other than carpenters and cement finishers, and to some extent, bricklayers and masons, there is no evidence that discriminatory practices were being noticeably relaxed.

2. In Manufacturing Industries which have Customarily Employed Negroes In certain manufacturing industries notably iron and steel, industrial chemicals, saw-mills, foundries, meat packing, and to some extent shipbuilding, where it is customary for many employers to hire Negroes for skilled and semiskilled as well as for unskilled work, the industrial expansion has opened many job opportunities to Negro workers. However, there is a good deal of evidence to show that even in industries in which Negroes have by custom been accepted, many establishments which have never employed Negroes in the past were refusing to employ Negroes for skilled and semiskilled work. This policy of exclusion to a large extent has helped accentuate the acute shortage which has developed in some areas in the foundry industry, for example.

In Delaware a shortage of skilled and semiskilled foundry workers developed early in the defense program. While Negro foundry labor could have been supplied, in part from local sources and in part by clearance from other areas, most employers who in the past had not made a practice of employing Negroes, continued to refuse to hire them for skilled and semiskilled work. A similar shortage in Wheeling, West Virginia, was reported beginning in November, when foundry workers were needed in steel mills, railroad shops and foundries. Here, too, the shortage was due in part to the failure of certain employers to use available Negro foundry workers. The owner of one large foundry in this city toured eastern Ohio communities offering higher wages to employed white foundry workers, if they would come to Wheeling. A shortage of skilled and semiskilled foundry workers was also reported in Louisiana, North Carolina, Georgia, Illinois, Michigan and Connecticut.

This shortage is not only directly traceable to the failure of certain employers to use Negro labor, but in part, it is the result of the failure of employers to upgrade and promote Negro labor to the skilled occupations. While the practice of upgrading workers is a common method of overcoming labor stringencies, this practice is frequently not extended to Negro workmen even in those industries in which they are extensively employed.

Apart from the industries listed above in which Negroes have more or less of a foothold (usually in the less desirable jobs or in the unskilled occupations), most of the opportunities for work have come from a handful of individual establishments whose policy it is to employ Negro labor in skilled and semiskilled as well as unskilled jobs.

3. In Defense Industries In certain of the most important industries associated with national defense, particularly aircraft, tank and armament manufacturing, powder

**Table 12.1**
Placements of workers in selected essential skilled and semiskilled occupations important to national defense, by color, October–December 1940 and January–March 1941[1]

| Selected Defense Occupations | October–December 1940 Total | Non-White[2] | Percent Non-White | January–March 1941 Total | Non-White[2] | Percent Non-White |
|---|---|---|---|---|---|---|
| Aircraft | 2,269 | 1 | .05 | 8,769 | 13 | .1 |
| Building construction | 13,121 | 722 | 5.5 | 69,637 | 2,061 | 3.0 |
| Electrical equipment | 1,716 | 18 | 1.0 | 1,066 | 5 | .5 |
| Metal trades[3] | 24,524 | 231 | .9 | 34,834 | 245 | .7 |
| Optical goods | 101 | 1 | 1.0 | 213[4] | 1 | .5 |
| Shipbuilding | 759 | 39 | 5.1 | 1,499 | 25 | 1.7 |

Source: Division of Research and Statistics, Bureau of Employment Security.
1. The two quarters are not comparable, since a considerable number of occupations were added beginning in January.
2. Nearly all "non-white" placements are Negro placements.
3. Includes "foundry and forging, machine shop and machine tool, metal processing and forming."
4. Includes instruments, watches, and clocks.

manufacturing, shell-loading, machine tool, and machine shop, in which acute shortages of skilled and semiskilled labor have developed, there is little evidence that employers who have in the past refused to hire Negro labor are now turning to this source of supply. A comparison of Negro and white placements made by the employment service between October 1941 and March 1941 in selected skilled and semiskilled occupations in important defense industries is given in table 12.1.

Of 11,000 skilled and semiskilled jobs in aircraft, only 14 went to non-white workers; of nearly 60,000 placements in the metal trades occupations, including foundry and forging, machine shops and machine tool, and metal processing less than 500 went to non-white workers. Only in building construction and shipbuilding did Negroes receive a somewhat higher proportion of the skilled and semiskilled placements.

Not only are non-white workers not receiving many skilled or semiskilled jobs in a great many defense establishments, but they are receiving very few jobs of any type, even unskilled. Table 12.2 shows all male placements[1] made by the public employment offices between October 1940 and March 1941 in 20 selected defense industries.

Only in shipbuilding, hardwood distillation, charcoal and naval stores, industrial chemicals, miscellaneous chemicals, and iron and steel products were non-white workers receiving a more or less proportionate share of the job opportunities. Of over

**Table 12.2**
Placements of males by public employment offices in twenty major defense industries, October 1940 through March 1941

| Industry | All Male Placements | | |
|---|---|---|---|
| | | Non-White | |
| | Total | Number | % of Total |
| Total | 154,673 | 6,151 | 4.0 |
| Aircraft and parts | 27,651 | 68 | .2 |
| Automobiles and automobile equipment | 10,889 | 272 | 2.5 |
| Clocks and watches | 677 | 0 | 0 |
| Electrical machinery | 13,100 | 75 | .6 |
| Harwood distillation, charcoal, and naval stores | 158 | 21 | 13.3 |
| Industrial chemicals | 10,322 | 1,218 | 11.8 |
| Industrial rubber goods | 515 | 2 | .4 |
| Iron and steel and their products | 30,103 | 1,472 | 4.9 |
| Lighting fixtures | 944 | 30 | 3.2 |
| Machinery (except electrical) | 29,674 | 624 | 2.1 |
| Miscellaneous chemical products | 1,665 | 89 | 5.3 |
| Motorcycles, bicycles, and parts | 100 | 2 | 2.0 |
| Nonferrous metals not elsewhere classified | 5,982 | 112 | 1.9 |
| Petroleum refining | 4,998 | 482 | 9.6 |
| Primary alloying, and rolling and drawing of nonferrous metals (except aluminum) | 1,672 | 19 | 1.1 |
| Professional and scientific instruments, photographic apparatus, and optical goods | 1,409 | 26 | 1.8 |
| Railroad equipment | 2,795 | 100 | 3.5 |
| Ship and boat building and repairing | 10,674 | 1,500 | 14.1 |
| Surgical, medical, and dental instruments, equipment, and supplies | 759 | 22 | 2.9 |
| Tires and inner tubes | 586 | 17 | 2.9 |

27,000 male placements in aircraft, only 68 or .2 percent were non-white. Of over 13,000 in electrical manufacturing, only 75 or .6 percent were non-white. Of nearly 30,000 in machinery (other than electrical) only 624 or 2.1 percent were non-white.

Compared with these percentages, it is estimated that Negro males in April 1941 formed about 12.4 percent of the total unemployed male reserve and about 12 percent of the skilled, semiskilled and unskilled workers, —the pool of manual labor on which manufacturing industries draw for workers. (See table 12.3.)

The extent to which anti-Negro discrimination in employment is operating and the type of jobs in which Negroes are finding opportunities for employment is revealed in table 12.4.

**Table 12.3**
Male job seekers registered in the public employment offices, by occupational group, by color, April 1941

| | | Non-White | |
|---|---|---|---|
| Occupational Group | Total | Number | % of Total |
| All occupations | 2,991,642 | 370,258 | 12.4 |
| Professional and managerial | 116,128 | 4,933 | 4.2 |
| Clerical and sales | 245,184 | 6,113 | 2.5 |
| Service | 229,233 | 68,689 | 30.0 |
| Agricultural, fishery, and forestry | 272,786 | 37,777 | 13.8 |
| Skilled | 671,086 | 26,515 | 4.0 |
| Semiskilled | 614,347 | 56,007 | 9.1 |
| Unskilled | 758,219 | 162,126 | 21.3 |
| Unassigned[1] | 84,659 | 8,098 | 9.6 |

1. Includes unemployables, recent students, persons without work experience, and unspecified.

In October 31.4 percent and in November 30.6 percent of the job openings went to Negroes, and thereafter the proportion of Negro placements in the service industries rose each month—a reflection of the exodus of white workers from the service field to the more desirable jobs and their replacement by Negroes, and of the greater reluctance of white workers to take jobs in the service industries.[2]

A study of the monthly State reports describing labor market developments from Sept., 1940, to June, 1941, revealed in spite of increasing difficulties in finding labor in certain occupations that employers were averse to hiring experienced and available Negro workers. The most recent reports, those for April, May, and June indicate that a somewhat less restrictive attitude on the part of some employers has been noticed in Pennsylvania, New York City, Buffalo, Northern New Jersey, Cleveland, Cincinnati, and to some extent in Michigan, but it still cannot be said that color restrictions have been substantially relaxed.

Prior to March the placement data in table 12.2 indicate that only a handful of Negroes were hired in all the aircraft plants in the country. Since March a slight change is evident. In most of the other defense industries or establishments in which it has not been customary in the past to employ Negroes, particularly machine shop and machine tool, electrical machinery, tanks, munitions and ordnance, and rubber, instances in which Negroes have been hired are still rare.

4. In Government Establishments Government-owned or operated shipyards are providing many employment opportunities for skilled and semiskilled as well as unskilled Negroes. In September 1940, the last date for which information is available, between 5,000 and 6,000 skilled and semiskilled Negroes were employed in the Navy yards, and

**Table 12.4**
Placements of job seekers

| Month | Selected Skilled and Semi-skilled Occupations[1] (I) | Twenty Selected Defense Industries[2] (II) | All Industries (III) | Construction (IV) | Manufacturing (V) | Service (VI) | All Other (VII) |
|---|---|---|---|---|---|---|---|
| Total | 215,427 | 22,600 | 2,676,976 | 565,600 | 539,355 | 881,380 | 690,641 |
| Non-white | 4,443 | 7,382 | 534,350 | 128,274 | 35,432 | 288,146 | 82,498 |
| Percent non-white | 2.1 | 3.3 | 20.0 | 22.7 | 6.6 | 32.7 | 11.9 |
| **1940** | | | | | | | |
| *October* | 16,856 | 27,647 | 407,494 | 82,354 | 74,818 | 129,269 | 121,053 |
| Non-white | 287 | 1,503 | 78,685 | 16,077 | 5,665 | 40,611 | 16,332 |
| Percent non-white | 1.7 | 5.4 | 19.3 | 19.5 | 7.6 | 31.4 | 13.9 |
| *November* | 16,575 | 25,015 | 364,799 | 80,554 | 73,458 | 113,518 | 97,269 |
| Non-white | 415 | 758 | 73,363 | 20,440 | 5,593 | 34,787 | 12,543 |
| Percent non-white | 2.5 | 3.0 | 20.1 | 25.4 | 7.6 | 30.6 | 12.9 |
| *December* | 14,183 | 26,126 | 377,697 | 82,336 | 68,866 | 114,639 | 111,856 |
| Non-white | 432 | 1,063 | 69,527 | 19,331 | 5,209 | 35,004 | 9,983 |
| Percent non-white | 3.0 | 4.1 | 18.4 | 23.5 | 7.6 | 30.5 | 8.9 |
| **1941** | | | | | | | |
| *January* | 44,606[3] | 30,743 | 363,163 | 93,182 | 69,993 | 124,447 | 75,541 |
| Non-white | 880 | 935 | 76,593 | 21,662 | 3,954 | 41,655 | 9,922 |
| Percent non-white | 2.0 | 3.0 | 21.1 | 23.2 | 5.6 | 33.0 | 13.1 |
| *February* | 44,546 | 31,278 | 344,335 | 88,096 | 72,647 | 111,434 | 72,158 |
| Non-white | 933 | 984 | 70,779 | 19,484 | 4,083 | 37,444 | 9,768 |
| Percent non-white | 2.1 | 3.1 | 20.6 | 22.1 | 5.6 | 33.6 | 13.5 |
| *March* | 39,305 | 37,085 | 376,308 | 68,849 | 84,411 | 127,061 | 95,987 |
| Non-white | 741 | 1,054 | 76,556 | 16,501 | 4,821 | 43,277 | 11,957 |
| Percent non-white | 1.9 | 2.8 | 20.3 | 24.0 | 5.7 | 34.1 | 12.4 |
| *April* | 39,256 | 43,106 | 443,180 | 70,299 | 95,162 | 161,012 | 116,777 |
| Non-white | 755 | 1,085 | 88,847 | 14,779 | 6,107 | 55,968 | 11,993 |
| Percent non-white | 1.9 | 2.5 | 20.0 | 21.0 | 6.4 | 34.8 | 10.3 |

1. These occupations include most of the occupations important to defense production, but some of these placements were made in industries other than the "twenty selected industries" in column II. For a list of the twenty industries, see table 13.2.
2. The quarterly totals here would differ slightly from those in table 13.2, which were preliminary.
3. The number of skilled and semi-skilled occupations was considerably increased in January and a few professional and technical occupations were added. This, in part, explains the great increase in placements beginning in January.

the number has increased considerably since September. Government arsenals are also hiring Negroes, but not as extensively as the Navy Department. Outside of Navy yards the practice of government establishments appears to be little better than that of many private establishments.

### Source of Discrimination

The failure to use Negro labor cannot be exclusively attributed to employers. Instances have come to the attention of the employment service in which employers were prepared to hire Negro workers, but the policy was opposed by the officials or members of certain unions. This has been true of both AF of L and CIO unions. Frequently, too, unorganized workers are opposed to the introduction of Negro labor. In some instances the failure to employ Negroes is due to the arbitrary action of foremen and subordinates, and company officials are unaware of the situation.

Some employers explain their refusal to hire Negroes on the grounds that their white workers will refuse to work with Negroes and will either lay down their tools or quit to work elsewhere. Instances in which both organized and unorganized white workers have protested against the introduction of Negroes are pointed to, and sometimes the position is taken that the loss of white workers may be greater than the number of Negroes who can be added, so that the consequences of employing Negroes may not increase production, but reduce it.

In answer it must be pointed out that there are many instances in which white and Negro labor work together in the same establishments, and frequently in the same shop and on the same job, and that employers who have consciously sought to enlist the active cooperation of white workers in the solution of this problem have often found it possible to overcome the traditional opposition of white workers. Often employers have met the problem by establishing separate departments or units for Negro workers.

To a large extent the opposition of white workers to Negroes in the past sprang from the fear of being displaced by Negroes who would work for lower wages. With the fixing of wage standards and the extension of unionism, Negroes can no longer be used to undermine labor standards, and this has removed the basis for much of the opposition to the introduction of Negroes. Union officials, particularly in a period of expanding job opportunities such as the present, have frequently cooperated with management in an attempt to work this problem out.

### Importance of Skilled and Semiskilled Negro Labor

It is true that Negroes have in the past found it almost impossible to obtain work in many skilled and semiskilled occupations and consequently there are only a few thousand Negroes available in the occupations now most in demand. However, other factors must be considered which throw a different light on their relative importance in the skilled and semiskilled labor supply.

First, a large number of Negroes have traditionally been employed as "helpers" in the skilled occupations, but the customary labor policy of not promoting Negroes has prevented them from rising to skilled and semiskilled jobs. As "helpers" many of them have gradually become qualified mechanics.

Second, the difficulty which skilled Negro workers have faced in finding employment at the skilled level has driven them to take employment as laborers and service workers. During the recent press and radio campaign conducted by the employment service, workers experienced in defense occupations who were currently employed in other trades were asked to register at the employment offices. The Harlem office in New York City has reported that in response to this plea a surprising number of employed Negro "laborers" have come in to be registered, and have revealed former experience in skilled and semiskilled occupations in which there is at present an acute shortage.

Third, the practice of some employment offices of registering Negroes, not according to occupational background and experience, but according to job opportunities (which are almost inevitably in unskilled or service occupations), further tends to obscure their importance in the skilled and semiskilled occupations.

Although Negroes constitute but 12.3 percent of the registered job-seeking reserve for the nation as a whole, they are vastly more important in many of the major urban-industrial areas of the country. Philadelphia and Baltimore offer typical examples of the significance of Negroes in the unemployed reserve. Philadelphia is typical of a northern city in which Negroes, while at first glance they do not loom large in the population, are found on close examination to constitute an extremely important part of the available local labor reserve. Baltimore, on the other hand, with a large proportion of Negroes in the labor force, illustrates how a drain on the supply of white workers soon leads to a shortage of white labor locally available for employment or for training and makes the employment of Negroes the only alternative to large-scale, otherwise unnecessary, importation of whites.

## Training of Negroes for Defense Work

The difficulty in placing Negro labor, has, of course, affected the vocational training programs. While Negroes form about 11 percent of the total force of gainful workers, and about 12 percent of the unemployed reserve, of 115,000 trainees in the preemployment and supplementary vocational courses established under the national defense program, in December 1940 only 1,900 or 1.6 percent were Negro. By March the training of Negroes was somewhat extended: of 175,000, about 4,600 or 2.6 percent were Negroes.[3] However, the training courses which are established for Negroes frequently do not offer training in the occupations for which there is the greatest demand, but in those in which Negroes traditionally find employment. It is not uncommon to find that while the local demand calls for machine shop, sheet metal, riveting and welding trainees, the courses opened for Negroes are in wood-working, bricklaying, and auto servicing, which may or may not be in demand.

There is abundant evidence that many local authorities if they are not dissuading Negroes from enrolling in defense training courses, are not encouraging them to enroll, in spite of the provision of the supplementary Appropriations Bill, which established the national defense vocational training program, that

No trainee under the foregoing appropriations shall be discriminated against because of sex, race, or color; and where separate schools are required by law for separate population groups, to the extent needed for trainees of each such group, equitable provision shall be made for facilities and training of like quality.

In their attemps to coordinate training with employers' demands, local training authorities have hesitated to use their limited facilities to train Negroes where employers would not hire them. More recently, in the face of threatened shortages of trainees, there has been some tendency to provide training for Negroes, with the expectation that employers will increasingly tend to relax their color specifications as the labor shortage becomes more acute. On the other hand, in some areas training facilities are not fully utilized because of the shortage of white applicants for training.

There are, however, some communitites where Negroes are being encouraged to enroll in the vocational courses. In Philadelphia, for example, the vocational authorities, the WPA and the NYA all favor a policy of encouraging Negroes to train in defense occupations, in spite of current difficulties of placement.

### Effect of Failure to Use Negro Labor

It has been said that the refusal to employ Negro skilled and semiskilled labor in manufacturing cannot seriously retard defense production, since traditional discrimination has excluded Negroes from many trades and consequently there are relatively few skilled and semiskilled Negroes available who are trained in the occupations most in demand. While it is true that only a relatively small proportion of skilled workers are Negroes, in many instances the failure to employ those who are available has not been without direct effect on the defense effort.

The situation in the foundry industry, so vital to national defense, has already been mentioned. A shortage of power sewing machine operators, needed to sew uniforms, would to a considerable extent be ameliorated if Negro operators were more widely employed. A Chicago plant manufacturing gas masks refused to hire Negro power sewing machine operators in spite of the fact that there were no white operators available. Meanwhile several hundred Negro operators registered with the Chicago employment service, many of them highly expert, continue to be unemployed.

In spite of a shortage of machinists in Ohio, machine shops continue to refuse to consider Negro machinists recommended by the employment service, and the same is true in other States. Aircraft manufacturers continue to comb the country for skilled and semiskilled workers, but most of them will not hire locally available experienced or trained Negroes whom the employment service could supply in small numbers. A vocational training instructor in New York City stated that white junior trainees with limited mechanical aptitude were being hired by aircraft manufacturers, who had refused employment to better qualified Negro trainees.

The metal trades division of the New York City Employment Service has repeatedly reported that the color restriction was one of the factors tending "to create what is generally referred to as a shortage." Until about April, Ohio reported that "there is no apparent increase in the hiring of skilled Negroes," but since April there has been a somewhat greater willingness on the part of some employers to hire Negro labor. Indiana reported that the shortage of labor in some sections has led employers to give consideration to the possibility of using Negro labor, but there has as yet been no considerable increase in Negro hiring.

Connecticut has reported that there is only a limited industrial market for Negro labor, but more recently the Connecticut Manufacturers' Association's Committee on Employment Problems is urging that opportunities be given "young trained Negroes" to prove their capacity and fitness to serve in such occupations as they have been trained for. Delaware reported that, in spite of a shortage of skilled and semiskilled workers, "white jobs" continue to be exclusively reserved for white workers. Maryland continues to report a shortage of white workers suitable for training. A long list of instances such as these could be supplied from the experience of many employment offices.

## Conclusion

It is clear that such discrimination tends to (1) retard production, (2) intensify the shortages of labor, (3) increase labor turnover, (4) contribute to the competition for labor, (5) cause unwarranted and unnecessary migration which is dislocating the labor market, and (6) accentuate the shortage of housing in certain areas, which in turn causes rents to rise, and compels government expenditure for defense housing.

If discrimination continues, it is likely to cause training facilities in certain areas to stand idle or compel the government at considerable expense to move large numbers of trainees and maintain them during the training period. Each of these manifestations has already appeared. That they will become even more extensive as the defense program develops cannot be doubted. The employment of Negro labor will not, of course, solve these problems, but it will help minimize them.

There are signs that the discrimination against Negroes is beginning to relax, but by and large the instances of relaxation are still isolated occurrences. Negroes are still regarded as a supplement to the supply of white labor rather than as an integral part of the labor reserve. If, in the end, employers will be compelled by circumstances to use Negro labor just as they were in the 1917–18 emergency, it would be rational policy to use such labor as the need develops, rather than wait until acute shortage compels it. Only in this way can the destructive effects of discrimination be minimized. Even under ideal conditions the problem of mobilizing the labor force on an orderly basis is an enormous task; in the present emergency, no unnecessary restrictions on the use of the labor supply ought to be allowed to impede it.

## Notes

1. The number of "non white" women placed in 20 selected defense industries between October 1940 and March 1941 was 146 compared with 23,675 white placements.

2. The placement figures in tables 12.1, 12.2, and 12.4 do not, of course, reveal the extent to which industry is hiring Negroes or non-white workers outside of the employment service. Many employers, particularly the large employers who have extensive personnel departments of their own, do not generally use the employment offices to any great extent, and many more Negroes (as well as white workers) have been hired than would appear from the data. Employment service placements, however, are a sufficiently large proportion of all job openings filled to be representative of general conditions. The disproportion of Negro placements in skilled and semiskilled jobs in most of the defense industries, and in all manufacturing industries, which is reflected in these figures is unquestionably typical of hiring outside the employment service.

3. The national defense vocational training program includes many types of training other than pre-employment and supplementary training. There are the CCC, NYA, OSY (Out of School Rural and Non-Rural Youth), the regular public school vocational program, apprenticeship programs, etc. In some of these, notably the CCC, NYA, and the OSY, the proportion of trainees who are Negro is higher.

## A Black Woman at Work in a Wartime Airplane Assembly Plant

**Figure 12.1**
Photo of a black woman working on the aircraft assembly line at the Eastern Aircraft Division of General Motors. reprinted from Deborah G. Douglas, *United States Women in Aviation, 1940–1985* (Washington, D.C.: Smithsonian Institution Press, 1990), 21.

# Graduate Technicians and National Technical Association (1944)

Most war workers, of all races, were new to the work they undertook and either needed training or to learn on the job how to weld Liberty ships, fill artillery shells, or stamp out helmets. Some, however, came to the work with technical skills. African-American engineers, with collegiate degrees, had been active in civilian work before the war and were, presumably, available for war work as well. The National Technical Association, formed after World War I, now prepared to see that trained African-American engineers were integrated into the nation's defense effort.

## Graduate Technicians

There was a total of 567 colored graduate technicians who were in active work, according to a survey made by the School of Engineering and Architecture of Howard University in 1939. This number was broken down into the following technical professions: architects, 150; civil engineers, 172; electrical engineers, 140; mechanical engineers, 123; chemical engineers, 20; aeronautical engineers, 4; and mining engineers, 3.

In the matter of employment, the following facts stood out:

The architects, besides attaining success in private fields, have recently been playing a conspicuous part in the design and construction of low-rent housing facilities under the U. S. Housing Authority. Some examples of this work include the Langston Terrace Housing Project, a low-rent project in Washington, D. C., and Washington Manor, a similar project in Charleston, West Va.

A colored architect drew up plans for the Army's air field near Tuskegee Institute where the first colored air corps Army unit is being trained; and an engineer constructed the airport on the institute's grounds for the Civil Aeronautics flying courses at the school.

A Negro civil engineer has to his credit important construction projects in the Middle West such as a municipal sewage treatment plant, railway bridges, highways and power plants; while another built a number of outstanding spans in Illinois and Indiana. Three Negroes have received Harmon awards for contributions in the field of technical science.

A colored contracting firm constructed the Army air base at Tuskegee, while another has to his credit the construction of a defense housing development for white

Reprinted from: "Graduate Technicians" and "National Technical Association," *The Negro Handbook, 1944*, ed. Florence Murray (New York: Current Reference Publications, 1944), 140–141.

citizens in Louisville, Ky., and was awarded a government contract amounting to more than a million dollars for the construction of a housing development in Baltimore.

A colored construction engineer built for the city of Camden, N. J., the first post office in the United States to have a roof designed for the landing of autogyro planes. A colored chemical engineer is chief chemist and plant superintendent for the H. and G. J. Caldwell Company, Massachusetts, while a colored mechanical engineer, who is consulting engineer for the C. A. Dunham Co., has more than twenty-five patents to his credit in the field of heating, ventilating and air conditioning, some of which have been widely used.

Another mechanical engineer has been granted a U. S. patent on his recent invention, "an optical device for determining the position of a tool." Two recent aeronautical engineering graduates were employed by the American Aircraft company.

The Howard University School of Engineering and Architecture is participating in the national defense training program. Special intensive short engineering courses are being given in engineering, drawing and tool designing and chemistry of powder and explosives.

---

### National Technical Association

The National Technical Association is an organization composed of colored technicians in various fields. During 1940 it set up a committee to handle matters pertaining to the participation of Negroes in engineering and architecture under the national defense program. Each chapter of the organization has its own local employment committee. These local committees are co-ordinated under a national committee on employment. As a result of their efforts, many Negroes now are working in the municipal, State and federal departments of engineering and housing, and in private industry. The association was organized in 1926.

The president of the association is John A. Lankford, Washington, D. C. and the secretary is James C. Evans, electrical engineer, State College, West Va.

## Defense and Wartime Employment (1946–1947)

By the end of the war, hundreds of thousands of African-Americans had served in the armed forces or worked in defense plants. Following these new war jobs, many had migrated to the West Coast, much as defense work in World War I had led to the Great Migration. The record of true integration and equal pay is, not surprisingly, mixed.

The defense and wartime civilian employment of Negroes increased by approximately 1,000,000 jobs between April, 1940, and April, 1944, it was reported by the United States Department of Labor in January, 1945.

The employment of Negro men rose from 2,900,000 to 3,200,000 during the four-year period, and the number of employed Negro women increased from 1,500,000 to 2,100,000. The Negroes' greatest employment advances were made in those occupations, industries, and areas in which the postwar adjustment will be most severe.

"The 700,000 Negroes in the Army have their civilian counterpart in the more than $5\frac{1}{2}$ million Negro workers in the United States," the Labor Department report stated.

This civilian labor force has experienced marked changes in both its occupational and industrial attributes, which are significant as indicators not only of wartime change but also of postwar employment opportunities.

The outstanding changes in Negro employment that occurred during the four-year period were a marked movement from the farms to the factories (particularly to those making munitions of war), a substantial amount of upgrading for Negro workers, but little change in the proportions occupied in unskilled jobs. As the Negroes' greatest employment advances have been made in precisely those occupations, industries, and areas in which the postwar adjustment will be most severe, the extent to which these gains can be retained will be largely dependent upon the maintenance of a high level of postwar employment.

### Changes in Occupations

The proportion of the employed male Negro labor force on farms declined from 47 percent in April, 1940, to 28 percent in April, 1944, or by 19 points; the proportion in industry increased by the same amount. The remainder of the major occupational groups showed changes of not more than about 1 point between 1940 and 1944.

Reprinted from: "Defense and Wartime Employment," *The Negro Handbook, 1946–1947*, ed. Florence Murray (New York: A. A. Wyn, Publisher, 1947), 99–101.

The shift from the farm to the factory is by far the most outstanding change that took place in the male Negro labor force during the war. Between 1940 and 1944, the number of Negroes employed as skilled craftsmen and foremen doubled, as did the number engaged as "operatives," i.e. performing basic semiskilled factory operations.

Altogether, the number in both categories rose from about 500,000 to a total of about 1,000,000 during the four years covering the national defense program and the entry of the United States into the war. In contrast, the number on farms, either as farm operators or laborers, decreased by about 300,000. In terms of the total numbers involved the other changes were small.

The number of Negro men working as proprietors, managers, and officials increased 50 percent in the four-year period, but in April, 1944, still had not reached 75,000.

Slightly over 7 of every 10 employed Negro women were in some service activity in April, 1940, and the great majority of these were domestic servants. After four years, the proportion in the services had decreased only slightly, although a significant internal shift had occurred.

The proportion working as domestic servants showed a marked decrease, while those engaged in the personal services, e.g. as beauticians, cooks, waitresses, et cetera, showed a corresponding increase. It is interesting to note, in this connection, that the actual number of Negro domestics showed a slight increase between 1940 and 1944 (about 50,000), but it was not enough to counterbalance the decline of 400,000 among white domestic servants.

As among the men, the most pronounced occupational shift among Negro women was the shift from the farm to the factory. In April, 1940, 16 percent of the entire female Negro labor force was on farms; four years later, that proportion had been halved. The total number of Negro women employed had increased by about a third; the number employed on farms had decreased by about 30 percent. On the other hand, Negro women employed as craftsmen and foremen and as factory operatives almost quadrupled during the same period.

No significant changes occurred in any of the other major occupational groups. Percentage increases were large; the number of Negro women working as proprietors, managers, and officials tripled; those working as saleswomen almost doubled, and those engaged as clerical workers rose to a number five times as great as in April, 1940. The actual numbers involved were very small, however, and made little difference in the occupational distribution of the employed Negro women.

### Position in Labor Force, 1940 and 1944

It is evident from the foregoing that Negro workers have experienced a considerable amount of upgrading; by April, 1944, both men and women were engaged in skilled and semiskilled factory operations which few had performed before the war. Nevertheless, a considerable proportion of the Negro labor force was still engaged in unskilled occupations and service activities.

Thus, 1 in every 5 Negro men was working as an unskilled laborer in April, 1940; after four years, the proportion engaged in that activity remained the same. The same situation was found in practically every other major occupational group.

### Shipyard Workers

By March, 1944, employment of nonwhite shipyard workers, predominantly Negroes, had increased more than fifteenfold since 1940, the Office of War Information reported on the basis of information furnished by the Maritime Commission, the Navy Department, the Bureau of the Census and the War Manpower Commission.

The number of nonwhite workers in 96 shipyards and naval establishments on March 1, 1944, exceeded the total peacetime employment of all workers in the industry.

The OWI report said that:

(1) Employment of nonwhite ship, boat building and repairing workers increased from 10,099 in the whole industry in 1940 to 157,874 in 96 shipyards on March 1, 1944. (Negroes constituted 96.9 percent of all nonwhite workers employed in the industry in 1940.)

(2) The 1944 employment of nonwhite workers exceeded by more than 4500 the total number of 153,364 persons listed as working in the whole shipbuilding industry by the 1940 census of the United States.

(3) While the phenomenal wartime growth of the shipbuilding industry increased total employment 888.9 percent, nonwhite employment during the same period increased 1463.2 percent, in 96 shipyards and naval establishments under contract to the Maritime Commission and the Navy Department. As a result, nonwhite shipyard workers constituted 10.9 percent of total employees in March, 1944, as compared with 6.5 percent in 1940.

A substantial numerical increase in the employment of nonwhite workers in West Coast shipyards was revealed in employer reports to the War Manpower Commission. Skilled and semiskilled shipyard employment there represents a new field for thousands of Negro workers who migrated from the East Coast and the South. The Kaiser shipyards offered the major employment to nonwhite workers on the West Coast.

Three Kaiser yards in the Portland-Vancouver area reported 4182 nonwhite workers on March 1, 1944, and four Kaiser yards in the San Francisco area listed 7102 nonwhite employees.

At the same time, 4922 nonwhite workers were employed at the California Shipbuilding Corporation in the Los Angeles area. Other West Coast yards employing sizable numbers of nonwhite workers included the Mare Island Navy Yard, San Francisco, 3934; Marinship, San Francisco, 2700; Naval Drydock, San Francisco, 1589; Puget Sound Navy Yard, Seattle, Wash., 1435; and Bethlehem-Alameda, San Francisco, 1199.

Nonwhite workers played an important part in the West Coast shipbuilding industry. Negro shipyard workers served on labor-management production committees, and the first Negro war worker to win a War Production Board Certificate of Individual Production Merit was an employee of a West Coast shipyard.

This worker, Charles H. Fletcher, a welder in the Moore Drydock yards at Oakland, Calif., worked out a device to use in welding insulation pins to deckheads. The device, which was recommended by the Board of Individual Awards for widest possible use throughout the shipbuilding industry, speeds up tack welding, used in temporarily joining units of the superstructure, by 400 percent.

The most rapid wartime expansion of nonwhite employment in the industry occurred in the United States Navy Yards in the early days of the defense and war emergency, employment records indicate.

## Postwar Outlook

The Negro gains have taken place in congested production areas where considerable readjustment of the labor force will be necessary. In general, the Negro has been able to get his war job in areas where a substantial proportion of the labor force was also engaged in war work. Information for four major congested production areas (Mobile, Charleston, Detroit and Willow Run, and Hampton Roads) shows that among the more than half a million in-migrants, about 1 in every 4 was a Negro. These cities will experience considerable labor turnover in the immediate postwar period.

In those occupations and industries in which the Negro has made his greatest employment advances, he was generally among the last to be hired. Therefore under seniority rules he is more likely to be laid off than the average worker in these occupations.

The war has given many Negroes their first opportunity to demonstrate ability to perform basic factory operations in a semiskilled and skilled capacity. The consolidation of the Negro's gains in the postwar period (and this is true, of course, for a sizable proportion of other workers as well) is dependent in large measure upon the volume of employment that then prevails.

## Bay Area Council against Discrimination, San Francisco (1942)

Discrimination against black skilled workers came not only from employers but also from white trade unions. During the war, the San Francisco Bay Area was a center of defense production, particularly in extensive shipyards. The Bay Area Council against Discrimination, which included some prominent liberals such as Carey McWilliams, long-time publisher of *The Nation*, took up the case of one skilled mechanic who had been rejected for membership by a local union.

**June 12, 1942**

***Officers and Members of the* International Association of Machinists, Local 68**
San Francisco, Calif.

Dear Sirs:
May we take this opportunity of requesting you to consider our views regarding the case of Charles Sullivan, Negro, qualified journeyman machinist, who has made application to your organization for membership.

Other Machinists' locals have admitted Negro workers to full status, the Lockheed local in Los Angeles, one of the largest in the country, and Local 79 in Seattle to mention two. Across the country millions of organized workers are struggling side by side, black and white, against the Hitler doctrine of racial discrimination.

Mr. Sullivan requested sixty-one of his fellow machinists to sign his application for membership in your organization. This application was signed by fifty-eight of these members. These are men who work with Mr. Sullivan day after day, week after week, who know him personally, and who know all the facts surrounding his ability and qualifications as a machinist. If these men who are in daily contact with Mr. Sullivan desire his membership in the union, it is a true indication of their willingness to work with him, and in view of his qualifications should be sufficient recommendation for his membership in your organization.

We are now in a war with the Axis powers, united with all democracy loving nations to defeat Hitler, the advocate of the vicious doctrine of racial supremacy. If we are sincere in this war to preserve humanity, we must not be guilty, either consciously or unconsciously of using any of Hitler's Nazi ideas. We cannot bend our best efforts in this war for liberty, freedom and equal opportunity and at the same time deny liberty, freedom and equality to anyone because of his race.

Reprinted from: Bay Area Council against Discrimination, San Francisco, to International Association of Machinists, Local 68 (June 12, 1942).

President Roosevelt, on June 25, 1941, issued an executive order in which he said: *"The policy of the United States is that there shall be no discrimination in employment in defense industries and government because of race, color, creed or national origin."*

This is a great deal different from Hitler's ideas. As far back as 1925, Adolf Hitler, still a petty gangster with ambitions, was writing things about the Negro people. Contrast his words with those of the President. Hitler's book, *Mein Kampf*, was published in 1925 and 1927. In it, Hitler has this to say: *"From time to time it is demonstrated ... in illustrated periodicals that for the first time here or there a Negro has become a lawyer, teacher, even clergyman, or even a leading opera tenor, or something of that kind ... which does not dawn upon this depraved bourgeoisie world that here one has actually to do with a sin against all reason, that it is a criminal absurdity to train a born half-ape until one believes a lawyer has been made of him ..."*

That is our enemy speaking. Anyone who refuses to recognize a fully qualified individual for any given craft or profession is either consciously or unconsciously following the above mentioned policies of Hitler. Contrast this with the remainder of what President Roosevelt has to say in his Executive Order 8802, 1941. *President Roosevelt said: "I do hereby declare that it is the duty of employers and of labor organizations in the furtherance of said policy and of this order to provide for the full and equitable participation of all workers in defense industries without discrimination because of race, creed, color or national origin."*

Full and equitable participation means not only to be employed but to participate in the organizations of and the organizing of the workers, for the purpose of choosing representatives and to bargain collectively. Division of the workers of any given craft creates confusion in collective bargaining, and it destroys organization, it does not provide full and equitable participation.

In regard to labor's official position, it was just a few weeks ago that the San Francisco Labor Council communicated with all of its affiliated unions to the effect that it was contrary to the policy of the American Federation of Labor to discriminate for racial reasons. The Labor Council requested and urged all unions to take into full membership all workers under their respective jurisdictions, regardless of race, color, creed, or national origin. The Labor Council further pointed out the damage that such discrimination was doing to the unity of the American people and the effect that such actions would have on our war effort; and may we call to your attention the fact that President William Green of the American Federation of Labor has appointed Frank P. Fenton to represent him on the Fair Employment Practice Committee. Mr. Fenton writes us, *"I am representing President Green on the President's Committee on Fair Employment Practice ... Our job is to eliminate any discrimination against minority groups because of race, creed or color or national origin. We have been holding meetings all over this country and have endeavored to do away with this discrimination both by employers and organizations. I hope that you will watch very carefully these complaints in your locality ..."*

If we are to win this war, we will need the best efforts of every single citizen. We cannot afford to hinder the war effort by refusing to recognize the equal rights, the liberty and the freedom of thirteen million Negro American citizens. These people are participating in this fight and must participate equally in all phases, both production and combat to preserve the liberty of all the people, including themselves. Discrimination, and those who practice it, destroys unity, helps Hitler. Let's get behind the President of the United States and carry on a real fight against Hitler's doctrine of discrimination.

We urge you and most respectfully request that your organization take into full dues paying membership Charles Sullivan, so that he may contribute his share to the organization, enjoy the benefits gained through organization, participate in the elections, the growth and progress of your union and its members.

Fraternally yours,
*Bay Area*
Council against Discrimination

# 13 After the War

**Elm City, A Negro Community in Action (1945)**
C. L. Spellman

**Mechanization in Agriculture (1941–1946)**

***The Negro in the Aerospace Industry* (1968)**
Herbert R. Northrup

# Elm City, A Negro Community in Action (1945)

C. L. Spellman

The economic opportunities of the war years and the consumer delights of the postwar era continued to bypass whole segments of the population and sections of the country. Those African-Americans who remained in the rural areas of the South during the post-war years enjoyed few of the benefits of prosperity and home appliances.

## The Community

The community is made up of Elm City, the center, and five surrounding neighborhoods, each of which is held together by a one- or two-teacher elementary school. For community purposes, the consolidated high school, the several businesses and the two important churches of the area, located in the village of Elm City focalize the interests of the Negroes in this one place. This means that for an area of approximately twenty-five square miles, they are drawn into a single community. Interestingly enough, this is not true for the white people in the same area, since for them there are three distinct communities, each having its consolidated school, church and minimum services. White and Negro communities in many cases, as this one, may not be coterminous. Some entities which for white people constitute complete community life, are for Negroes only parts of neighborhoods.

## Races and Sexes

White people and Negroes make up practically the entire population of Wilson County generally. In 1940 there were only seven people in the county classed as "other races," and none of them lived in the community area. The population of the community is 48 percent Negro, a figure which is substantially higher than the one for the county as a whole, which is 41.9 percent. . . .

## Occupational Classification

The 1940 census shows that of the 6,480 employed Negroes in Wilson County 14 years old and over, 2,391 or 33 percent of the total, engage in agricultural occupations; 1,116 workers or 17 percent engage in domestic service, and 1,610 workers or 24 percent engage in service and labor other than domestic service. This means that about 75 percent

Reprinted from: C. L. Spellman, "Elm City, A Negro Community in Action," *Rural Sociology* 10 (June 1945): 174–179, 186–187. 

of the Negroes of the county engage in agriculture, domestic service and a few other common labor occupations. Since the community is patterned about like the county, the same kind of occupational distribution runs also through the community. Both have an urban "nucleus" in which are found the usual urban occupations, and an agricultural hinterland in which are found the usual agricultural occupations. If the community should vary from the pattern of the county, it would vary in the direction of more Negroes participating in the agricultural occupations due to the relatively higher rate of land ownership here than in the county as a whole. Incidentally, absence of more other industries here reflects the county's favorable resources for growing tobacco, one of the South's leading cash crops....

### Mobility

The generally high rate of tenancy here predisposes the Negroes to a high rate of mobility. Much of it is intra-community mobility, but there is considerable intercommunity movement. Late depression years saw many actual cases of replacement of Negroes by white tenants. In one neighborhood, so many Negroes have been replaced that the school is now in the center of an extensive white settlement. The distance away of the Negro families now renders this school almost non-functional for their social and communal needs and activities.

### Communication and Interaction

The community is crossed from north to south by U. S. Highway 301 on which the Carolina Trailway Buses run. They make stops in the community at Elm City and Sharpsburg. The area has no other hard surfaced highways, but it does have an adequate network of well kept secondary roads belonging to the State Highway System. Distances between places are not too great for the energetic to walk from one point to another.

There are no telephones in Negro homes here, and only a few receive daily or weekly newspapers, although papers are available from Wilson seven miles away, or Rocky Mount eleven miles away, if the people wanted them. Lack of enthusiasm for newspapers probably reflects the generally low educational status of the Negro family heads. The fathers averaged only 4.28 grades completed and the mothers 5.34 grades completed.

Mail facilities are made available to the area from the post office at Elm City and two rural routes from Wilson. Mail posted at either office will reach the addressee the following day. In regard to receipt of mail, many open-country families handicap themselves by not owning a private mail box. It is not unusual to find a half dozen families receiving mail in the same box, but this has its complications, especially when the smaller children go to the box to get the mail and lose some pieces belonging to other families.

Battery radios play a great part in keeping these people in touch with the world. They listen very alertly during tobacco marketing season to keep up with the trend of the market, and also each Sunday evening to a special radio hour sponsored by a Negro undertaker of Wilson. This program, especially since it is directed and announced by the Negro county agent, who works extensively in the community, is a unifying influence of considerable importance....

## The Economic Situation

Soil is about the only natural resource the community now has, and it is rapidly becoming depleted in some instances. The area economy is based upon tobacco as the main cash crop, followed by cotton. Other crops that grow well are potatoes, peanuts, cowpeas, soybeans, etc. Those who wish industrial work must go to Wilson, or Rocky Mount, the two nearest towns of consequence. The farm economy is built upon tenant labor, with most of the desirable land being farmed by white landlords or corporations. Negro land ownership in the area is low, although it is probably higher here than in the county as a whole. Housing facilities compare favorably with other areas of the county, but are poor in aspects such as toilet facilities, screens, provision for safe drinking water and the like.

## Conclusion

Here we come to the end of the description of the Elm City Negro community. A definite effort has been made to feel the pulse and catch the tempo of life and social processes as expressed in the development of a community from its beginning. The brief sketches of personalities and movements, joys and sorrows, inter-racial and intra-racial relationships constitute the warp and woof of an emerging community fabric. From such a description, social workers, ministers, teachers and others have a valuable source of basic information about the community by which to orient their efforts toward further progress. Upon this as a base, future efforts toward community improvement may be made in perfect knowledge of the resources to be utilized, the deficiencies to be overcome and the pitfalls that must be avoided.

## Mechanization in Agriculture (1941–1946)

The mechanization of agriculture, particularly in the South, had been long overdue. Picking up speed during the Depression, when wealthy white planters often used government subsidy checks to buy tractors, the true industrialization of farming began right after the war. Its paradigmatic form was that of specialized tractors, designed to pick cotton, corn, tomatoes, and a host of other crops. But the rural electrification begun in 1935 by the New Deal's Rural Electrification Administration was important too, as were improved irrigation and new kinds of chemical fertilizers and pesticides. Even in its early years, this coming industrialization boded ill for the small, impoverished, tenant farmers and sharecroppers of the South.

### Mechanization of Farming in the South

The mechanization of farming in several areas of the South is ushering in changes of great importance. Texas and Oklahoma, even before Pearl Harbor, had seen the invasion of tractors transforming the countryside into multiple-sized farms. The Mississippi Delta and the better lands of the Old Southeast have, during recent years, been cultivating larger crops with fewer laborers than in pre-tractor days. In two of the thirteen Southern States the percentage of farm operators using tractors in 1945 approximated the national average. These two, Oklahoma 30.3 and Texas 29.1, were followed by Florida with a percentage of 14.4. The national average is 30.5 (see table 13.1.)

### Hand Labor Cannot Compete with Machines

Dr. Arthur Raper has pointed out in the booklet, *Machines in the Cotton Fields*, published by the Southern Regional Council, September, 1946, that "we need first of all to recall that cotton and tobacco, the farmers' main sources of cash in the South, are two of the least mechanized crops in the nation." The agricultural South using hand labor cannot without change prosper in an age of mechanized production. However, well intentioned and industrious, the man with a hoe and a one-mule primitive plow cannot maintain respectable standards of living in a country where other men use labor-saving machines. Likewise, a region characterized by primitive methods of production must remain economically backward. As long as most everything bought by the southern farmer is machine-made while everything that he sells is hand-made, the differentials in living standards will be to his disadvantage.

"Mechanization in Agriculture," in *Negro Year Book, A Review of Events Affecting Negro Life, 1941–1946*, ed. Jessie Parkhurst Guzman, 180–183 (Tuskegee: Tuskegee Institute, 1946).

**Table 13.1**
Per cent of farm operators reporting tractors on farms in 1930, 1940, and 1945,* and per cent increase from 1930 to 1940, and 1940 to 1945 for selected states**

| | Farm Operators Reporting Tractors on Farms | | | | |
|---|---|---|---|---|---|
| | Per cent Reporting Tractors | | | Per cent Change | |
| State | 1930 | 1940 | 1945 | 1930–40 | 1940–45 |
| *United States* | 13.5 | 23.1 | 30.5 | 70 | 32 |
| *13 Southern states* | 3.9 | 7.8 | 11.0 | 90 | 44 |
| *8 cotton states**** | 3.9 | 9.3 | 13.2 | 112 | 43 |
| *South Atlantic***** | 3.6 | 4.8 | 7.3 | 30 | 52 |
| Virginia | 5.4 | 6.2 | 8.4 | 22 | 36 |
| North Carolina | 3.9 | 4.3 | 6.4 | 12 | 63 |
| South Carolina | 2.0 | 3.1 | 5.4 | 38 | 75 |
| Georgia | 2.1 | 3.8 | 5.9 | 59 | 57 |
| Florida | 7.4 | 10.2 | 14.4 | 47 | 42 |
| *East South Central* | 2.1 | 3.6 | 5.3 | 72 | 50 |
| Kentucky | 2.8 | 4.4 | 6.2 | 63 | 41 |
| Tennessee | 2.7 | 4.4 | 6.7 | 72 | 53 |
| Alabama | 1.7 | 2.9 | 4.5 | 64 | 56 |
| Mississippi | 1.5 | 2.7 | 4.1 | 91 | 50 |
| *West South Central* | 5.7 | 14.9 | 20.9 | 125 | 49 |
| Arkansas | 1.8 | 4.3 | 6.6 | 121 | 53 |
| Louisiana | 2.4 | 4.6 | 6.9 | 89 | 50 |
| Oklahoma | 11.4 | 22.9 | 30.3 | 75 | 32 |
| Texas | 6.4 | 20.6 | 29.1 | 165 | 42 |

*Figures for 1930 and 1940 from U.S. Census; figures for January 1, 1945, are taken from estimates made by Bureau of Agricultural Economics as shown in "Number and Duty of Principal Farm Machines," by A. P. Brodell and M. R. Cooper, F. M. 46, Washington, D.C., November 1944.

**The percentage of farm operators using tractors in January 1945 was arrived at by showing a percentage gain in operators from 1940 to 1945 equal to the percentage gain in number of tractors during the five-year period.

***South Carolina, Georgia, Alabama, Mississippi, Louisiana. Arkansas, Texas, and Oklahoma.

****Excludes Delaware, Maryland, D.C., and West Virginia.

### Not All Parts of the South Can Be Agriculturally Mechanized

Not all areas of the South can be transformed into mechanized, large-scale farms. Professor Peter F. Drucker, Bennington College economist, after an extensive study of the Southern Region, has drawn up a map indicating the geographic areas which lend themselves best to mechanization of cotton production. (See *Exit King Cotton, Harper's Magazine*, May, 1946). These areas, in brief, are the rich lands of the Mississippi Valley extending from the Gulf to upper Arkansas and Tennessee; the Gulf Coast, especially around Corpus Christi; some of the low-lying hill counties of Alabama and Mississippi—wherever a yield of more than one bale an acre is obtainable; at least two-thirds of Texas; and the new, irrigated cotton lands of New Mexico, Arizona, and California—where cotton can be grown for as little as four or five cents a pound on mechanized farms. Not included, however, are the low-yield, high-cost regions—all of South Carolina, Georgia, and eastern Texas, most of Arkansas, Mississippi and Alabama—which will, in all probability, be forced out of cotton production. In these low-yield, high-cost areas more than half of all the cotton farmers of the United States live—"primarily small, poor farmers," writes Drucker, "who have no alternative cash crop and neither the capital nor the training to develop one."

### New Inventions Increase Probability of Mechanized Farming

While the industrial revolution has tardily reached the South, a number of recent inventions are hastening the movement toward mechanization. The cotton-picking robots, for example, each doing the work of a half-hundred human harvesters, have already demonstrated their worth. While the number of these machines installed at present is small and relatively insignificant, there is the probability of increasingly large numbers ahead. Already, three of the largest manufacturers of farm machinery—International Harvester, Allis Chalmers, and John Deere—are in the race to supply the market. The mass production of these cotton harvesters is definitely in the plans for the future.

Complementing the extraordinary efficiency of these machines, the Graham Page fire-spitting cultivator and the McLemore "Sizz Weeder" will reduce to a minimum the human labor formerly required to keep the cotton fields free from weeds. This type of cultivator will render the fields practically weedless for a considerable period of time. Tests with the "Sizz Weeder" at the Stoneville, Mississippi, Experiment Station, have disclosed an overall operating cost of 48 cents an acre. This is small indeed when compared with $4 to $12 an acre usually paid hand laborers to chop out weeds with a hoe. Also it is estimated that a laborer with a hoe can "chop" only a half-acre in a ten-hour day; while a two row "Sizz" can cover 25 acres in a day. A four row cultivator can cover 46 acres in the same length of time. Furthermore, squirting a band of intense heat just above the ground, the flame cultivator has another value. It not only sears the weeds and weed seeds, but it kills insects as they are knocked to the ground by the moving machine. The cotton plants are unscathed because of the toughness of their stalks.

In the judgment of Colonel A. J. McLemore, inventor of the "Sizz", his weed-destroyer will bring about complete mechanization of thousands of farms which produce cotton, corn, sugar cane, vegetables and other crops. The chief reason why the cotton robot has not been used more extensively in the past is because the farmer, even when possessing a tractor, cultivator and a cotton picker, still had to retain a sizable force of laborers just to hoe the crops.

#### Mechanized Farm of the McLemore Brothers, Montgomery County, Ala.

On the farms of the McLemore Brothers, white farmers, totaling 7,700 acres in Montgomery County, Alabama, there is a 150-acre tract of cotton land that, in the 1946 season, was prepared, planted, fertilized, chopped, weeded, defoliated, and picked entirely by mechanical means. This was probably the first time that the human hand rarely touched the cotton from the time plans were made until the burlap-wrapped bale of cotton was delivered from the gin press. What is even more extraordinary is the fact that a single man did the entire series of operations on this 150-acre field.

#### Social and Economic Problems Involved in Agricultural Mechanization

Needless to say, the human effects of a complete mechanization of the cotton industry alone would upset the equilibrium of production and of life far and wide. Approximately 10,000,000 human beings in the South would be directly affected. Probably 2,000,000, it is estimated, would be occupationally displaced. Of this number, perhaps 1,200,000 would be white and at least 800,000 Negro. Whether these displacements would create serious economic and social problems depends upon the period of time consumed in the changes. Some experts in the field of southern economics believe the changes would be sudden and chaotic; others, that the changes would be distributed over a considerable period of time. Dr. Raper believes that each mechanical picker will, however, displace more workers at one time than the tractor, and will displace them more completely, especially since cotton picking is the one remaining big hand process in cotton production. Hand workers will commonly be thought of as surplus labor only after a mechanical picker has been put into operation on a particular farm. Other nearby planters may continue with traditional hand methods of operations for another year or two, while some few growers may continue hand methods of production because of the ease with which they can secure from among the families already displaced by mechanized farms the very kind of workers they like to use. "Workers will be displaced farm by farm, year by year. Operators still relying on hand methods of production will remain as dependent as ever upon the availability of workers. In short, hand workers on any given cotton plantation are indispensable as workers right up to the time that they are displaced by machine pickers when most of them will not be needed at all."

Still further is the possibility that the effects of displacement may be mollified by the reduced necessity for women and children working in the fields. Colonel McLemore made a pertinent remark on this aspect of the situation. "The majority of hired pickers," he said, "are Negroes. They include all members of the family. When mechanized cotton farming comes into its own, it will not be necessary for the women and children to be hoeing and picking cotton. The women can stay in the home where they are needed more, and the children will have more time for schooling."

## Organized Efforts Necessary to Meet Displacement Problems

"What new activities should be launched by the vocational agricultural people, the Agricultural Extension Service, the Farmers' Home Administration and other agricultural agencies, the churches, and the farmers' organizations to help as many families as practicable to make a good living on farms, and to help those who leave the farms to get ready to do something else?... Small operators can be served by their neighbor's machinery when custom work is done at equitable rates. Also a group of small farmers can own and operate machinery jointly. Cooperatives might prove helpful to the small, independent farmer in securing the advantage of machinery without being saddled with uneconomic equipment. It is not implied here that the present farms in the poor land areas are large enough if properly managed, but it is important to remember that the increase of the size of the farm is but one of the ways to develop an adequate farms unit."

## *The Negro in the Aerospace Industry* (1968)

Herbert R. Northrup

While new technology was destroying small-scale agriculture, especially in the southern states, other areas, especially on the West Coast, saw a new and heavily subsidized aerospace industry emerging from the foundations of aircraft production laid during the recent war. Coinciding with a slowly growing commitment on the part of the federal government to fair employment and even, in some cases, affirmative action, this new aerospace industry became a test case for a national commitment to equal access to the new technologies spawned by the war.

### The General Picture and "Affirmative Action" since 1966

Table 13.2 shows employment in the aerospace industry by occupational group, race, and sex for 21 companies and 127 establishments located in all regions of the country. The sample includes all major aerospace corporations and about 60 percent of the industry's labor force. It includes all major facilities covered by the data in table 13.2, omitting only some of the smaller plants reported therein.

The first significant fact revealed by the data in table 13.2 is the continued expansion of Negro employment—almost 5 percent of total aerospace employees in 1966, and nearly twice the percentage revealed by the similar, but not completely comparable, data in table 13.2. This is the highest percentage of Negroes found in any analysis of the industry's racial data heretofore. Moreover, it is likely that the percentage of Negroes in the industry has expanded since 1966. Continued expansion of aircraft production, and considerably less expansion in missile and space production, a tight labor market, and pressure by the government for "affirmative action" have all contributed to the expansion of Negro employment since 1964 and after 1966.

As already noted, the great boom in commercial aviation and the needs of the armed services for helicopters, jet fighters, and bombers in Vietnam have all contributed to the expansion of the aircraft segment of the aerospace industry. The fact that aircraft production utilizes large numbers of semiskilled personnel whereas missile and space vehicle production requires a much higher complement of skilled workers and technical and professional personnel has been reiterated earlier. The production shift to aircraft and the tight labor market which has existed since 1965 have combined both to enhance Negro employment opportunities, and to give aerospace (as well as

Excerpted from: Herbert R. Northrup, *The Negro in the Aerospace Industry*, Report No. 2, The Racial Policies of American Industry (Philadelphia: University of Pennsylvania Press, 1968), 34–53.

**Table 13.2**
Aerospace industry, employment by race, sex, and occupational group, total United States (21 companies, 127 establishments, 1966)

| | All Employees | | | Male | | | Female | | |
|---|---|---|---|---|---|---|---|---|---|
| Occupational Group | Total | Negro | Percent Negro | Total | Negro | Percent Negro | Total | Negro | Percent Negro |
| Officials and managers | 71,328 | 292 | 0.4 | 70,638 | 289 | 0.4 | 690 | 3 | 0.4 |
| Professionals | 179,436 | 1,435 | 0.8 | 175,513 | 1,375 | 0.8 | 3,923 | 60 | 1.5 |
| Technicians | 63,999 | 1,209 | 1.9 | 57,284 | 1,128 | 2.0 | 6,715 | 81 | 1.2 |
| Sales workers | 720 | 2 | 0.3 | 673 | 2 | 0.3 | 47 | — | — |
| Office and clerical | 130,261 | 3,692 | 2.8 | 51,289 | 1,986 | 3.9 | 78,972 | 1,706 | 2.2 |
| Craftsmen (skilled) | 164,991 | 7,595 | 4.6 | 158,623 | 7,050 | 4.4 | 6,368 | 545 | 8.6 |
| Operatives (semiskilled) | 155,167 | 18,417 | 11.9 | 122,869 | 13,566 | 11.0 | 32,298 | 4,851 | 15.0 |
| Laborers (unskilled) | 8,065 | 1,804 | 22.4 | 6,344 | 1,619 | 25.5 | 1,721 | 185 | 10.7 |
| Service workers | 14,055 | 3,124 | 22.2 | 12,015 | 2,792 | 23.2 | 2,040 | 332 | 16.3 |
| Total | 788,022 | 37,570 | 4.8 | 655,248 | 29,807 | 4.5 | 132,774 | 7,763 | 5.8 |

Source: Data in author's possession.

other) employers added incentive to find ways and means of utilizing Negroes and others whose training and background are below what was available in the looser labor markets of the late 1950's and early 1960's. Government and public policy to enhance employment opportunities for Negroes has thus been operating in what is fortuitously a favorable climate for success.

Affirmative action takes many forms. It involves such programs as recruiting in depressed areas like the Watts section of Los Angeles, developing special training programs for the disadvantaged, or simply giving preference in fact to Negroes. All the major Southern California aerospace firms, for example, set up special recruiting efforts in Watts after the 1965 riot there. Aerojet-General went further by establishing a tent-making subsidiary operated and manned by Negroes in the heart of Watts. Other companies financed training and motivational centers there. In addition, in late 1966, and early 1967, nearly 8,000 on-the-job trainees were enrolled in Southern California aerospace firms under the provisions of the federal Manpower Development and Training Act. Nearly all these trainees were classified as disadvantaged, and virtually all were Negroes or Mexicans.

In other areas, these special programs have also been pursued. United Aircraft's Pratt & Whitney division in Connecticut has made notable strides in rescuing dropouts and converting them through an elaborate training program into semiskilled or even skilled personnel. Pratt & Whitney has over 200 men and women in the training section of the personnel department, and had 7,000 persons in formal training in 1967 plus another 5,000 in a special short course.

Boeing's Vertol division in Philadelphia has developed a careful, elaborate training program which has put many persons, including large numbers from the heavily Negro Chester, Pennsylvania, area into useful jobs, especially in its machine shop. McDonnell in St. Louis has among the highest proportion of Negroes in the industry in part because of its willingness to train those who might ordinarily be considered unemployable. Many other examples could be given of aerospace industry training and of special efforts to enroll Negroes in training courses.

Also in Philadelphia, General Electric's missile and space division has been a leading supporter of the Opportunities Industrialization Center, the much publicized self-help motivational and training institute, which has now been set up in over thirty other localities. Avco is building a plant in a Boston slum which will be operated and manned entirely by Negroes. General Dynamics, which is building the F-111 plane in Fort Worth, Texas, has established a satellite plant at San Antonio to train and to provide jobs for persons considered heretofore to be unemployable. Both the Avco facility which will employ 250 at full production, and that of General Dynamics which expects to have 200 jobs, are being established in collaboration with federal programs seeking to reduce hard-core, inner city unemployment.

All major aerospace companies are members of Plans for Progress, with Lockheed being the first company enrolled. Plans for Progress concerns are, in effect, committed to go beyond nondiscrimination, and to develop programs and activities designed to further the employment of Negroes and other minorities. This has involved such already noted programs as hiring and training hard-core unemployed, establishing work areas or plants in Negro slum areas, special recruitment drives at Negro schools and colleges, and a host of other programs.

As the leading and largest government contractors, aerospace concerns are constantly being inspected by representatives of the various government procurement agencies, the Office of Federal Contract Compliance which now coordinates enforcement of the various Presidential Executive Orders relating to nondiscrimination by government contractors, the Equal Employment Opportunity Commission, and in many cases by state agencies as well. The obligations to take "affirmative action" are constantly before the aerospace companies, and they are usually the first to be "invited" to take part in new programs.

Pressure by the government has also resulted in companies literally giving preference to Negro applicants. Many leading aerospace concerns have told their employment interviewers that every effort should be made to hire Negroes who come even close to meeting minimum qualifications. Some have adopted an unwritten policy of outright preference: give the job to the Negro if he is available. One midwestern plant, where Negro employment was "disappointingly low" to the company's headquarters officials and to the government inspectors, hired 13.4 percent of its Negro applicants in the first seven months of 1967 as compared with 11.4 percent of its total applicants. Moreover, although total employment in this plant declined during this period, Negro employment increased even beyond the percentage increase resulting from the higher proportion of Negro recruits. This probably indicates some selection in layoffs to improve the percentage of Negroes on the payroll.

In one Southern California plant, which is located in a community where very few Negroes or Mexican-Americans dwell, the company moved to better its minority group employment (Mexican-Americans as well as Negroes) even though 10 percent of its 21,000 employees were from these groups. From November 1966 to August 1967, minority groups comprised 16.3 percent of all hires and were involved in 11.5 percent of all upgradings and promotions. Sixty percent of Negroes who applied were employed, but only 6 percent of the white persons who sought work were employed!

Numerous other examples could be cited of various types of "affirmative action." It is, for example, not uncommon for personnel executives to counsel with Negroes who have poor attendance or excessive tardiness records; or to approve several in-plant transfers for those who do not seem to be able to "get adjusted" to a supervisor; or otherwise to attempt to understand and to meet the problems of new Negro employees who seem unable to conform to the rules, regulations, or mores of the factory.

The executives of some companies have, of course, a greater commitment to equal employment opportunity than do those of others. For some it is a necessary order of business, to others, it is that plus a moral commitment. Within this framework, the author has not found any company of the twenty-one surveyed which was not making a real effort to expand Negro employment. Yet total Negro employment was less than 5 percent in 1966. This is approximately one-third of the proportion in the automobile industry and considerably less than that in many other industries. The reasons for this seeming paradox will be made clear after the ensuing discussion of occupational differences and intraplant movement of workers.

### Officials and Managers

A feature of the present affirmative action campaign in the aerospace industry is the search for Negro managerial talent. The great bulk of the 292 Negroes listed as "officials and managers" in table 13.2 are first-line supervisors. Aerospace companies have diligently searched their ranks for supervisory talent, and continue to do so. Some Negro supervisors preside over all Negro labor or janitorial gangs, but most are out on the line managing mixed crews. Despite their small number and percentage, there have been a sufficient number of breakthroughs so that Negroes deserving of promotion out of the ranks now obtain full and fair consideration, or occasionally even preference, for supervisory appointments.

Few Negroes now have middle or top management jobs in the aerospace industry—or in industry generally. Traditionally Negroes with talent, education, and motivation have not sought careers in business. The doors have been much more open in certain professions, particularly teaching, and to a lesser degree, medicine and the law. Now the barriers are lowered. The potential Negro executive is today assiduously pursued by business recruiters, and the aerospace companies are in the forefront of the pursuers. But the availability is very small. The prestigious business schools, conscious of their need to attract and to interest Negroes in graduate business study, have developed several programs to do so, but the success is limited. The number of Negroes trained, or training for, executive positions continues to remain very small. It will be many years before the Negro manager is a common occurrence in aerospace or in any other major industry. Few today are found above middle management; this author knows of none in the top ranks of the aerospace industry. Several are in the personnel function; some in the engineering middle management. The fact that the Negro middle manager is becoming more common, if still relatively rare, presages a continued upward movement, qualitatively and quantitatively, albeit slowly.

### Professionals, Technicians, and Sales Workers

The professional group is the largest occupational category in the industry. Aerospace companies have scoured the country looking for professional and technical employees. Unfortunately, Negroes seeking professional education and attainment have only recently been welcomed into engineering work, and consequently few Negroes even today seek engineering degrees. Moreover, the engineering taught at many of the traditionally Negro colleges is often substandard in general, but particularly inferior for the aerospace industry whose professionals are so often asked to develop products never heretofore made.

Nevertheless, most aerospace companies have their Negro engineer success stories (usually shown in pictures in the annual report!). Aerospace may well have as large, if not the largest, supply of Negro engineers of any industry, as it has of white engineers. Many companies in the industry have been recruiting Negro engineers for a long period, some as far back as World War II. A number do indeed have Negroes who are key members of their professional group. A Negro engineer who is available today can count on attractive job offers, both in terms of money and of the character of the work, from almost every major aerospace concern that learns of his availability. The unfortunate effect, of course, of years of discrimination is that so few Negroes can qualify as engineers.

Technicians are somewhat similarly situated in so far as Negro employment is concerned. Their technical training is the result either of school work somewhat below the engineering level, or they have advanced by experience and/or special training up from the ranks of shop craftsmen. In either case, they have backgrounds of work or training from which few Negroes were once admitted and for which few Negroes are now trained. Nevertheless, real progress has been made in this area. There are quite a few first-rate Negro technicians in aerospace plants, some of whom came up through the ranks over the years. But here again, future prospects are good only if the Negro population can become convinced of the potential which technical education promises. The jobs are there, but the obstacles to achieve the jobs remain formidable, as our discussion of the craftsmen situation, below, will again emphasize.

As has already been noted, few employees in the aerospace industry are classified as sales personnel, and those few are usually private plane salesmen or other specialty marketing personnel. A major distributor of one of the largest private plane manufacturing companies is a Negro. His salesmen and sales agencies are largely white and he is reputed to be very successful. Selling to the government, or to the major airlines, is a job calling for the designation of an executive, manager, or professional in so far as the aerospace industry is concerned. Whatever the designation, few Negroes are so employed. This is the practice throughout American industry, a practice that must change with the times. It will be instructive indeed, as the aerospace industry moves with the times, as it must, and utilizes Negroes in sales, marketing, sales engineering, and customer contact work, to observe how such changes are accepted by the industry's prime contractor, the government.

## Office and Clerical

Despite some major progress, Negroes comprised only 2.8 percent of the clerical employees for 1966 in the table 13.2 data. This reflects, among other aspects, the late start at which Negroes were accepted as office workers in the industry. The office and clerical group is the fourth largest in the industry, comprising 16.5 percent of the total work force. It not only offers a multitude of jobs, but includes many such as purchasing clerks, bookkeepers, personnel assistants, etc., from which promotion into lower-management ranks is customary.

In recent years, aerospace companies have made strenuous efforts to expand their Negro clerical work forces. They have sponsored special training programs, visited and recruited at predominantly Negro high schools, and taken graduates of special training groups, such as those sponsored by community human relations commissions, Urban Leagues, or the Opportunities Industrialization Centers. Two prime reasons for their inability to expand their Negro clerical percentage are lack of training and background and the locational factor.

Just as the segregated schools of the South and those in the inner city core are relatively deficient in their mathematics training, so they are in English grammar, punctuation, and spelling. The girl who graduates from such an institution and is told that she is a typist receives a rude shock when she is tested in an industrial setting. Her speed is likely to be substandard, but practice can overcome that. What is more serious is that she is likely to be a poor speller, have little idea of proper punctuation, and have had inferior training in how to set up a memorandum or business letter.

Aerospace industry officials—and most others—would often prefer to train high school graduates initially, rather than have to retrain them. The girl who thinks that she has learned stenography often resents being retaught and feels discriminated against because she is unaware how lacking in skills she actually is. As one personnel officer told the author: "When we suggest that we are willing to train a girl at our expense, she is sometimes outraged and files a case of discrimination with the state or the federal Equal Opportunity Commission. Inevitably, the record of her typing and English test exonerates us. But in the meantime, we have spent hours proving that we did not discriminate, lost a potential employee whom we need, and hurt our image with the friends and neighbors of the girl."

The locational problem is a serious one which will be discussed in greater detail below. Suffice it here to point out that most aerospace facilities are located on the outskirts of cities and cannot be reached except by private car. Even, however, where public transportation exists, the travel problem remains a significant factor in reducing Negro participation in the industry and especially, the participation of Negro women. Studies of the author in other industries have confirmed the fact that, even where men are willing to travel long distances to work, women seeking work in an office are not. Neighborhoods in which Negroes are concentrated are usually long distances from aerospace facilities. As a result, qualified Negro women tend to gravitate toward office and clerical work in the nearby center cities. Until Negroes can or do find housing in the outer city and suburbs, the aerospace industry's search for more Negro office and clerical employees will have only limited success.

### The Craftsmen Problem

Craftsmen is a broad designation in the aerospace industry. It includes maintenance and machine setup men, tool and die workers, machinists, electricians, millwrights, and others typically found in this category. It also includes some persons engaged in what is essentially assembly work. And it includes mechanics with great capability and versatility in electro-mechanical, hydraulics, pneumatics, or combinations of all three, as well as many other highly or even uniquely skilled workers.

During the past two decades, there has been a tremendous upgrading of skills in the industry. According to the Vice-President of Industrial Relations of Lockheed, "an amazing upward ... shift in job skills occurred.... In 1944, what might be called the 'normal labor market' could fill more than 80 percent of the workforce needs (operatives and low-skilled). Today less than half these jobs can be filled from that market. Even more critical is the sharp cutback of low-skilled jobs—those that might normally be filled by 'almost anyone.' A generation ago more than half the jobs were in this category. Today less than one-quarter are, and the number of workers at the craft level has about tripled."

If this change in skill composition is critical for labor and management, it is even more so for Negro labor. Obtaining skilled blue collar work has always been difficult for Negroes in all industry. The craft unions in the building and metal trades have been historically, continually, and almost universally antagonistic to acceptance of Negroes as apprentices in areas of work under their control. Small businessmen in these trades are frequently former craftsmen who share the racial outlook of their onetime colleagues, but who, in any case, are in no position to oppose the unions on such questions or often, even to train many apprentices.

Negroes desirous of seeking employment leading to skilled blue collar work also cannot as easily short-circuit formal training as can many white workers. The latter often have the opportunity to work as aids or helpers to friends, neighbors, or relatives who work at a trade. Except in the southern trowel trades (bricklaying, plastering, or cement finishing), where a tradition of Negro participation has existed since slavery days, the Negro aspiring to skilled work finds few members of his race already in the trade and therefore able to aid him, and few whites willing to do so.

As a result, Negroes aiming at craftsmen's work find their best opportunities with those large employers, who have the need and financial resources to train and to upgrade their work force and who are expected as a social obligation, and now as a matter of law or public policy, to make certain that Negroes participate in that training and upgrading. Of course, the aerospace industry fits this description as well as, if not better than, most others. Moreover, the aerospace industry has probably done as good a job, if not a better one, than most industries in recruiting, training, and upgrading Negroes to the craftsmen level. The percentage of Negro craftsmen shown in table 13.2 is approximately equal to the percentage of Negroes in the industry—4.6 percent craftsmen, 4.8 percent total Negro. The author knows of no other major industry where Negro representation at the skilled level equals total representation.

Of course one may look at these data in an adverse way—that the overall recruitment level of Negroes is poor. The thrust of this analysis, however, is that the relatively low participation of Negroes in the aerospace industry is, above all, a function of the skill requirements of the industry. The rate of Negro participation in the industry is substantially reduced by the low percentage of Negroes in the salaried areas. In this aerospace is not unique, but rather typical of American industry. Since, however, professionals are the largest occupational category in the industry, comprising 22.8 percent of the total industry labor force, and since white collar and salaried employees make up 56.6 percent of the industry labor force, total Negro participation in the industry is affected in a major way by the lack of Negro representation in these occupational groups. Under the circumstances, Negro participation in the craftsmen group is relatively high, although certainly less than one-half the Negro population ratio in the country.

Among the other factors which keep Negro participation in aerospace craftsmen work from rising at a more rapid rate is of course educational deficiencies among Negro youth. Segregated southern schools and those in the inner city slums of northern cities provide inferior educational backgrounds particularly in communication skills and mathematics than do the white southern, or outer city or suburban schools which few Negroes attend. The large school drop-out rates of Negroes accelerate this problem. The lack of Negro family industrial background in industry and skilled craft work, the demoralized Negro family structure with its missing father, and until recently, the predominant feeling that aspiration to skilled industrial work was not a practical hope, have all reduced the availability of Negroes for skilled work.

To offset these problems, the aerospace companies have made special efforts to recruit Negro apprenticeship candidates. The results have not been encouraging. Close contacts with high schools have uncovered far too few Negroes with the interest and mathematics background or capability. Those on the margin of mathematics competence have been offered special courses or tutoring by several companies, with very limited success in either enrollment or course completion. The Negro who qualifies in mathematics is now firmly college-oriented and is likely also to be uninterested in an apprentice course.

As a result of the historical problems of discrimination, the Negro aerospace craftsman, like his professional counterpart, is disproportionately found in the lower echelons of his group. Proportionately many fewer Negroes are tool and die workers, top-rated machinists, electromechanical, hydraulic, or pneumatic mechanics than are whites. Negro craftsmen are more heavily concentrated in less sophisticated sheet metal, pipe, millwright, or bench mechanic operations. Many of the latter have worked up from semiskilled operatives.

Such upgrading is the prime source of Negro craftsmen today. Many have come into the industry as unskilled employees and now are craftsmen with high-rated jobs. The aerospace industry is probably the most training conscious industry in America. Given the fantastically changing technology which underlies its products, it can do little without continuous training. Most companies post training notices throughout plants and personnel departments constantly urging employees to take training courses, and thereby helping companies solve the skill shortages. In recent years, special efforts have been made to encourage Negroes to take advantage of training programs. . . .

Unfortunately, Negro participation in training is almost universally below the proportion of Negroes in a plant. Motivational factors appear very important.

Willingness to contribute one's time to train for a better future depends on background, expectations, and genuine belief in opportunity. That all three are lacking to some degree in the Negro community is not difficult to understand. There is also undoubtedly some reluctance on the part of Negroes in plants to do anything which would transfer them from a job situation which appears acceptable to white fellow workers to one which may not be. Some Negroes (and of course whites, too) are making more money than they ever expected. They see no need to spend time learning new skills, especially in a volatile industry where the end of a contract, or of hostilities in Vietnam, could bring a layoff. Yet until training opportunities are fully grasped, Negro upgrading will not achieve its great potential in the aerospace industry.

## Operatives

The operative jobs vary widely from some rather close tolerance assembly work to essentially entry jobs. The operative group was traditionally the largest occupational category in the aircraft and parts industry, and still is in the manufacture of engines and aircraft. In the overall aerospace industry, however, the number of craftsmen now exceeds the number of operatives. In missile and space work, the number of craftsmen greatly exceeds that of operatives.

Negroes hold about the same proportion of operatives jobs in the aerospace industry as their proportion to the general population. Their operative representation is also more than twice that of their representation in the industry as a whole. Again, within the operative group, there is a tendency for Negroes to be disproportionately concentrated in the lower-rated sectors.

The aerospace companies which have both missile and space and aircraft work have a much higher percentage of Negroes in the latter. Aircraft engine factories also usually feature a higher proportion of Negroes than do missile and space facilities operating in the same area or region. The reasons, of course, are the character of work and the training and educational backgrounds of Negro applicants.

Assembly line work exists in aircraft factories, but it is far different than automobile assembly line operations. An aircraft company that produces one plane per day for a year or two is operating volume production by the standards of its industry. Automobile assembly plants are geared to produce up to sixty vehicles per hour. The aircraft jobs are not nearly as broken down as are those in the automobile industry. The latter industry can take unskilled persons off the street, and with little or no training start them on the assembly line. This cannot be done in aerospace. The assembly line work is much more complicated, and the cost of mistakes is too great to risk inadequate training. Before new applicants are put onto the line, they must have careful training, varying from a few weeks to several months. Life is involved, and the quality of workmanship cannot be compromised.

All major companies which have lower-rated semiskilled work have elaborate introductory or vestibule training programs. Some now even teach shop mores and manners—how to dress, use sanitary facilities, and live under shop regimentation—as well as the basic fundamentals of workmanship before taking up the rudimentary components of work to be done. From such programs, there is usually a high initial dropout rate, but those that remain often become useful workers, and many advance up the occupational ladder.

The future of Negroes in the aerospace operative category is largely tied to the expansion of aircraft production. A combination of tremendous commercial expansion and the needs of the Vietnam war for combat aircraft and helicopters has created a tight labor market. Companies now have to train workers for entry jobs in most areas in order to obtain any labor, and Negroes have profited immensely from this situation. Layoffs would hit them hard as their seniority is less as a group than whites. Until, however, aircraft production slows down, one may expect continued expansion of Negro representation among aerospace industry operatives.

### Laborers and Service Workers

Not many employees in aerospace are found in the "laborer" category—about 1 percent of the total labor force. They are low-rated employees who do odd jobs around the plants and offices, but do not work in production. Negroes make up a far disproportionate share of this small group in aerospace as they do in all industry. For the Negro (or white) with little education, ability, or indeed prospects, these jobs offer a livelihood. Their insignificant number in aerospace, however, attests both to the nature of work in the industry, and the declining opportunity for those with little skills or potential. The unskilled person with reasonable potential and motivation today in aerospace need not even start as a laborer, but in most cases, can move directly into training for work in the operative category.

Service workers entail a wider occupational grouping and make up about 2 percent of the labor force in the industry, with Negroes again highly disproprotionately represented. They include porters, messengers, cafeteria workers, and if not classified as laborers, groundskeepers and janitors—all jobs traditionally open to Negroes, and usually dead-end jobs. Also included in this grouping are plant guards, a larger group in aerospace than in most industries because of classified work and security problems. In many areas, Negroes were denied the opportunity to work as plant guards until very recently. The rationale seems to have been that Negro guards would have difficulty handling problems affecting white personnel. This reasoning was maintained long after Negro police officers and officers in the armed services lost their novelty. There is still underrepresentation of Negroes among plant guards, but no longer are they denied opportunities in this classification as a matter of course.

### Negro Women in Aerospace

The aerospace inudstry became a large employer of Negro women during World War II, and the tradition has continued, especially in Southern California. Substantial gains have been made in the utilization of Negro women in the shop in recent years. They comprise in the table 13.2 sample 8.6 percent of the 6,368 women classified as craftsmen and 15 percent of the 32,298 female operatives.

One reason for the increased use of women, and therefore also of Negro women, is the large amount of electrical and electronic assembly work in modern aerospace technology. Thousands of wires and cirucits wend their way through the structures of aircraft, missiles, and space vehicles. Women are especially adept at such small assembly work and are extensively used for it. As the labor market tightened and civil rights pressures increased, Negro women increased their share of such work.

Recent years have also seen an increased utilization of Negro women in sheet metal assembly in aircraft plants and as grinders, welders, and small machine tool operators in engine plants. Several companies have had better experience with Negro women than Negro men in these jobs, particularly women with families. The tradition of female family support seems to generate a strong determination to take advantage of opportunities in the Negro women employed by these aerospace firms. In one such case, a plant manager bitterly opposed using Negro women until forced to do so by the company president. Later he was so pleased with the results that he had to be stopped from issuing an order requesting preference for Negro women.

There is still some reluctance to use Negro women in the shop—indeed a few plants have no women—but if the labor market continues to be tight, this will probably change. Perhaps since shop work is not available in the inner city, as is office work, Negro women who work in blue collar jobs are apparently more willing to travel than are female office workers. It is nevertheless quite likely that the distance of most aerospace plants from Negro communities tends to reduce considerably the availability of Negro women for the industry's plant labor force.

# VII THE MOVEMENT AND BEYOND

Each of the smaller movements within The Movement of the 1960s and 1970s—feminist, peace, environmental, civil rights, even "black power"—contained a tension between the struggle to become a fully accepted part of a perhaps radically reformed society, and the desire to tear it up, tear it down, and start over with a better society than could ever be formed out of the detritus of the old. Most of the issues that the various constituents of The Movement had to wrestle with involved technology: Was it out of control or simply under the control of the wrong people? Was it too modern, or not modern enough? Did higher education in a technical field represent a betrayal of racial or gender solidarity and power, or did it form an essential basis for the assertion of that solidarity and power?

Enslaved Africans had brought with them from their homelands a complex culture, an integral part of which was a knowledge of the tools and techniques necessary to do the work of the world, and these had greatly enriched the technological resources of the new Americas. Now after World War II, with the destruction of the great European empires in Africa as well as Asia, the deliberate underdevelopment of those continents had left dual economies that had served the imperial powers well but had devastated the cultures and economies of their colonies. American foreign aid to the newly independent Third World began in 1948 with President Harry Truman's Point 4 program. This was largely geared to providing large-scale engineering infrastructures such as hydroelectric dams and airports, highways, and port facilities. These official efforts were gradually augmented by a host of private ventures formed by missionary groups or, less traditionally, groups of scientists and engineers who sought to redevelop formerly colonial economies along technological lines more appropriate to their time and place. Some African-Americans explored the possibility that understanding and having access to modern technology was as important for the liberation of Africans as it was for their own people in America.

By the mid-1970s a concerted effort was underway to bring under-represented groups into the engineering professions through intense recruitment by engineering schools. Both women and African-Americans were targeted, but a generation later the results were disappointing for both; the white, masculinist culture of engineering yielded only little, and grudgingly, to diversity. Heavily committed to mathematics as the proper path to practice, typical engineering students had begun that study early in their education. Both the culture and social reality of being black or female militated against the kind of education that would prove successful. Nevertheless, the numbers did rise slowly and a number of new focused societies, like the National Society of Black Engineers, joined the older National Technical Association (1926) in encouraging young African-Americans to become engineers and in celebrating their successes.

By the end of the twentieth century the world "technology" has, in popular usage, become almost synonymous with the computer. Whether or not this is, somehow, uniquely the Information Age, computers and the Internet have become the technological environment for an increasing proportion of the things we do and value. As noted in previous sections, technologies always have the potential of redistributing power, and the computer is no exception; indeed the early years of the personal computer (PC) were shrouded in visions of a liberating, democratizing technology that could dissolve inequalities and create a "global village." By the end of the century however, it became increasingly clear that while computer skills may not make people independent, the absence of those skills can be a considerable handicap to full participation in the country's social and economic life.

And finally, at the beginning of the twenty-first century the costs of technological "progress" are becoming ever more obvious and insistent. If African-Americans are being left behind disproportionately in the dash to computerize all of society, they are also demonstrably being forced to pay the environmental price for those who are moving quickly forward. Part of the power technology gives to those who control it is the ability to pass on its costs to others. The nation's environmental movement grew out of an almost exclusive concern for the welfare of "nature," defined as the exotic opposite of all that was human. Rachel Carson's epic challenge to modern technology, *Silent Spring,* warned of the devastating effects of DDT on the bird populations, but people as well had suffered from environmental contamination. By the end of the twentieth century, economically disadvantaged, and most often African-American, communities were making it abundantly clear that the burdens of hazardous-waste incinerators and unclean air, industrial pollutants and environmental lead, are consistently and deliberately foisted upon them. The fight for "environmental justice" is a reminder that technologies give benefits but also, and inevitably, have costs and that, in a society steeped in racial injustice, neither is distributed fairly. Black Americans are the best prepared to realize that while historically they have been blocked from the full enjoyment of the benefits, they have had more than their share of the costs; in either case, they have been deeply involved with the technological history of America.

# 14 Setting a Political Agenda

## Revolution in a Technological Society (1971)

Samuel D. Proctor

## Revolution in a Technological Society (1971)

Samuel D. Proctor

One demand of the black power movement on American campuses during the 1960s and 1970s was for the creation of black studies, as a collection of courses, a department, a program, and most importantly as a recognition that black culture, the black experience, and a black future were not only legitimate but valuable. However, black studies, like any other component of the humanities and even the social sciences, seemed a somewhat tangential subject in the context of the modern technological world, a subject not even close to the center of power and progress in the country. The question facing young African-Americans could be put as starkly as this: Should the politically aware and active student major in black studies or in engineering to gain the kind of power she or he would need to change the world?

---

There is no precedent anywhere for the situation facing the Black people of America. In terms of our origins here as a people, our survival in a cruel and forced demographic transposition, our adaptation to an alien culture, and our sustained protest, we are without precedent. It is nothing short of amazing that so few observers recognize the complete novelty of our situation in this country. All sorts of analogies are made to characterize our struggle but it has subtleties and nuances to it that defy every analogue.

First, we were not slaves by conquest in war, nor slaves taken as vengeance, nor slaves by virtue of voluntary contracts, nor slaves by virtue of human failure and degeneracy. We are the result of the largest market in human cargo known to the world. Our situation is unique.

We survived two and a half centuries of the most massive artificial transplanting of living organisms out of their native habitation the world has ever seen. But we have survived! There were only 20 of us in 1619 and today there are 20,000,000. We endured the change in temperature, the change in climate, the change in ecological balance, the change in barometric pressures, change in flora and fauna, change in ultra violet rays, and changes in language, religion and customs. By all the laws of Darwin, Spencer, and Malthus, we should be extinct by now. But, there were just 20 of us in 1619 and there are 20,000,000 today.

We are unique in another sense: We were denied a chance to live with our own heritage, our own ancestral legacy. We had to abandon all that made us human and assume the role of cattle; our super-ego, our cultural personhood, the "thou" in

---

Reprinted from: Samuel D. Proctor, "Revolution in a Technological Society," *Negro History Bulletin*, 34 (January 1971), 6–9.

us versus the "It," in Martin Buber's terms, were all crushed and we were treated as though we had none of these human attributes. So, despite the physical jeopardy, the social alienation, and the psychological paranoia that were thrust upon us we survived.

This makes the black community in America unique. We are unmatched in the world.

Now, here we stand with our remarkable survival test behind us and, before us lies this massive technological society.

Before we could do anything about it, it had a hold on us. None of us can judge it in purely objective terms because we have been nurtured in its matrix.

A. Whatever education that we have was training to come to terms with this society. All of the mathematics illustrations, the poetry, the novels, and the skills were grooming us for "Main Street."
B. Our heroes are those who have made it here. We are now finding other heroes worldwide, but our real heroes, close-up, were crowned and regaled in Americas arenas, on her stages, and by her cameras.
C. We have such a long career incubation period in America that before one can choose whether or not to join the establishment he is on his way.
D. The values that technology supports have eclipsed other values, and success today can appear in only one form: namely, success in the technology race. The values that belong to a more simplistic, pastoral society are associated with dullness, lack of initiative or lack of capital.

In order to dampen our emotional enthusiasm for America's gadgetry, her plumbing and her swimming pools, her Eldorados and her playthings, her credit cards and her shopping centers, we would need to be hypnotized for a generation.

The profit motive in the economy surrounds us like the air we breathe. Our retirement plans are interwoven with the stock market. The foundation grants that we court are products of a market economy. The jobs we plead for—if they come tomorrow or next year—will be provided out of the market economy.

In other words, just as soon as our survival of slavery was accomplished we found ourselves with hardly any choice but to become participants in a rapidly expanding, urbanized, technological society that has a built-in reward system, an inhibition to criticism, severe penalties for those whose participation is lukewarm, and highly visible rewards for the faithful. Nevertheless we have some severe judgments to pass on this urbanized technology.

(1) It has made people insensitive to the weak and the dependent person. It applauds competition and success, not cooperation and supportiveness.
(2) It is racist. It keeps authority in white hands. It has a tribalism about it that makes only gestures at meritorious achievement.
(3) It thrives on colonialism. It demands a favorable trade balance with weak countries in Latin America, Asia and Africa.

(4) It has to sacrifice moral principles for economic concerns. It can lose 50,000 American boys in Southeast Asia to contain Communism but *not one* in South Africa to contain white supremacy.
(5) It fails to distribute its benefits where they are deserving; those who work the hardest enjoy the least. The poor through their consumption of goods subsidize the rich who don't work at all.
(6) It is devoid of a self-corrective mechanism. The political power resides where economic power dwells. And social passivity is adorned as patriotism. Dissent is assailed as treason. Liberty is an empty slogan. Justice is wrenched from the poor until it distorts itself and becomes oppression.
(7) It's military superiority has become an end in itself and it defeats moral initiatives that might otherwise be taken in the interest of peace.
(8) It has learned to live comfortably with chronic poverty.
(9) It has lost the admiration and the respect of smaller, newer nations who have hardly anywhere else to turn without being swallowed up in the Communist camp, which is just as materialistic with less freedom.
(10) It stands afraid and irresolute before clear choices that mankind must make, because it will not risk any of its apparent advantages.

Yet, it is our country. We have none other. We have earned our place here with our blood and tears, our honor and our lives. Is there anything that we can do to strengthen the moral fibre of the nation or to bring about a shift in the way she uses power and wealth. How does a modern, technological country experience a revolution? How do we prepare to shift the locus of power? How can we behave so that our efforts will serve to diffuse the concentration of power around materialism and militarism and redirect it in support of a more humanizing social order, the spread of justice and the protection and well-being of the weak?

Let us learn something from our own past in this country. Let us not be deterred by doctrinaire rhetoric about the system and fail to see what the real possibilities are. We have no precedent for our situation anywhere. There is nothing to go on. The capitalism of the early 19th century is not the capitalism of today. The rural setting of China or Cuba or Algeria is not our setting. The black numerical majority of W. Africa or the Carribean is not our situation. The apartheid of South Africa is not our situation. The strategies used to transfer power must be related to the stern realities of the American community. Moreover, too many blacks have too much at stake in America to be sanguine about her destruction.

This is our dilemma. We see her hypocrisy and her failure, her rigidities and her moral inertia, but the nation and her institutions are a trust to her people and there are many blacks who share in that trust.

We have contributed to her financial strength, we have supported her military might, we have striven to develop and reform her institutions, and every fibre of her being is strung with the lives of black people. It is not her destruction that we desire. It is her redemption.

First we must find a way of combining the new thrust called "black power" with the older efforts that were called "racial pride." We must celebrate our togetherness and our continuum and forsake those divisive spirits who want to make careers on fragmenting black people.

When T. C. Walker roamed all over Gloucester County, Virginia, begging black people to keep their land and buy some more he called it "race pride." Today it would be called "black power." When Dr. Luther P. Jackson gave his life to get black Virginians to vote he called it "race pride." Today it is called "black power." When black folk in Atlanta, Durham and Richmond; Philadelphia, New York and Nashville built their own banks, churches, lodges and restaurants, marched in their own parades and performed on their own stages, this was called "race pride." Today it is black power.

Of course, there is inevitably a generation gap. There has always been a generation gap. But it is shameful, ludicrous and absurd for older blacks to try to interpret the manifestations of pride for today's black youth; and, for today's black youth to look with contempt upon those who did "their thing" in their historical context before the sixties.

Before our full force can be applied to the task of displacing and diffusing power we need to focus our attention on the real enemy and refrain from internecine sniping. There is no need for blacks, young or old, to entertain and amuse patronizing white audiences, large or small, in hotel banquet halls or in East Side cocktail parties, by interpreting each other out of existence.

Indeed there are many life-styles among us. The black from small isolated Yankee communities who grew up with white playmates sees things a bit different from a black born in South Philadelphia or in Richmond's Churchill. Those who were reared in Methodist parsonages with praying parents will see things differently than those reared around bars and cussing parents. There is variety in our talk, in our stride, in our perceptions and in our *Weltanschauung*. But this variety need not breed contempt. It should breed a deeper, richer and longer lasting spirit of genuine community.

So, before we face the issue of the power that is lodged at the center of this urban technological society we will have to find a way to stay in the same room together, with honest dialogue, mutual respect, a warm fellow—feeling and an awareness of the potency of our enemies.

What a pity that so many talented, resourceful and dedicated black minds have been read out of the movement. To be sure some read themselves out, but the problems we face are so well known, the evidence is so clear, and our suffering so common to us all that we need not have to unify around particularistic and esoteric black themes. Those who have strong views that give them a unique identity should nurture these among like-minded brothers and sisters, but let us get our thing together on terms that unify rather than divide.

This has particular relevance to a society that is cybernated, keypunched, televised and system-analyzed. It has capacity. It has momentum. It has swift reflexes. Whatever you do evokes a rapid and tell-tale response. A strike of TWA stewardesses stops contractors, breaks up board meetings, holds up stock transactions, delays marriages, fouls up connecting flights, closes down the airport in Columbus, Dayton, Denver, Kansas City, and points in between; a Con-Ed brown-out stops elevators with millionaires locked in, shuts down computers, closes La Guardia, jams the D. C. airport because the tower grounds northbound flights. Ralph Nader finds a flaw in Ford's connecting rods and 400,000 cars have to go in for check-up. How then can we make our best approach to challenge entrenched power? We must have 22,000,000 black voices chanting themes upon which the great masses of them can harmonize. We can sing different parts, but the melody must be supported by harmonic chord progression, thunderous cadences and ringing Amens. I repeat, the story of black oppression is not that complicated; the reality of the denial of civil rights is not that incomprehensible; our estrangement from the flow of benefits in America is not that enigmatic, the options open to us are not that varied. Genius is not the capacity to obfuscate issues that have an innate transparency. It is the capacity to see novel approaches to issues that have eluded men of lesser talent. So let that black genius come forth—now—and show us those themes upon which the humble and the gifted, the poor and the sufficient, the young and the old, the college bred and the unlettered can all embrace with a whole heart.

I have been to Dachau and to Auschwitz. Into those chambers were marched the humble, the gifted, the poor, the sufficient, the young, the old, the college bred, and the unlettered.

Not only do we need a continuum among us, linking new black power with old race pride, but before we become so apocalyptic and eschatalogical about revolution, we need a further reapporchment between black identity and the acquisition of those skills that transcend race and culture.

In order for us to embrace our own "nationhood" we need not embrace the concomitant of contempt for that residium of scientific fact, scientific methodology and scientific objectivity that will make all of the difference between success and failure.

There is an increase of 110% of black college enrollment in the past 5 years. But these students have not given themselves to those disciplines that African students, for example, find so necessary for their struggle. Our students are still agonizing over identity crises in a painful and prolonged process. In fact many of the counselors who serve them have ushered them into this paranoid condition and have left them there.

One does not find out who he is or what he is by isolated introspection and cogitation. Your dignity is not defined in a corner. You know that dignity that fills the chest and throws the shoulders back by coming on out among people with your head high. Dignity is a social attribute, won in context of social variableness, not defined in privatized withdrawal. Likewise, one's identity is assured when his selfhood stays intact while in the crossroads of life.

It is important to study black history, black poets and playwrights, black art and music and to ground oneself in the black experience. But this is the beginning of things. It is the putting of the person together. Then it is time to move out and attach that person to a chemistry problem, an algebraic equation, a French novel, a physics experiment, a research topic in history or a riddle in political science.

These are not exclusive alternatives. But I fear greatly that the importance of keeping these parallel emphases is lost. There is some sneaking notion abroad that suggests that too much truck with the sciences is like too much concern for this capitalistic system, too much rehearsal for a place in it. There is not nearly the enthusiasm for those disciplines that the future will require that there should be.

Counselors to black students will have a lot to answer for in the future when this enlarged crop of black graduates all end up in low level civil service jobs, with no capacity to work where power is created and where power is exercised.

We simply cannot make the mistake on our own that the British imposed upon the Africans; namely, breed a crop of paper pushers who need to wait a generation for engineers, metallurgists, meteorologists, surgeons and architects.

I began by remarking the uniqueness of our situation. Part of it is the fact that we can see so clearly what kinds of skills and capacities and instrumentalities it will take to dislodge the concentration of power. Now we have the orators and the rhetoric. We need the accountants and lawyers, the technicians and analysts; we have the ball players and singers; now give us some engineers and surgeons.

Being black and beautiful does not exclude one on logical, ideological or historical grounds from organic chemistry. There may be educational deficits to be overcome but we will succeed in spite of them rather than fail because of them.

There must be an ingredient in our black pride that represents a new interest in the mastery of those skills that Hindus, Moslems, Wasps, Communists, and the Silent Majority—even—must have if they are going to talk about dislodging or redistributing power in a technological society.

I want to live long enough to see every black boy today who raps about the system in a position to pick up one of the pieces of the system and turn it from an instrument of greed and oppression to an instrument of compassion and justice.

Finally, with a new togetherness and with a new breed of young blacks who will include in their identity a well-mastered discipline that the nation will need, let us develop an unprecedented interest in the political machinery of the country that will make effective use of our new strength in the cities.

No one knows what direction change will take. There is wisdom in the notion that from the top of the mountains there is a peripheral vision that one cannot have looking down from half-way up one side. We cannot dictate the terms of power redistribution until that power is real to us—in our hands. Today, the black masses in the cities constitute a field for sociology dissertations, educational experiments and economic dilettantism. I am crushed by the equating of the black masses with urban blight. Let the black masses be equated with black political strength. This is happening in Newark, in Atlanta, in Norfolk, in Richmond and in Washington. Only 39% of the eligible voters went to the polls in 1968. For every one who voted, two were asleep, standing on a corner, leaning on a bar, or just pacing up and down. Whenever reference is made to anything urban it is a reference to black pathology!

All of this must change. Before we get caught up in the rhetoric of revolution let us comprehend the reality of politics. The prelude to revolution the kind of revolution that redeems rather than destroys, that results in black power rather than black incarceration—let us translate the black urban masses to black political articulation.

The heavens are not going to divide and lower a new Jerusalem in our midst. But if we get together, acquire some critical skills and make our presence felt, the waters will part. We'll march across them like dry land, scale high mountains and make the desert blossom like a rose.

# 15 Ties to Africa

**Mickey Leland and the USAID Bureau for Africa**

**Solar Cooking Demonstration in Akwasiho Village, Ghana**
Hattie Carwell

## Mickey Leland and the USAID Bureau for Africa

Mickey Leland (1944–1989), who was trained as a pharmacist, had a record of public service even before being elected to Congress in 1979. In Washington he concentrated especially on the problems of homelessness and hunger in America and, by extension, in Africa as well. After his untimely death, the Office of Sustainable Development of the U.S. Agency for International Development's Bureau for Africa, received responsibility for a new program, called the Leland Initiative, to extend Internet services to African nations.

### Mickey Leland

Born November 27, 1944, Lubbock, Texas
Died August 7, 1989, Gambela, Ethiopia

Mickey Leland was a singularly effective spokesman for hungry people in the United States and throughout the world. During six terms in the Congress, five years as a Texas state legislator and Democratic Party official, he focused much needed attention on issues of health and hunger and rallied support that resulted in both public and private action.

Growing up in a primarily Black and Hispanic neighborhood, he attended segregated public schools. He became a civil rights activist while still in college, bringing national leaders of the movement to Houston. After his graduation, while serving as a clinical pharmacy instructor at Texas Southern University, he helped set up a "door to door" outreach campaign in low-income neighborhoods to inform people about available medical care and to do preliminary screenings. The patterns of a lifetime of advocacy and action to help poor people were already forming.

His Congressional district included the neighborhood where he had grown up, and he again became an advocate for health, children and the elderly as he had in the legislature. His leadership abilities were soon recognized and he was chosen to be Freshman Majority Whip in his first term, and later served twice as At-Large Majority Whip.

As he visited soup kitchens and makeshift shelters, he became increasingly concerned about the hungry and homeless. The work for which he is best remembered began when Leland co-authored legislation with Rep. Ben Gilman (R-NY) to establish the House Select Committee on Hunger. Speaker Thomas P. "Tip" O'Neil named

Reprinted from: "Mickey Leland" and "USAID Bureau for Africa, Office of Sustainable Development. Leland Initiative: Africa GII Gateway Project," available at http://www.info.usaid.gov/regions/afr/leland/project.htm.

Leland chairman when it was enacted in 1984. The Select Committee's mandate was to "conduct a continuing, comprehensive study and review of the problems of hunger and malnutrition."

Although it had no legislative jurisdiction, the committee, for the first time, provided a single focus for hunger-related issues. The committee's impact and influence would stem largely from Congressman Leland's ability to generate awareness of complex hunger alleviation issues and exert his personal moral leadership. In addition to focusing attention on issues of hunger, his legislative initiatives would create the National Commission on Infant Mortality, better access to fresh food for at-risk women, children and infants, and the first comprehensive services for the homeless.

Leland's sensitivity to the immediate needs of poor and hungry people would soon make him a spokesman for hungry people on a far broader scale. Reports of acute famine in sub-Saharan Africa prompted Speaker O'Neil to ask Leland to lead a bipartisan Congressional delegation to assess conditions and relief requirements. When Leland returned, he brought together entertainment personalities, religious leaders and private voluntary agencies to generate public support for the Africa Famine Relief and Recovery Act of 1985. That legislation provided $800 million in food and humanitarian relief supplies, and the attention Leland had focused brought additional support for non-governmental efforts, saving thousands of lives.

Leland's ability to reach out to others with innovative ideas and to gain support from unlikely sources was a key to his success in effectively addressing the problems of the poor and minorities. He met with both Pope John Paul II about food aid to Africa and with Fidel Castro about reuniting Cuban families. In Moscow as part of the first Congressional delegation led by a Speaker of the House in the post-Cold War era, he proposed a joint U.S.-Soviet food initiative to Mozambique. As chairman of the Black Caucus, Leland proudly presented the first award the Caucus had ever given to a non-black—to rock musician Bob Geldof, honored for his Band Aid concert and fund-raising efforts for African famine victims.

Mickey Leland was a powerful advocate on other causes as well. He was a very effective promotor of responsible and realistic broadcast television and cable programming, seeking to reduce violence and assure that the characters realistically portrayed the rich diversity that makes up the American public. In addition, he frequently put his stamp on postal legislation, seeking at every turn more efficient delivery of postal services while protecting the rights of postal employees.

Mickey Leland fought hard to prevent food aid from being used as a political tool. He advocated communication with all governments, even those considered enemies, in order to further humanitarian goals and supported the right of U.S. citizens to provide humanitarian assistance to civilians of any nation. His successful initiatives expanded funding of primary health care in developing countries, including UNICEF's child survival activities and Vitamin A programs supported by the U.S. Agency for International Development (USAID). Under Leland's aegis, the Select Committee reemphasized the priority of hunger and the alleviation of poverty within the foreign assistance program of the United States. The committee successfully championed expanding credit to the poorest individuals in low-income countries and the use of proceeds from the sale of donated U.S. commodities for health, education and other grass-roots development activities.

Mickey Leland died as he had lived, on a mission seeking to help those most in need. He was leading a mission to an isolated refugee camp, Fugnido, in Ethiopia, which sheltered thousands of unaccompanied children fleeing the civil conflict in neighboring Sudan. Fifteen people died on the mission, including two members of his staff, philanthropist Ivan Tillem, State Department and USAID personnel, as well as Ethiopian nationals.

---

**USAID Bureau for Africa, Office of Sustainable Development**

**Leland Initiative: Africa GII Gateway, Project Description and Frequently Asked Questions**

Q: Exactly what does the Leland Initiative try to do?
A: The Leland Initiative is a five-year $15 million US Government effort to extend full Internet connectivity to approximately twenty African countries in order to promote sustainable development. The Leland Initiative (LI) seeks to bring the benefits of the global information revolution to people of Africa, through connection to the Internet and other Global Information Infrastructure (GII) technologies. The Internet is emerging as a low cost pathway that allows information to be more accessible, transferable and manageable; ready access to information is becoming the catalyst that transforms economic and social structures around the world and supports fast-paced sustainable development Even as African countries move toward more open economies and societies, there remain formidable constraints on sustainable development in such areas as the environment, disease prevention, literacy and private sector development. Africa needs access to the powerful information and communication tools of the Internet in order to obtain the resources and efficiency essential for sustainable development.
Q: What are Leland's three strategic objectives and expected results?
A: The project seeks to be flexible, ready to address the obstacles and opportunities within any given country, including support for policy reform, facilitating low-cost, high-speed access to the Internet, and the introducing proven mechanisms to build networks of active users. Its Strategic Objectives are to:

a. Create an Enabling Policy Environment (Objective 1): The project will promote policy reform to reduce barriers to open connectivity. The expected results:
1. Affordable prices, based upon costs plus profit, conducive to broad expansion of the user base;
2. Delivery of Internet services by private sector providers; and,
3. Free and open access to information available through the Internet, in conformance with host country laws.

b. Create a Sustainable Supply of Internet Services (Objective 2): The project will identify appropriate hardware, assist with full Internet connectivity, and assist private sector ISPs to develop their industry. The expected results:
1. Indigenous ISPs, trained in marketing and business plan development, offering full Internet access and better communication between counterparts in Africa and the world;
2. Country-wide access, with special attention to extension (rural) issues; and
3. Internet Society Chapters serving as advocacy and support organizations.
c. Enhance Internet Use for Sustainable Development (Objective 3): The project will increase the ability of African societies to use the communication and information tools of the Internet. The expected results:
1. Local and international partnerships sharing information related to sustainable development in manufacturing, business, the environment, health, democracy, education, and other sectors;
2. Indigenous partnerships to create and maintain new information resources based in the African experience which feed the GII;
3. Increased African capacity to use information in decision-making and in managing scarce resources;
4. Broadened user base for information systems and telematics services; and
5. Indigenous training capacity for users and ISPs.

The blend of activities addressing these Strategic Objectives will vary from country to country, depending on the level of e-mail and/or Internet connectivity. The development portfolio of the USAID Mission in each country will be very important, especially in implementing a strategy for growing the user base in priority sectors.

# Solar Cooking Demonstration in Akwasiho Village, Ghana

Hattie Carwell

Enslaved Africans had brought with them to the Americas a range of technologies that proved valuable in the New World. As the twentieth century drew to a close, private efforts and "appropriate technologies" as well as government aid for "high tech" infrastructure were attempting to aid in African redevelopment after centuries of colonial mis-development. The National Technical Association, founded by African-American technologists in 1926, reported one such project in 1997.

## Introduction

The Solar Cooking Demonstration and Solar Energy Project is a collaboration between Isaac Moore, electrical engineer; Hattie Carwell, chairperson of the International Committee of the Northern California Council of Black Professional Engineers (NCCBPE); and Dr. Wade Nobles, the newly-selected Nkosuohene (Chief of Development) for Akwasiho-Kwahu, Ghana, and immediate past president of the National Association of Psychologists (all living in Oakland, Calif.). This collaboration provides a good combination of technical problem solving capability with sociological and cultural considerations deliberated from the beginning of planning. They serve as the technical advisory team for this project.

The purpose of the initial visit to Akwasiho Village was to introduce the usefulness of solar energy. Arrangements for the Solar Cooking Demonstration and Solar Energy Project in this village were made with the Village Chief and Council of Elders. The demonstration was designed to show how solar energy can reduce the need to find firewood for cooking, water pasteurization and food preservation. The Solar Cooking Demonstration was the first of a series of activities and part of a long-term development project to create a Solar Village Living Center. Accordingly, only a cooking demonstration was scheduled for the first visit.

The Project is a simple, cost-efficient way to improve the living conditions of the people of Akwasiho by introducing solar energy technology as a source of sustainable development. The objective is to train villagers to construct and utilize solar products and to distribute and market the products not only in their own village, but in neighboring villages and beyond. Solar technology is introduced as an ecologically friendly means of cooking. It also improves health conditions by pasteurizing the water supply to provide safe drinking water and for cleaning items to destroy germs

Reprinted from: Hattie Carwell, "Solar Cooking Demonstration in Akwasiho Village, Ghana," *NTA Journal* 71 (Summer 1997), 34–35. 

and hence control diseases that result from them. Preliminary planning and development of the concept started in October 1995.

### Preparation

At the invitation of Nana Keseku II (Chief of the Village) and with the permission of the Elders, the Nkosuohene (Chief of Development) arranged the visit. Before going to the village a meeting was held with Nana Keseku II to develop an implementation plan. We reviewed the proposal that had been sent earlier to the Chief to explain the concept of using solar technology to develop economic opportunities in the village where unemployment is high. It was decided that a committee of six villagers (two sub-chiefs, two men and two women) would be established to promote and guide the development and use of solar technology. The members would be chosen because of their leadership capability and status. Their education would be at least at the high school level and both men and women would be included. The Village Chief, Nana Keseku II, along with Nana Berku I (Dr. Wade Nobles), the Development Chief of the village, would serve as ex-officio members and serve as links to the U.S.

In preparation for the cooking demonstrations, I developed a list of materials needed for the villagers to build their cookers—cardboard box, string, plain glass, aluminum, black pot, scissors—to make the parabolic shaped cooker which was jointly designed and constructed by Isaac Moore and myself. In the village, we used recycled boxes and string, but the other items had to be purchased. It cost approximately U.S. $4.00 for these materials. The glass was purchased and cut in Accra. The other items were readily available in a store in a town close to the village.

### Demonstration

The Solar Cooking Demonstration was conducted July 25, 1996, from 12:30 p.m. to about 4:00 p.m. July is still the rainy season in Ghana. We started the three-hour drive to Accra to Akwasiho at 7 a.m. with an overcast sky looming overhead.

The village is at the foothills of the Kwahu mountain range. The people in this area are known for their entrepreneurial spirit. This agricultural community relies primarily on selling and trading family crops.

Although Nana Keseku II had initially suggested that the demonstration be conducted just for the committee members, who will be the ones who will promote the use of solar technology, the Village Elders requested a public demonstration to be held in the Chief's palace courtyard where all important events are held.

During the visit, the solar cooker was assembled, and it was demonstrated how the sun could be used to cook local foodstuffs. In this particular case, rice, which is a dietary staple for the villagers, was chosen for the demonstration. The rice used was obtained from the villagers. It took approximately 20 minutes to assemble the cooker, since the parts had been precut prior to leaving California. It was transported on the airplane in a backpack in its collapsed and completely flattened state. The cooker was constructed from a 16-inch by 20-inch cardboard box. The basic solar energy concepts and principles of the cooker design (see appendix) were explained to the villagers as the rice cooked. The explanation was presented by Hattie Carwell in English, and one of the villagers provided simultaneous translation into Twi. Approximately 100 villagers watched the demonstration with great interest. The age range of the audience was from four years old to over 70.

Some other uses of solar energy also were described. It was explained that the sun could also be used to preserve and can food, to heat water and to help kill germs that cause illness. The villagers were told that electricity could be generated from solar energy with another type of equipment.

It was clear that the adult attendees easily grasped the solar energy concepts and the advantages and disadvantages of using the sun. Their questions indicated that they understood the limitations. For example, one asked, "What do you do when there is no sun?" Others asked, "Can you use wood rather than cardboard or tin rather than aluminum foil?" They realized that solar cooking would mean that they would have to change the time at which they normally prepare meals. They were curious about why the cardboard box did not catch fire. I allowed them to touch the glass cover and the sides of the box to answer their questions. I invited them to see how the sun was focused on the black pot. They watched as I moved the box and adjusted its angle of tilt to maintain the same focus. Children were asked to chant to encourage the sun to come out from behind the clouds and to make the same observations as the adults. Each one watched the temperature rise and fall as the sun went in and out of the clouds. Due to the overcast weather conditions, the maximum temperature obtained was 155°F, whereas temperatures of 250°F were easily reached during tests on a sunny day in California.

The rice did not completely cook because the crowd tended to block the sun as they observed the process; but the villagers remarked that they could smell the rice cooking and could see the steam condensate on the glass cover over the pot. The difference between the ambient temperature of 70–75°F and the 155°F inside the box was very convincing.

## Construction by Villagers

After the explanation and as the rice continued to cook, volunteers among the villagers were selected and divided into six teams of four each to build their own parabolic shaped solar cookers. Both men and women built the cookers with enthusiasm. Spontaneously, the teams modified the design that I had demonstrated. They were able to obtain the same shape with one box (one of my earlier prototypes). A boy, approximately 15 years old whose family recently moved to the village, indicated that he had built a solar cooker before where he used to live. He was very supportive of the use of solar energy technology.

## Conclusions

The introduction of solar cooking was successful and has sparked an interest in the possibilities for the use of solar technology. Nana Keseku II is very supportive of the possibilities and with the permission of the Village Elders, he has set aside land for other demonstrations of solar energy applications. Despite the enthusiasm expressed during the day, it will be the responsibility of the Village Solar Energy/Cooker Project Committee to assure that the villagers continue to use the cookers and to guarantee that the innovation of cooking with the sun is permanently infused within the community. Supplementing the firewood method of cooking will mean an adjustment to existing lifestyles. The Committee must come up with techniques to persuade people to make the change.

As a follow-up to the demonstration, the Committee will be asked to encourage villagers to use the cookers built that day and to keep records of the results. Several promotional activities also are planned. The training guide, which provided information on how to design and build solar cookers and provides helpful cooking hints, will be distributed to the schools in the village so that classroom demonstrations can be conducted with children and design contests held. Solar-powered lights will be placed outside the library which has recently been established by Nana Kwaku Berku I (Dr. Wade Nobles). Solar information will be placed in the library for those who would like to learn more about solar energy technology.

## Appendix: The Physical Parameters That Affect Solar Cooking

Sunlight is composed primarily of visible light and at somewhat longer wave lengths, near-infrared (heat) radiation. When this sunlight enters an enclosure with a front window and hits a dark surface, such as that of a cooking vessel, it is absorbed and converted to heat. Some of this heat is reradiated as far-infrared radiation at even longer wavelengths. Glass and many clear plastics transmit the incoming visible and near-infrared solar radiation but do not transmit the back-emitted far-infrared radiation, which is trapped in the enclosure. In addition, the enclosure traps convection currents of heated air, which also can carry away heat. The combination of these constitutes the "greenhouse effect," which allows objects within the enclosure to attain significantly higher temperatures than objects in the open.

The heating effect can be further enhanced by the use of external reflective surfaces, which collect sunlight over a large area and concentrate it into a smaller area, as shown in figure 1 (not reprinted here). The most efficient heating occurs when sunlight is incident from the front, along the axis of the cooker, as shown. The best cooking conditions are with sunlight unobstructed by clouds and when the sun is high in the sky, between about 10 a.m. and 2 p.m. standard time. The cooker should be shielded from wind, which tends to cool it.

Aluminum foil, covering the flaps of a cardboard box, constitutes an efficient collecting surface, readily available and with a minimum of cost and complexity (particularly important factors in developing countries). Triangular pieces of cardboard, also covered with aluminum foil, are used to hold the four main flaps in the proper angular configuration (see photograph), and also contribute to the collecting area. Thin sheet plastic can be used instead of costly and fragile glass for the front window of the enclosure.

To maximize heating efficiency, it is also important to insulate the sides and bottom of the enclosure to prevent heat loss by conduction, convection or radiation. Effective but low-cost insulation can be obtained with spaced layers of cardboard, the innermost layer of which is covered with aluminum foil. Crumpled newspaper also can be used as insulation material. It is also important for the enclosure to be air-tight, especially around the edges of the front window, to avoid heat loss by convection.

To maximize heating efficiency and minimize the required cooking time, the cooking vessel should be black in color and supported by a black metal tray. The tray should be in good thermal contact with the vessel, but insulated from the outside.

# 16 Engineering Careers

*Women and Minorities in Science and Engineering*

**History of the National Society of Black Engineers (1997)**
Jerry Good

**Black Women Engineers and Technologists**
Valerie L. Thomas

*Careers in Science and Technology* **(1993)**

## *Women and Minorities in Science and Engineering*

The National Science Foundation, as a part of its responsibility for the health of the nation's science and engineering activities, tracks the statistical record of participation in the technical professions. Its studies have consistently shown large disparities between races and between sexes.

---

Data for 1978 show black, Asian, and white scientists and engineers concentrated in different fields of science and engineering (figure 16.1). Blacks were more likely than whites to be scientists than engineers; over half of the whites but only 27 percent of the blacks were engineers, and blacks represented less than 1 percent of all engineers (figure 16.2). In science, blacks were more likely than whites to be social scientists or psychologists; 37 percent of the blacks were social scientists and almost 12 percent were psychologists. For whites, the comparable figures were 15 and 10 percent.

The field distribution of Asian scientists and engineers is similar to that for whites (figure 16.1). Over half of both whites and Asians were engineers rather than scientists.

The index of dissimilarity can be used to summarize overall field differences among racial groups. The index of dissimilarity between whites and blacks at all degree levels in 1978 was 38. This figure means that roughly 38 percent of the blacks would have to change fields or occupations to have a distribution identical to that for whites. The index of dissimilarity between Asians and whites was 13.

Regardless of race, about one-fourth of S/E doctorates in 1979 were life scientists (text table 16.1). This was the only field for which such similarity existed. The various racial groups were distributed quite differently between engineering and science and across fields of science. A larger proportion of blacks and native Americans were social scientists and psychologists, while a larger share of the Asians were engineers and physical scientists. At the doctoral level, the index of dissimilarity between blacks and whites in 1979 was 20; between Asians and whites, it was 26.

The relatively high proportion of women among black doctoral scientists and engineers (23 percent in 1979) does not appear to affect the field distribution of blacks. Although black men were more likely than black women to be physical, mathematical, and environmental scientists and engineers, over 70 percent of the black male doctoral scientists and engineers were still in the life and social sciences and psychology. Among whites, 54 percent were in these fields.

---

Excerpted from: National Science Foundation, *Women and Minorities in Science and Engineering* (Washington, DC: NSF, January 1982), 12–14. Reprinted by courtesy of the National Science Foundation.

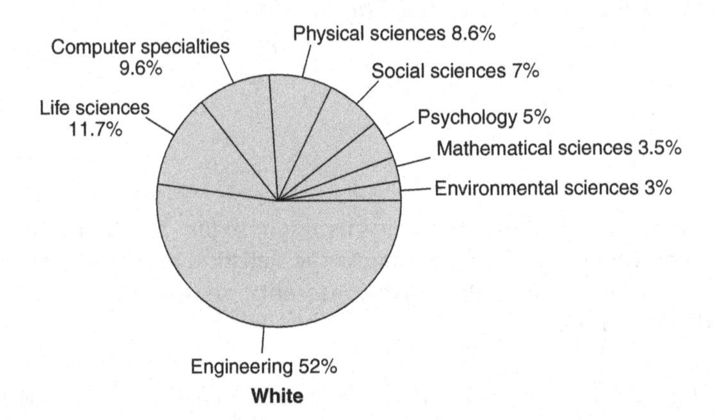


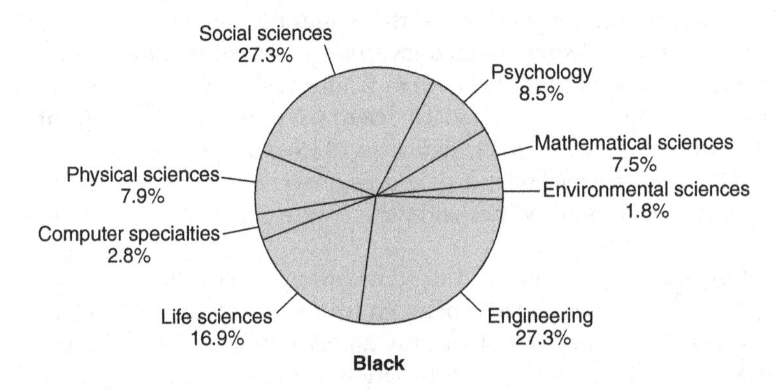


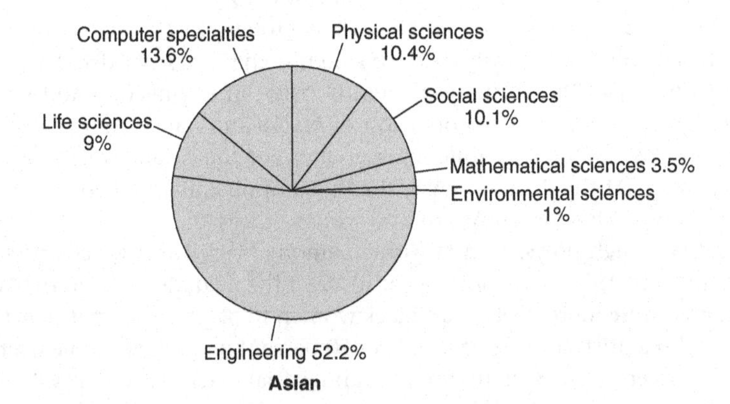


**Figure 16.1**
Employed scientists and engineers by race and field, 1978. From data in appendix table 6b (not reprinted here).

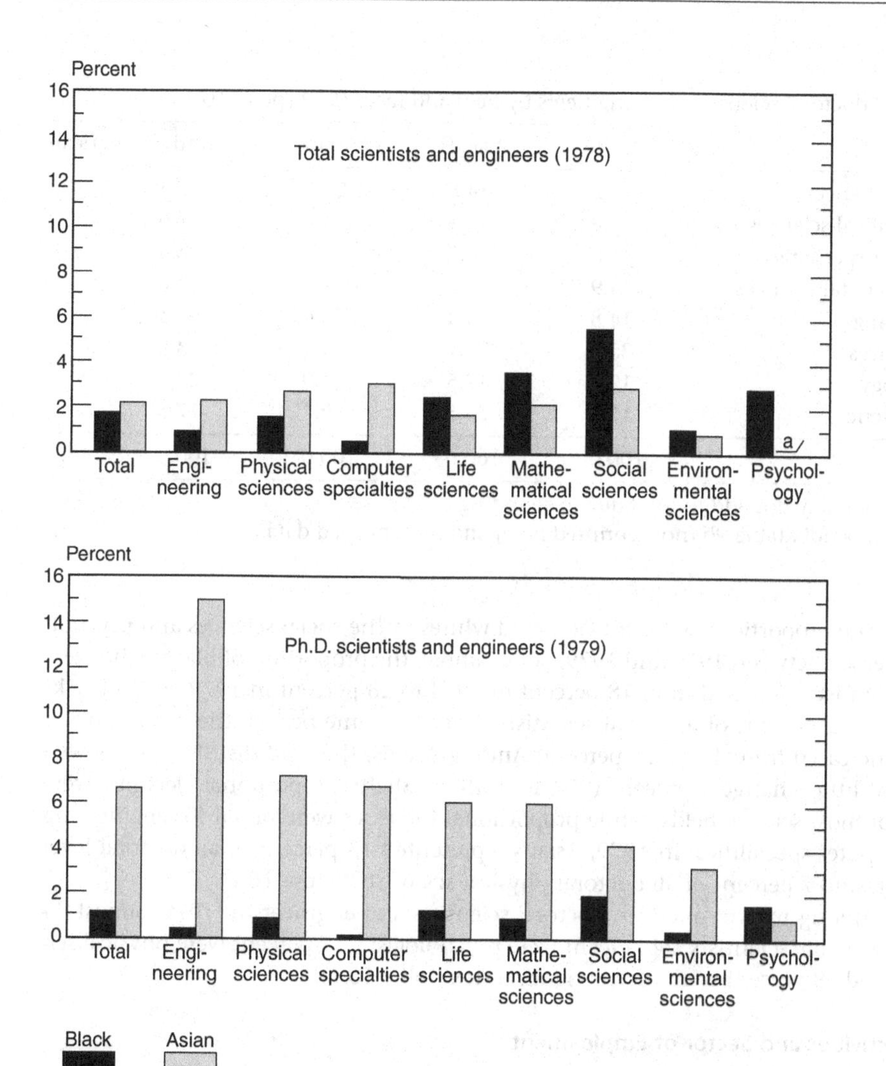


**Figure 16.2**
Racial minorities as a percent of all employed scientists and engineers by field. ([a] Too few cases to estimate.) From data in appendix tables 6b and 9b (not reprinted here).

**Table 16.1**
Employed doctoral scientists and engineers by field and race: 1979 (percent)

| Field | White | Black | Asian | Native American |
|---|---|---|---|---|
| Physical sciences | 19.2 | 14.8 | 20.7 | 17.3 |
| Mathematical sciences | 4.8 | 4.5 | 4.5 | 5.6 |
| Computer specialties | 1.0 | 0.2 | 2.2 | 2.4 |
| Environmental sciences | 4.9 | 1.9 | 2.3 | 3.0 |
| Engineering | 14.8 | 5.4 | 35.6 | 7.1 |
| Life sciences | 25.7 | 27.1 | 23.1 | 23.8 |
| Psychology | 12.7 | 17.5 | 1.9 | 23.3 |
| Social sciences | 15.7 | 28.5 | 9.7 | 17.5 |
| Total | 100.0 | 100.0 | 100.0 | 100.0 |

Note: Detail may not add to total due to rounding.
Source: Appendix table 9b (not reprinted here) and unpublished data.

The proportions of both blacks and whites in the social sciences and psychology increased between 1973 and 1979. For example, the proportion of blacks who were social scientists increased from 18 percent in 1973 to 28 percent in 1979, when blacks represented 2 percent of all social scientists. Over the same period, the proportion of whites increased from 13 to 16 percent. Among Asians, the field distribution showed relatively little change between 1973 and 1979. Slight proportional declines were noted for most science fields, while proportional increases were noted for engineering and computer specialities. In 1979, Asians represented 15 percent of all doctoral level engineers and 7 percent of all doctoral physical scientists (figure 16.2).

Among native American doctoral scientists and engineers in 1979, almost 24 percent were life scientists, 24 percent were psychologists, 17 percent were physical scientists, and 18 percent were social scientists (text table 16.1).

### Work Activities and Sector of Employment

Data on work activities by race are not available for all scientists and engineers. However, some identification of differences in work activities by race can be gained by examining the activities of experienced scientists and engineers (those in the labor force at the time of the 1970 Census of the Population), recent graduates, an doctoral scientists and engineers.

Among experienced scientists and engineers, 30 percent of both blacks and whites were likely to work in some aspect of management. Asians, however, did not participate in management to the same extent as their white or black colleagues; only 19 percent held management positions.

Among recent graduates at the bachelor's level, however, the findings have been mixed. Whites more often reported management as their primary work activity

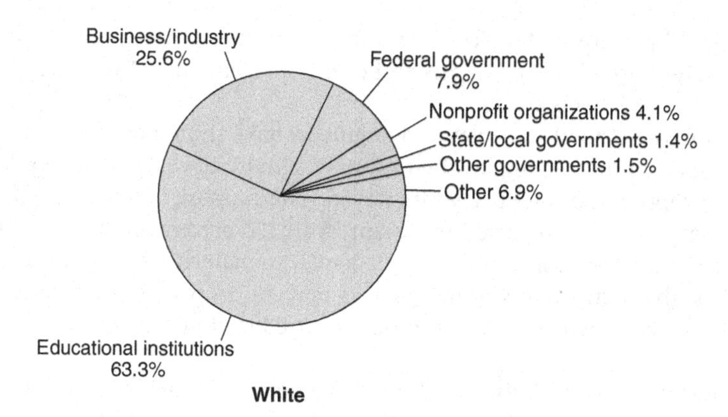


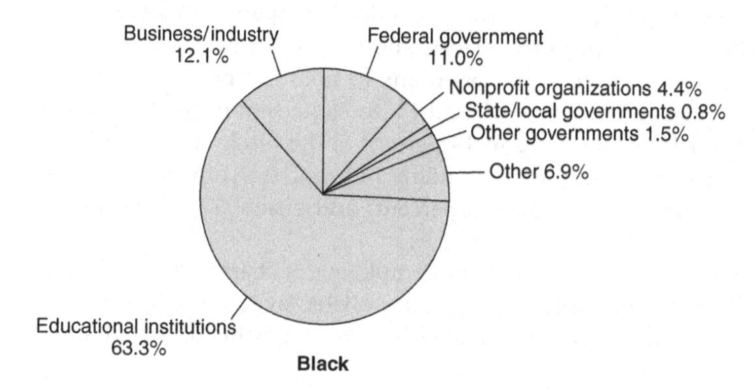


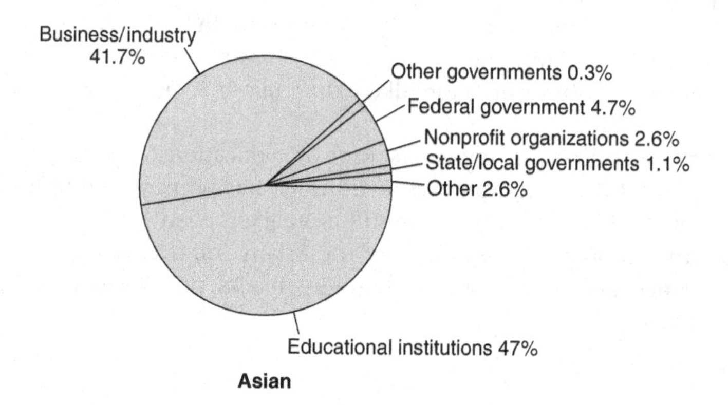


**Figure 16.3**
Employed Ph.D. scientists and engineers by race and type of employer, 1979. From data in appendix table 29 (not reprinted here).

than blacks or Asians (13 percent vs. 9 percent). Among recent master's degree holders, blacks were more likely than whites or Asians to be in management (17 percent, 12 percent, and 6 percent, respectively).

Work activities of doctoral scientists and engineers have shifted over time. For all races, the proportions citing teaching as their primary activity have declined, while the proportions reporting management have increased. The most significant proportional gains in management were reported by Asians, from 12 percent in 1973 to 24 percent in 1979. For whites, the proportion in management remained stable at around 23 percent; for blacks, the increase was from 24 to 28 percent. In part, these changes reflect sectoral shifts in employment opportunities from educational institutions to business and industry.

Within educational institutions, however, whites were more likely than blacks or Hispanics to be tenured. Of those who received their doctorates in science and engineering between 1960 and 1978 and who were academically employed in 1979, about 62 percent of the whites, 47 percent of the blacks, and 64 percent of the Asians were tenured. Blacks were less likely than whites or Asians to hold full professorships. Of those who earned their degrees between 1960 and 1978, 28 percent of the whites and Asians were at this rank in 1979 compared to 19 percent of the blacks. It is interesting to note that most (90 percent) of the Asians holding full professorships were foreign born. Much smaller percentages of whites (13 percent) and blacks (27 percent) were foreign born.

Sector of employment affects a number of employment characteristics, including work activities and salaries. Reliable data are not available by race for all scientists and engineers. Data by race, however, are available for some segments of the S/E work force.

Among experienced scientists and engineers in 1978, almost two-thirds of the whites, one-third of the blacks, and over half of the Asians were in business and industry. Among recent graduates at both the bachelor's and master's levels, Asians were more likely than whites and whites were more likely than blacks to be in business and industry.

Most doctoral scientists and engineers were in educational institutions in 1979, although the proportion in educational institutions has been declining for all races since the early 1970's. Blacks, however, are still more likely than whites or Asians to be in educational institutions. Over two-fifths of the Asians and over one-quarter of the whites were in business and industry in 1979. Among blacks, only 12 percent were in this sector (figure 16.3).

## History of the National Society of Black Engineers (1997)

Jerry Good

In an attempt to support African-Americans in the technical professions and to attract more young people into careers in science and engineering, in recent years a number of national organizations have formed, such as the National Action Council for Minorities in Engineering (NACME), the Black Data Processing Associates (founded in 1975), and the National Association of Precollege Directors (founded in 1978). One of the most active is the National Society of Black Engineers.

---

The Society of Black Engineers was the concept of two Purdue undergraduate students, Edward Barnett and Fred Cooper, and Faculty Advisor, Arthur Bond, an electrical Engineering Ph.D. candidate and Purdue Staff Member. The idea was to establish a student organization to help improve the recruitment and retention of Black engineering students at Purdue. In the late 1960s, a devastating 80% of the Black freshman entering the engineering curriculum at Purdue were dropping out. Ed Barnett and Fred Cooper approached the Dean of Students with the idea of starting the Black Society of Engineers (BSE). The Dean agreed and asked the only Black faculty on Staff, Arthur Bond, as advisor to the new Society. Later in 1971 Arthur Bond wrote the Black Society of Engineer's constitution, which he modeled after other engineering societies.

On October 11, 1971, the Black Society of Engineers met for the first time. At that time, Edward Barnett, a senior in industrial engineering, was elected President; Fred Cooper was elected Vice President; and Cynthhia Gunthner (now Cynthia Milton) was elected Secretary. About 15 of the entering 25 Black freshman class and 7 of the 12 upper classman attended the first meeting. The primary aim of the fledgling organization was to increase the number of Blacks attending Purdue's School of Engineering and, most importantly, to make sure those who entered the engineering school graduated.

The objective was accomplished by study groups led by upper classmen and exam files donated by all members. The grade point average of the entering class of 1971 was tracked and proved to be a full percentage point higher than prior classes of Blacks, proving the success of the program.

In 1974, Tony Harris, a senior in mechanical engineering was elected President; John Logan, Vice President; Rudy Nichols, Secretary (who was later replaced by George Smith as Secretary); and Brian Harris was elected Treasurer. The name of the organization was changed to the Society of Black Engineers (SBE) to be more in line with

---

Reprinted from: Jerry Good, "History of the National Society of Black Engineers," typescript, April 24, 1997.

traditional engineering organizations on campus. At the beginning of the 1974 school year there were 104 Black Engineering students on Purdue's campus. The Society of Black Engineers had about 60 members, approximately 30 of which were active and contributing members. Meetings were held in the Library of the Black Cultural Center.

Mr. Harris was also President of the Purdue Chapter of the American Society of Mechanical Engineers, the first Black to hold that office. That organization proved to serve later as a role model for several precepts of NSBE. Tony Harris fashioned a letterhead and purchased stationary. Harris drew the now familiar crossed lightning bolts to represent the energy of the Black engineering students and the torch to symbolize the hopes that they shared. The letters SBE were superimposed across the symbol. At the bottom of each page of the letterhead he had printed "Dedicated to a Better Tomorrow".

The four officers, Tony Harris, John Logan, George Smith, and Brian Harris, who were also roommates, served as kind of "think tank" for the events that followed. A Resume book listing all the Black engineering students was compiled by Kevin Mason and Jerry Good and sold to companies interviewing at Purdue in the Fall of 1974. Also Brian Harris and his projects committee coordinated and held the first company reception and banquet in the February 1975, attracting over 60 companies and their representatives to the campus, as well as many faculty and Black Engineering alumni. This banquet and corporate reception was the predecessor to what has become a mainstay to the national organization and local chapters nationwide. It was at this banquet that the concept of a national Black engineering organization, an idea that the think tank had been developing for some time, was first made public. Dr. Arthur Hansen, President of Purdue University was present and immediately liked the idea. He pledged his support and later made good on that pledge.

With the help of Dr. Bond, now an engineering professor and still the faculty advisor, and the assistance of Mrs. Saunie Taylor who was responsible for minority affairs at the Office of Freshman Engineering, detailed plans were laid for going national. Proceeds from the banquet left SBE with working capital of almost $2,000. It was agreed by the membership to use this money to establish a national organization rather than for scholarships for members of SBE. Agreement was unanimous.

While preparing the budget, it became obvious that the $2,000 would not cover the kind of meeting the Think Tank group envisioned. Anthony Harris called President Hansen and explained the situation and also met with Dean Hancock. After explaining to the Dean how "supportive" and "eager" (only mild embellishments) the President had been to receive his help, the Dean directed Mr. Harris to contact Harold Amrine, the head of freshman engineering, and state his case. In early December, 1974, Harris met with Dr. Bond, Harold Amrine, and Thomas X. Fletcher. After hearing the proposal, Mr. Amrine was extremely impressed. He said it was the first time a student organization had come to him with funds of it's own requesting assistance. He agreed on the spot to match their money from his budget.

Determining who would attend became the major problem. How do you reach Black engineering students nationwide? They now understood the problem some companies were having finding Blacks to hire. Mrs. Saunia Taylor furnished Smith and Tony Harris a list of every accredited engineering program in the country, including Hawaii. Mr. Harris then wrote a letter to each Dean or President (288 in all) explaining what they were attempting to do, and asking their help to identify Black student leaders, organizations or faculty members. Although many could not see the need for such a group and considered it divisive (why purposely separate the Blacks from the mainstream?), about 80 schools answered affirmatively.

After making personal contacts with the 80 schools, the SBE members were pleasantly surprised to find that many universities had Black student organizations of engineering, science or math students very similar to theirs. The names were all different, but the objectives were basically the same. Although some were older, none were more active, with exception to the group at the California Polytechnical, Pomona, CA.

A date was set, April 11–12, 1975, for the first national convention. Students that could not get support from their schools or could not afford to attend were sponsored by Purdue's SBE. With Mrs. Taylor's help and sanction by the office of Freshman Engineering, all the resources of Purdue were made available to the group. Now all that remained was the agenda.

Mr. Harris divided the country into 6 regions, again patterning them after the 12 regions of the ASME. All of the schools that had agreed to send representatives were placed into regions. Those six regions still exists today. The Think Tank expanded to include John Cason, Junior in Chemical Engineering (elected 1st National Chairperson), and Virginia Booth, Sophomore in Industrial Engineering (and later elected 4th National NSBE Chairperson). Together these six decided on what would occur at the conference: 1) Select a name; 2) Choose a national symbol; 3) Rough draft a charter; 4) Choose a national headquarters; and 5) Elect a chairperson. With these as the conference objectives, the first national conference of the National Society of Black Engineers began.

Forty-eight students representing 32 schools began arriving Friday, April 11, 1975. Most visitors were housed at the Campus Inn in Lafayette, Indiana. Several stayed with Purdue Students. In addition, many of the Purdue SBE members attended and participated in the conference. The first evening a welcoming dinner was held with remarks by Tony Harris and Dr. Bond. Later that night, a party was held at the Shreve dormitory.

The following morning after breakfast, the opening session began at the Purdue Memorial Union. Mr. Harris opened the meeting and established ground rules. Five splinter groups were formed to contemplate each of the five decisions to be made at the end of the day. At the end of the splinter sessions, a representative would summarize the discussion with an oral report. It was agreed that each school would receive two votes on each issue. A simple majority would decide. Ed Coleman, Brian Harris, John Logan, Stanley Kirtley, and George Smith led the splinter groups. After lunch, the entire group reconvened and the voting began. Several names were identified, with the one submitted by the Purdue contingent, the National Society of Black Engineers, being selected by majority vote. The fact that it resembled other engineering organizations and already had some national exposure as SBE, were determining factors. Next a symbol was chosen. The Purdue SBE symbol had been modified by Jerry Good and several others. The letter "N" was added and the SBE letters were shown horizontally. This was submitted along with several and the arguments were heard for each. Finally the NSBE symbol was selected for much the same reason as the name. Selection of an official charter proved to be much more difficult. It was agreed that it would be the first order of business at the subsequent conference.

The delegates also agreed that Purdue would become the National Headquarters for NSBE. Finally came to electing the first national chairman. After hearing brief speeches from contenders (Mr. Harris could not run since he was a graduating Senior), John Cason, junior in Chemical Engineering at Purdue was elected. John Cason later left Purdue for personal reasons and William A. Johnson II, California Polytechnical, assumed the office as National Chairperson. Harris thanked all participants and received a standing ovation. This concluded the formal portion of the conference, but everyone remained to view a film brought along by William Johnson of California Polytechnical, Pomona contingent which was prepared with the help of the Los Angeles Council of Black Professional Engineers. After viewing the film, it was agreed that Los Angeles Cal-Poly, Pomona Campus would be the site of the next NSBE Conference.

## Black Women Engineers and Technologists

Valerie L. Thomas

As in most other fields, African-American women in technical fields have struggled against the double handicaps of being both black and female. Their record of accomplishment can, like most other measures, be seen as a glass half full or still half empty. At the time she wrote this, Valerie Thomas was Assistant Director for the National Space Science Data Center, NASA/Goddard Space Flight Center. (Text citations for notes 6, 10, 11, and 30 are missing in the original article.)

Black women in the U.S. who have chosen careers in engineering and technology have found themselves in a very unique position throughout history because they were Black and/or women. At times this dual identity has caused doors of opportunity to be closed to them, and at other times it has caused doors to open for them. Despite the difficulty of the technical subject matter and the many obstacles that these women have had to overcome, many of them survived and went on to make significant contributions in their respective fields. This article is an assessment of the role played by Black women in engineering and technology historically, in the present, and with a view toward the future. Whenever possible, role models will be highlighted and the factors which influenced their decision to pursue a technical career are identified.

### Historical Perspective on the Engineering Discipline and Black Women in the Workforce

Before delving into the early contributions of Black women, it is important to consider the development of the field of engineering as a discipline and then look at the general picture of Black women in the workforce in order to establish the appropriate perspective.

Dr. Eugene Deloatch has done a very excellent chronology of the history of engineering, the highlights of which are summarized here.[1] Formal engineering education began in 1802 with the establishment of a civil engineering program at the U.S. Miliary Academy at West Point and Rensselear Polytechnic Institute followed by offering engineering studies in 1824. By 1845, other schools established programs for technical education and studies and a dual track developed: programs at traditional colleges and universities such as Yale (1846), Harvard (1847) and Cornell (1868); and "institutes" were established expressly for providing technical education, e.g., MIT (1862),

Reprinted from: Valerie L. Thomas, "Black Women Engineers and Technologists," *SAGE* 6 (Fall 1989), 24–32. Reprinted by permission of the author.

Stevens Institute of Technology (1867), etc. Because of the prohibitive cost of an engineering education, the Morrill Act was passed in 1863 to provide for a nationwide federally subsidized system of agriculture and mechanical or "land grant" colleges to increase the affordability of technical education. Even though the number of U.S. engineering schools increased from 6 to 70 between 1862 and 1872, this did not help Blacks because of their high concentration in the South. A second Morrill Act was passed in 1890 to correct the inequities; however, the emphasis was on equipping Blacks with manual rather than leadership skills in engineering. In 1910 an engineering school was established at Howard University. Until that time, even though Blacks could go to engineering schools in the North, they could not be admitted to the schools in the South, which were federally and state subsidized.

By 1870, engineering professors began putting more technical content into their curricula and the field of electrical engineering emerged. Electrical engineering was stimulated by the ideas generated by Alexander Graham Bell, Thomas Edison, Louis Latimer and George Westinghouse. The post World War II (WWII) period was a boom in engineering. Prior to 1942, Howard University's enrollment never exceeded 100 students. After WWII, Howard's enrollment increased rapidly and schools of engineering were established at 6 other historically Black colleges and universities. In 1816 there were 100 engineers in the U.S. and by 1900 the number of engineers was second to the number of educators in the country.

### Women in the Workforce

According to Debra Lynn Newman, in her article on "Black Women Workers in the Twentieth Century," domestic work was the least desirable of the occupations available to Black women at the beginning of the 20th century because of the long hours (from 6 AM to 9 or 10 PM, with little time off), low wages and the sheer drudgery of the chores.[2] During this period in history, one-third of all Black females were employed: 95% evenly divided between agriculture and domestic work; 2% as seamstresses, tobacco and factory workers, and 1% as teachers. Many Black women endured the sacrifices associated with working in "white people's homes" to earn money to educate their daughters so that they could have the opportunity for greater occupational accomplishments.

With the outbreak of World War I (WWI), industrial job opportunities that were previously unavailable to Black women opened up.[3] This was due to the lack of sufficient numbers of men because of the draft and the cessation of the influx of immigrants by legislation. These job opportunities were temporary and had to be relinquished with the return of the solders to civilian life. Of all the Black women in the labor force, 5.4% were in industry in 1920 compared to 3.7% in 1910; however, this does not reflect the women who lost their jobs after the war.[4] During WWII, Blacks brought more pressure on the federal government than was done during WWI. A planned March on Washington in July 1941 (100,000 marchers were expected), organized by A. Phillip Randolph to protest discrimination and demand employment opportunities for Blacks in the defense industries, was aborted when President Roosevelt reached an agreement with Randolph that resulted in an Executive Order prohibiting discrimination in the federal government and in the defense industries.[5] This did not lead to an effective penetration of industry by Black women but they became an important part of the nation's clerical segment and many domestics moved to "service workers" for institutions such as hotels, schools, cafeterias, hospitals, etc. with better hours and wages. Black female professionals increased from 1.5% in 1910 to 7.7% in 1960 and 10% in 1970, with a slight increase by the time of the 1980 census.

### Black Women as Technical Pioneers

Black women have been and continue to be pioneers in the fields of engineering and technology. With the first Ph.D. in physics awarded by any American school in 1866, Black male achievements quickly followed with the first Black to earn Ph.D.'s in the following areas: Edward Bouchet (also the first Black elected to Phi Beta Kappa), physics, Yale University, 1876; Alfred O. Coffin, biology, Illinois Wesleyan, 1889; St. Elmo Brady, chemistry, University of Illinois, 1916; no Black held a Ph.D. in mathematics until 1925 when Elbert Cox received one from Cornell University.[7] Black women did not earn Ph.D.s in the sciences until many years later when Evelyn Boyd Collins[8] earned a Ph.D. in pure mathematics from Yale University in 1946 and Marjorie Lee Browne[9] received a Ph.D. in mathematics from the University of Michigan in 1949. Women were also distinguishing themselves in the areas of geology and chemistry with Margurite Thomas—being the first to earn a doctorate in geology in 1942 from Catholic University and Marie Maynard—earning one in chemistry from Columbia University in 1947. In the area of physics, it took over one hundred years after the first doctorate was conferred by any American school before a Black woman received one. This accomplishment was achieved by Shirley Jackson[12] at MIT in 1976. Also in the 1970s, the first Black women received Ph.D.s in chemical engineering. Jenny R. Patrick[13] who received her degree from MIT in 1979 has been credited with the first; however, another Black woman, Lilia Abron,[14] received a doctorate in chemical engineering from the University of Iowa in 1971. The author did not find a citation for the first Black woman to earn a Ph.D. in engineering. However, there is at least one Black woman, Christine Darden,[15] who has a Ph.D. in mechanical engineering from George Washington University in 1983. The first Black woman to obtain degrees in engineering from Howard University, Michigan State University and Tuskegee are listed in table 16.2.

### Women in the National Technical Association

Prior to WWII, despite Black women's academic achievements in the technical fields, job opportunities and membership in professional technical organizations in their fields were nearly non-existent. This was also true for Black men. In 1926 a group of Black men formed the National Technical Association (NTA) for the advancement of

**Table 16.2**
Black women firsts in science and technology

| Name | Distinction | Year | Institution |
|---|---|---|---|
| Madame C. J. Walker | Built research and production laboratory | 1910 | Private enterprise (millionnaire and philanthropist) |
| Marguerite Thomas | Ph.D. in geology | 1942 | Catholic University |
| Evelyn Boyd Collins | Ph.D. in pure mathematics | 1946 | Yale University |
| Hattie T. Scott | Graduate in civil engineering | 1946 | Howard University |
| Marie Maynard | Ph.D. in chemistry | 1947 | Columbia University |
| Marjorie L. Browne | Ph.D. in mathematics | 1949 | University of Michigan |
| Yvonne Y. Clark | B.S. in mechanical engineering | 1952 | Howard University |
| Hildreth G. Florant | Female professional member | 1954 | NTA |
| Maxine R. Rosborough | Female national officer (sec.) | 1960 | NTA |
| Delores R. Brown | Engineering graduate | 1967 | Tuskegee |
| Lilia Abron | Ph.D. in chemical engineering | 1971 | University of Iowa |
| Zella N. Jackson | Mechanical engineering graduate | 1974 | University of Michigan |
| Shirley Jackson | Ph.D. in physics | 1976 | MIT |
| Zella N. Jackson | Black systems operations manager—divison level | 1978 | IBM |
| Jennie R. Patrick | Ph.D. in chemical engineering | 1979 | MIT |
| Zella Jackson | Career Woman of the Year | 1982 | Hawaii |
| Christine Darden | Ph.D. in mechanical engineering | 1983 | George Washington University |
| Irene Long | Chief of NASA's Medical Operations & Human Research Branch, Biomedical Office | 1982 | NASA/Kennedy Space Center |
| Valerie L. Thomas | National president | 1984 | NTA |
| LaBonnie Bianchi | EE graduate | 1961 | Howard Univ. |
| Bertha Carmichael | Aboard sealift command oceanographic research ship | | Military |
| Mae Jemison | Astronaut | 1987 | NASA |

science and engineering by Blacks and for breaking down the barriers in the profession due to race prejudice.[16] Through its annual meetings which included presentations of technical papers and scientific and social discourse, NTA became a very effective network for obtaining job opportunities in the members' respective fields. Black women joined NTA and shared in the leadership of the organization. Hildreth Griffin Florant,[17] a metalurgist, became its first female professional member in 1954 and Maxine Kernodle Rosboraugh,[18] an architect, became the first female national officer in 1960 when she served as National Secretary. In the mid 1970s there was a tremendous influx of women who were very active at the local and national levels as members and leaders. In 1984 Valerie L. Thomas, a mathematician, became the first National President of NTA. NTA has survived the test of time, making it the oldest predominantly Black technical organization. Its objectives now include the motivation of minority youth to pursue technical fields.

## Early Opportunities

During the early 1940s, the only professional jobs open to Blacks with degrees in physics and engineering (most of which were earned at historically Black colleges) were teaching, preaching and the legal or medical profession.[19] Most of the Blacks with technical degrees found doors shut to them in private industry. Therefore, they gravitated to the U.S. Post Office, which was known to have the largest concentration of educated Blacks in any agency. At the other agencies,[20] only menial and janitorial jobs were open to the Blacks.

Around the time of WWII, the Black scientist in industry ceased to be an outstanding rarity and trained Blacks could consider industrial jobs as areas of normal employment. In addition, research competence and facilities in Black institutions began receiving the recognition that brought in grants and contracts to conduct research for industry and government agencies. The integration of Black scientists into professional societies and organizations placed them in the mainstream of development in the associated areas of science.

WWI provided opportunities for subprofessionals such as technicians and draftpersons. The shortage of men at Fort Monmouth in 1942 caused Corryne Godwin and Muriel Robinson Baldwin and two other women from Brooklyn College, all graduates with technical degrees, to be hired as junior professional assistants.[21] Corryne went on to become one of only two Black women at Fort Monmouth to become grade GS-13 (senior engineer). Helen Hayes was hired as a chemist in 1942 and dismissed in 1946 during a reduction-in-force. Corleze Holiman, a Black woman electrical engineer who reached the GS-13 grade level, started in 1943 in the engineer-in-training program. Mary Tate, a computer specialist GS-13 in the Communications Research and Development Command (CORADCOM) when she retired, started as a laboratory technician in 1945, later became a mathematician doing scientific and engineering calculations, and in 1948 became professional as a computer analyst. All of these women were college graduates.

During the late 1940s, a small select group of Black scientists worked at the Argonne National Laboratories in the Chicago suburbs, under a heavy veil of secrecy, helping to convert atomic materials for useful purposes in medicine and industry.[22] Two of the Black women working on that project were Ella Tyres, a biologist, and Blanche Lawrence, a junior biochemist. Ella Tyres is a Spelman graduate who first managed the animal farm and later conducted research to determine the effects of radiation on humans. Blanche Lawrence is a Tuskegee graduate and started out as a lab technician before becoming a junior biochemist.

During the early years of the National Aeronautics and Space Administration's space program, a Black woman, Dr. Evelyn Boyd, worked as a mathematician with the Space Computing Center in Washington, D.C. and supervised three people who formulated calculations for tracking unmanned and manned capsules that were launched for the Mercury Project.[23] Dr. Boyd was graduated *suma cum laude* from Smith College in 1945 and received advanced degrees from Yale University. She worked three and a half years with the Department of the Army before joining the Space Center.

## Aerospace Physician

When she was a very young girl, Dr. Irene Long had a fascination for watching the airplanes take off and land on Sunday afternoons at the Cleveland-Hopkins Airport which is next to NASA's Lewis Research Center and when she was not doing that, she tagged along with her father for flying lessons.[24] When she was 9 years old, she told her mother that when she grew up she was going to become a doctor for the nation's space program and go to work at the Kennedy Space Center (KSC) in Florida. She actively pursued her dream, receiving a B.A. in pre-medicine/biology at Northwestern University in 1973; earning a Doctor of Medicine in 1977; and completing a three year residency in aerospace medicine. During her aerospace medicine residency, Dr. Long had rotational assignments at NASA's Ames Research Center (ARC) in California and Kennedy Space Center in Florida.

Her dream was realized in July 1982 with her appointment as KSC's first Black woman Chief of the Medical and Environmental Health Office in the Biomedical Operations and Research Office. Some of the activities for which she and her staff are responsible include: providing and planning for emergency medical services in support of space shuttle launch and landing activities; day to day shuttle activities; coordination of human Life Sciences Flight Experiment requirements; operational management of the Baseline Data Collection facility used for pre and post flight physiological data collection; and screening and monitoring of research laboratory subjects and participating in research protocol development and implementation.[25] She sits in the Firing Room above the Launch Control Center when the space shuttle is launched and is one of a team of doctors at the landing site collecting post landing data on the medical condition of the astronauts when the shuttle returns safely.

Dr. Long's ultimate goal is to deliver medical care for people in the space environment.[26] She wants to treat people who are ill but who have been exposed to the zero-gravity of space. In that environment, people's bodies will have different responses and their treatments will have to be different. She wants to fly aboard the space shuttle and gain first-hand knowledge of how the changes in the astronauts' bodies due to weightlessness affect their medical care.

### Black Woman Astronaut

Dr. Mae Jemison has one of the most challenging of the technical careers as the first Black woman astronaut.[27] Dr. Jemison, a native of Decatur, Alabama, grew up in and claims Chicago as home, and attended Stanford University on a 4-year National Achievement Scholarship. She received a B.S. in chemical engineering and a B.A. in African and Afro-American Studies from Stanford in 1977. She then went to Cornell University where she received the doctor of medicine degree in 1981. After completing her internship at Los Angeles/USC Medical Center, she went to Sierra Leone and Liberia as Area Peace Corps Officer. While in Liberia, she received very valuable experience in the provision of medical care, management of the pharmacy and laboratory records, medical administration and supervision of personnel. On returning to the U.S., she joined the CIGNA Healthplans of California and was a general practitioner when she was selected by NASA for the astronaut program in 1987. She completed her training in August 1988.

### Overcoming Polio

At the age of six months, Patricia Richardson was diagnosed as having polio in her right leg. She spent her early years living with her grandmother, brother and cousin in rural Mississippi and did not think of herself as being handicapped, even though she maneuvered around with the support of leg braces and crutches.[28] When her brother and cousin went to the fields to pick cotton, she was one step behind them and when they returned home with their sacks filled with cotton, she carried her own sack of picked cotton. At age 7, accompanied by her brother, she was put on a train to Chicago to join their mother who had migrated there. It was there that she was treated like a handicapped child: called cripple and handicapped by the taunting children, sent to a special school for the physically disabled and forced to get on the "big bus" to ride to school. However, she did not think of herself as handicapped and convinced her mother to send her to a regular school. Even though her mother had been told by the doctors that she would never walk without the leg braces and crutches, she made a decision that she would walk without having to use them. When she visited her father during the summers, she would spend long hours practicing walking without the braces, releasing one part at a time while her leg grew stronger and she was finally able to walk without them.[29] At age 15, she underwent major surgery to correct the problem of one leg being longer than the other, a common result of polio, and spent the next nine months lying helplessly on her living room sofa while 40% of her body was in a cast. With that much time on her hands, she did a considerable amount of introspection. Realizing that there were many parts of her that she did not like, she finally stopped feeling sorry for herself and created a new self-concept which is best summed up by the expression, "There is a winner in you."

She went on to earn a B.S. in mathematics from Chicago State University, spent 12 years working as a systems analyst in Fortune 100 companies and today she is President of her own consulting firm, R&R Systems Consultants, Inc., a computer consultant firm with a strong commitment to excellence and service to a society faced with the task of assimilating and utilizing the rapidly advancing information technology. In addition, she is the Board of Directors Chair for Firman Community Services, one of the premiere social agencies in the nation; and she serves as the 2nd female National President of the National Technical Association.

## Black Women Engineers and Technologists Become Sought After

The fate of Black women engineers and technologists changed drastically by the mid to late 1960s. The Civil Rights Act of 1963 declared discrimination in hiring to be illegal and Executive Order 11246 required affirmative action plans from companies with more than 10,000 employees as a condition for receiving federal contracts. This meant that companies that wanted government contracts had to develop a plan which showed how they would actively locate and hire minority engineers and technologists. In 1960, there were 862,000 engineers in the U.S. and 1/2 of 1% (4174) were Black.[31] By 1974, there were only 4300 women engineers in the U.S. and the number of minority women engineers in the U.S. was too small to count (according to the National Science Foundation). Therefore, it was very difficult for the government and corporations to find Black women engineers and the competition for the small pool that existed was very stiff. This resulted in Black women engineers being very much sought after and heavily courted. This continues to be true.

In 1988, engineering disciplines were among the most lucrative for minorities and all graduates. According to the College Placement Council, the average starting salaries for electrical, civil, computer, industrial and mechanical engineers were $29,688, $25,428, $30,408, $27,420 and $29,412 respectively.[32] One graduating Black woman received $200/month more than the female average because of her major, chemical engineering.[33] However, there are some disadvantages of being Black and female in a technical environment. There is a high probability that there will be few peers to rely on who are minority women and whose judgement can be trusted for feedback.[34] This can have an adverse effect on Black females' self-confidence and performance. In addition, their upbringing as females and minorities may not have prepared them with lessons of social aggressiveness to form relationships that will advance them on the job.

## Entrepreneurs in Technical Fields

Of these three entrepreneurs, Mary Alicia Roach Paige has taken the most unusual path to become the owner of an electronic data processing/telecommunications firm,[35]

Computer Engineering Association, Inc. (CEA). She was graduated from the New England Conservatory of Music with the hope of becoming a music teacher. Instead she spent 12 years working as serial librarian for the federal government and 11 years as head cataloger for the town of Randolph, Massachusetts. In 1978, at the age of 57 after losing her only sister to cancer and ending a 14 year marriage, she began to put her life back together by starting her own business. Her conceptualization of CEA was influenced by her experience working with library information systems which she found to be very chaotic and spent a lot of time straightening out. Her initial staffing for CEA included a secretary and two of the three people who responded to the ad that she placed in a newspaper for an office manager for a female owned startup company. The two men that she hired had the expertise that she lacked and together they complemented each other.

CEA grew from three local employees and a deficit in 1979 to 33 employees nationwide and $500,000 in revenue in 1981.[36] The staff includes physicists, engineers, mathematicians, statisticians, meteorologists, computer scientists, programmers and system analysts. CEA offers services to government, industrial and commercial organizations and has completed projects in analyzing, designing, developing and implementing a wide variety of management and business information systems. A high technology division was added to the organizational structure to handle over the horizon ERS radar, plasma physics, particular orbit analysis, laser radar, space spectroscopy and magnetic field studies.

Zella Jackson's motivation for becoming an entrepreneur was her need to utilize her God-given talents and at the same time bring some balance to her life.[37] Her decision in selecting a field to pursue in college was a very important one for her. Because of her poor background, she decided to pursue a career that "guaranteed" professional success. She felt that she could not afford to gamble on liberal arts. Therefore, she became the first Black woman to earn a B.S. in mechanical engineering from Michigan State University in 1974. She continued at Michigan State and received an MBA in operations research in 1975. She has had a very impressive list of work experiences: first Black degreed person to work for Lansing Board of Water and Light in Michigan; technical marketing specialist for Dow Chemical Company in California; first Black person to hold the position of systems operations manager with Divisional level responsibilities for IBM in the General Products Division in San Jose, California; business consultant; co-founder (with her husband) of Novasearch Consultants in 1981; and the first Black woman to become Career Woman of the Year (1982) for the state of Hawaii.

Jackson has been so many firsts and onlys that she has stopped counting. Even though they represent outstanding achievements, there have been some down sides. When she started and finished engineering school at Michigan State, there were no female rest rooms in the engineering school and all of her mail was addressed to Mr. Zella Jackson. She has indicated that the overt and subtle discrimination to which she has been subjected was just short of being cruel. Though she is not bitter, she can now look back and see it for what it was. She has concluded that there is discrimination against women which is so far reaching that nothing short of a full revolution could change the corporate and the business world. Consequently, she has rewritten the rules for herself because she feels that "the white man's business rules don't work!" By creating a business and family life around her talents and needs (instead of ignoring or subjugating them), she is now happy and has a balanced life. She earns an income well into the six figures; works three to four days a week (which leaves extra time for her $2\frac{1}{2}$ year old son); enjoys a serious hobby (cycling); travels; has national recognition for her professional achievements; and has a husband of 10 years. So she asks, "What else is there?"

The last of the three businesswomen, Dr. Lilia Abron, developed PEER (Pollution Engineering Environmental Resource) as a logical extension of her counseling work that she was doing on the side while teaching at Howard University.[38] Dr. Abron was born in Memphis, Tennessee, to parents who were both high school teachers and into a home where education had a very high priority. As long as Dr. Abron and her siblings stayed in school, her parents committed to providing them with everything that they needed and half of what they wanted. Her parents and the parents of her friends shielded them from the harsh realities of segregation. After graduation from high school, she enrolled in LeMoyne College in Memphis on a four-year scholarship to pursue her interest in medicine and later become a pediatrician. However, after her second year she came to the realization that she did not like dissecting dead things.

In the 1960s, engineering was not a field in which Black women were highly visible and, from what Abron knew about it, was not very appealing to her. LeMoyne, a Black college, provided a nurturing environment; the professors, mostly white and from the North, took great care in directing the students toward success. Since there was no engineering major at LeMoyne, they created one and tailored a curriculum for her.[39] She received her engineering degree from LeMoyne, and went on to Washington University in St. Louis, MO, on a fellowship in sanitary engineering, which combined her love for medicine and science. She was one of two women and the only Black in the program of six students. Obtaining a job after receiving her M.S. from Washington University was very difficult, but she did finally get one which she held onto until she enrolled in a doctoral program at the University of Massachusetts in Amherst. She completed her requirements at the University of Iowa, receiving a Ph.D. in chemical engineering in 1971. By then the job hunting situation had changed: her credentials, coupled with being a Black woman, were very valuable; she was heavily courted by industry; but she decided to return to the South to teach, taking assignments at Tennessee State and Vanderbilt University. In 1974, after three years of teaching in Tennessee, she went to Howard University.

While at Howard she started doing consulting in 1978, and after the workload increased, it became difficult to juggle with her teaching responsibilities. President of PEER became her full-time job in 1981. Since that time, the company has continued to grow. The types of assignments that it handles are: pollution control and abatement; occupational medical services; and medical monitoring. In 1988, it employed more than 115 people in 9 offices and did over $1.5 million worth of business.[40] In addition to her technical and business accomplishments, Dr. Abron has three sons for whom she is providing the same kind of example and nurturing that were provided by her parents.

### Young Black Female Role Models

So far the discussion has focused on Black women who are well established in their careers and have already made their mark in their respective fields. I will now discuss young Black women engineers and technologists who are not too many years away from their school days and who have most of their career still ahead of them. A questionnaire was designed and distributed to gather information to give us a snapshot about some of these women. Twenty-three women, who live in ten states and the District of Columbia, responded. Of the women who responded, 35% had at least two college degrees; 39% of their degrees were in engineering, while physics and mathematics represented 19% and 23% respectively; 6% were MBAs and 13% were other (architecture, medicine and biology); the schools that they attended ranged from historically Black colleges and universities to technical institutes such as MIT and Rensselaer Polytechnic Institute; their first degree was earned from 1969 to 1989, with 78% of those earned from 1980 to 1986. This is by no means a statistical sample; however, it is interesting to note that 65% of these women are classified as engineers on their job and the other 35% are identified as computer scientist, manager, operations research analyst, physicist, architect, aerospace physician and business consultant. This group has had experience working at 42 different installations, 55% of which were in industry, 26% in the government and 7% in other (hospitals and a not for profit organization). The average number of job changes is 1.7. All but two of the respondents belong to professional organizations; 65% belong to at least two organizations. These professional organizations include: discipline specific (49%), general technical (8%), sorority (8%), labor (2%), honor societies (15%) and other (18%). Although the general technical category represented only 8% of the total number of organizations to which the women belong, eleven responses from them indicated membership in these organizations. These women have already begun to distinguish themselves: 13% were valedictorians and salutorians in their high schools; 22% received math and science department awards in college or participated in the honors program; 15 scholarships/fellowships were received by members of the group; 13% have presented technical papers at a symposium or annual meeting; seven job performance awards were received by women in the group; and 26% have been designated as Woman of the Year, Outstanding Young Woman of the Year, Outstanding Careerist of the Year, and Outstanding Young Woman of America.

When asked for the factors which influenced their decisions to pursue a career in engineering or other areas of technology, the two most frequent answers given by 39% of the women for each answer were: 1) the respondent was good at and/or enjoyed mathematics and science and; 2) participation in special programs (e.g., introduction to engineering programs for high school students; programs that provided technical job experience to college students; and a seminar on Black women in engineering). Jennifer M. Sims, operations research analyst with the Department of the Army, indicated that her interest in mathematics was the biggest influence in her decision. Some of the women participated in two special programs, one while in high school and the other while in college. For example, Anita Alexander, electrical engineer at NASA/Lewis Research Center, attended an early engineering exposure program at Case Western Reserve University and then worked in the summers at E.I. DuPont during her four year college career. One of the women, Janet C. Rutledge, electrical engineer at Georgia Tech, did not know what an engineer was until she participated in a National Science Foundation sponsored program at Georgetown University during the summer before her senior year in high school. Before that experience, she thought that engineers drove trains. She now has a B.S. and an M.S. in engineering and expects to have her Ph.D. in engineering in 1989. Another one of the women, Darlene Brummel, attended a seminar on Black Women in Engineering at the University of Maryland, Princess Anne Campus, and was so impressed by the statistics on Black women in engineering that she decided to major in engineering. Two years later, after taking a drafting course as an elective and getting straight A's on her assignments, she realized that it was demanding but interesting and the creativity was there. Her work was compared with "the guy's" work in her class and it was better. It was then that she decided to become an architect and she now works at NASA/Goddard Space Flight Center as an architect.

The next highest ranked factor (which was indicated by 30% of the women) that influenced their decision was their families. When the family was mentioned as an influencing factor, there was usually an indication of a specific person in the family who had the most direct influence. In over half of the cases, this person was the father who was a technical role model and/or provided direct training or counseling. For most of the other cases, the mother supplied the daughter with support and encouragement and demonstrated a belief in her daughter's capabilities. Rhonda Marie Lewis' father is an excellent example of the former. He is a self-employed electrician who taught himself the principles of electricity and spent a lot of time with her as a child, explaining the laws and theories of physics. He emphasized "how essential mathematics and science were in making a better tomorrow." She now works as quality assurance engineer at NASA/Stennis Space Center. Caroline L. Ledbetter, physicist at the Naval Weapons Station, is an example of the latter. The next two factors, each indicated by 13% of the women, were guidance from the high school counselor and good employment potential. It is interesting to note that even though the special programs are among the highest factors (indicated by 39% of the group) that influenced the career decision for these women, the guidance counselor as an influencing factor was rather low. This tends to imply that the students were not finding out about the special programs from their counselors nor were the counselors having much of an impact on their career decisions.

When these women were asked about ways to increase the number of Black women in engineering and technology, they identified what could be done in four broad categories: role models and mentors; outreach activities; early introduction to engineering careers; and early education which emphasizes mathematics and science. The most favored of these were the role models and mentors, recommended by 47% of the women. The other three categories were tied, with 35% of the women suggesting each of them. In the area of role models, two types of uses were identified; higher visibility of the Black woman currently working in the technical fields and an interactive role model/mentor relationship with Black girls. Cynthia Kovice Adams has suggested that we make the public more aware of Black females in the technical fields and this will invoke the "word of mouth" concept. We also must be accessible to Black girls, for children learn by example. She goes on to say that we must discuss with them our jobs, responsibilities, rewards for doing our jobs well, personal satisfaction, material gains and upward mobility. Terri Wood, software engineer at NASA/Goddard Space Flight Center, recognizes the importance of the participation of mentors in special programs sponsored by GSFC for providing technical job experience to high school and college students. She says that "all of these programs would not work without the mentors ... they should be given rewards of recognition by their organizations for participating in these programs."

The women who stressed education and early introduction to engineering careers agreed that this should start early; however, they had different ideas about how early. Some said junior and senior high school, others said elementary school and there were some suggestions for pre-school. Zella Jackson feels that, "It starts the moment that the baby is in your arms!" Regardless of the exact level at which the training starts, there is a consensus that there needs to be an emphasis on excellence, specifically in mathematics and science. We must "plant" a notion of excellence and achievement at an early age, preferably grade school, according to Joan Higginbotham, electrical engineer at NASA/Kennedy Space Center. She continues, "Give them a push in the right direction ... Young children need the frame of mind to know that they can succeed with lots of hard work." Darlene Brummell thinks that we should encourage girls as early as the third grade to develop mathematics and analytical skills because this is the age for them to build their confidence as well as their character, to know that they are "good with numbers" and have the ability to find solutions to difficult problems, and to realize that they have the option of pursuing a nontraditional profession with confidence. There were suggestions for students in the higher grades also. At those levels, mathematics and science emphasis should be complemented with technology fairs, site visits and paid summer experiences. Another twist to the education was offered by Anita Alexander, who suggested workshops and seminars to educate parents in motivating youngsters to seek engineering and the sciences as a profession.

A discussion of outreach programs includes much of what has already been mentioned; however, there were some additional activities cited. These included making introductions to engineering programs such as MITE[41] and Preface more available to Black girls; tutoring assistance; networking; providing cultural experiences to broaden their horizons; not limiting participation to lower income children; seminars and a sister in engineering program.[42] In all the recommendations for increasing the numbers of Black women in engineering and technology, there was a common theme of involvement. Ledbetter summed it up very well when she said, "There must be a commitment to help others and not to rest on our laurels."

As the women who were surveyed look toward the future, they are very optimistic. Eighty three percent of the women used very positive adjectives to describe the future for Black women while 13% used the words dim and bleak and 4% could not comment because of not knowing any other Black women engineers. Even those who used negative descriptors qualified them with a comment about the need to increase the number of women going into these fields. The extent of the problem of the lack of visibility and small numbers of Black women in technical fields, and the even smaller number of managerial and supervisory roles becomes clear when 35% of the respondents envision more involvement by Black women in leadership roles. They also look forward to seeing Black women owning their own companies and becoming astronauts. Anita Alexander is very pragmatic when she says that Black women still have to contend with the double standard, unlike their white counterparts; and characteristics such as being articulate, enthusiastic, independent, aggressive and forthcoming are integral elements in the survival and success of the Black female in the engineering profession. Higginbotham pulled it all together very succinctly when she said, "There is nowhere to go but up."

## The Year 2000

The composition of the workforce in year 2000 is projected to be quite different from what we have today. There will be a larger segment of minorities and women: 23% more Blacks, 70% more Asians and other races (American Indians, Alaska natives and Pacific Islanders), 74% more Hispanics and 25% more women adding 3.6 million, 2.4 million, 6.0 million and 13.0 million more workers respectively.[43] Altogether, the minorities and women will make up 90% of the workforce growth and 23% of the new employees will be immigrants. This increase in the number of Blacks and Hispanics could pose a problem because both of these groups have traditionally had higher unemployment rates than whites and the occupations that will have the greatest growth rate are not those typically filled by Blacks and Hispanics.[44] Occupations requiring the greatest amount of education will have the most rapid growth rate. People with less than a high school education will find it harder to find a job with good pay and chances for advancement. If nothing is done to change this trend, the problems will be exacerbated. If serious changes and made in the educational preparation of Blacks and Hispanics, tremendous opportunities await them in the high growth occupational areas such as engineering and technology.

The technological progress and rapid rate of technological change will quickly make equipment and human skills obsolete.[45] These changes will also make products and services obsolete. Therefore, workers in year 2000 will have to be continually learning new techniques and jobs will be very demanding. Large corporations will find it difficult to respond very quickly to rapid changes: thus, there will be great opportunities for entrepreneurs.

When answering the questions about the future of Black women engineers and technologists, the surveyed group was right on target. If the suggestions for increasing the pool of Black women with the necessary technical skills are successfully implemented, then Black women will be well prepared to fill the positions that will be available in year 2000. They envisioned women forming their own companies and filling leadership positions in industry and the government. Because of the unique makeup of the workforce, skills necessary for managing a diverse and multi-cultural workforce will be crucial. According to Floyd and Jacqueline Dickens, authors of *The Black Manager: Making it in the Corporate World,* managing diversity requires one to recognize the cultural differences the employees bring to their jobs. In order to manage a diverse workforce and extract added value from it, you must acknowledge and be sensitive to those differences.[46] Well prepared Black women engineers and technologists could be ideal for managing the diverse workforce in the 21st century.

## Notes

1. Eugene M. DeLoatch, "Electrical Engineering—A Historical Perspective," *NTA Journal,* 60 (October 1986), pp. 8–9.

2. Debra Lynn Newman, "Black Women Workers in the Twentieth Century," *SAGE: A Scholarly Journal on Black Women, III* (Spring 1986), pp. 10–15.

3. *Ibid,* p. 10.

4. *Ibid,* p. 11.

5. *Ibid,* p. 13.

6. *Ibid,* p. 13.

7. *Negro Year Book,* 1952, p. 96.

8. *Ebony Magazine,* August 1960, p. 7.

9. *Negroes in Science, Natural Science Doctorates, 1876–1969,* p. 60.

10. *Holders of Doctorates among American Negroes,* pp. 159–160.

11. *American Men and Women of Science,* 15th ed., p. 488.

12. *Ibid,* p. 10.

13. *Ebony,* May 1981, p. 6.

14. Victoria Hendrickson, "Minorities Who Made It," *Graduating Engineer,* Oct. 1988, pp. 144–145.

15. Ivan Van Sertima, *Blacks in Science* (New Brunswick: Transaction Books, 1986), pp. 255–257.

16. "NTA Historical Highlights", *NTA Journal*, 56 (1982), p. 9.

17. "60 Years of Progress, The Story of NTA," *NTA Journal*, 60 (Oct. 1986), p. 28.

18. "Our Women in Scientific and Technical Fields," *NTA Journal*, 56 (1982), p. 40.

19. "High Level Achievers," *NTA Journal*, 61 (Jan. 1988), p. 14.

20. *Negro Year Book*, 1952, p. 96.

21. "High Achievers," *NTA Journal*, 61 p. 16.

22. *Ebony*, 1949.

23. *Ebony*, Aug. 1960, p. 7.

24. "A Pioneer in Aerospace Medicine," *Ebony*, Sept. 1984, pp. 61–62.

25. "NASA Biography," NASA/John F. Kennedy Space Center.

26. "A Pioneer in Aerospace Medicine," *Ebony*, Sept. 1984, p. 64.

27. "Dr. Mae C. Jemison: The World's First Black Woman Astronaut," *NTA Journal*, 61, Oct. 1986, p. 27.

28. "A Winner In You," Chicago P.M., 1 (1988), p. 5.

29. *Ibid*, p. 46.

30. Fatina Shaik, "Double Blessing or Double Whammy?" *NTA Journal* 56 (1982), p. 41.

31. M. Lucius Walker, Jr., "Statistics on Blacks in Engineering," *NTA Journal*, 1976, p. 41.

32. Nicholas Basta, "Job Prospects for Minority Engineers," *Graduating Engineer*, Oct. 1988, p. 164.

33. Fatima Shaik.

34. *Ibid*, p. 42.

35. "Computer Engineering Assocs., Inc., A Minority-Woman-Owned Business," *NTA Journal*, 56 (1982), p. 23.

36. *Ibid*, p. 23.

37. Responses from questionnaire and telephone interview.

38. Victoria Hendrickson, "Minorities Who Made It," *Graduating Engineer*, Oct. 1988, p. 144.

39. *Ibid*, p. 144.

40. *Ibid*, p. 144.

41. Minority Introduction to Engineering (MITE) Program; Pre-Engineering Program (Preface).

42. A program suggested by Sabra Lynn Townsend, industrial engineer, for reaching back in a supportive, positive way to encourage better grades in high school, higher self-esteem and better work habits in our "little sisters." She suggested that we should develop an "each one, teach one" attitude.

43. Ronald E. Kutscher, "An Overview of the Year 2000," *Occupational Outlook Quarterly*, Spring 1988, p. 4.

44. *Ibid*, p. 4.

45. "Getting Ready for Work in the 21st Century," *USA Today*, Aug. 1988, p. 10.

46. Walter M. Perkins, "Straight Talk from Smart Black Managers," *Graduating Engineer*, Oct. 1988, p. 72.

## *Careers in Science and Technology* (1993)

To help raise the expectations of minority youth and attract more of them to technical careers, the National Technical Association, the Department of Energy, and NASA compiled biographical sketches of successful black, Hispanic, and Native American engineers and scientists. One important thread connecting most of these careers is work for or in the federal government, the combination of lavish spending for military and space-related technologies and an affirmative policy of nondiscrimination made many of these careers possible.

---

**Col. Guion S. Bluford, Jr., Ph.D.**
Aerospace Engineer and Former NASA Astronaut

Guion S. Bluford, Jr., was the first African American to fly in space, aboard the eighth flight of NASA's Space Shuttle (STS-8) in August 1983. He participated in this five-astronaut flight crew as a Mission Specialist. He is a Colonel in the U.S. Air Force. He was selected as a Mission Specialist Astronaut by NASA in 1978.

Guion S. Bluford, Jr., was born in Philadelphia, Pennsylvania. He earned a B.S. degree in Aerospace Engineering from Pennsylvania State University in 1964. He joined the Air Force and attended pilot training at Williams Air Force Base in Arizona and received his pilot's wings in January 1965. He then went to F-4C combat crew training in Arizona and Florida and was assigned to the 557th Tactical Fighter Squadron in Cam Ranh Bay, Vietnam. He flew 144 combat missions, 65 over North Vietnam.

In July 1967, he was assigned to the 3630th Flying Training Wing at Sheppard Air Force Base in Texas, as a T-38A instructor pilot. He served as a standardization/ evaluation officer and as an assistant flight commander. In early 1971, he attended Squadron Officers School and returned as an executive support officer to the Deputy Commander of Operations and as School Secretary for the Wing.

In August 1972, he entered the Air Force Institute of Technology residency school at Wright-Patterson Air Force Base in Ohio. In 1974 he received an M.S. degree with distinction in aerospace engineering, after which he was assigned to the Air Force Flight Dynamics Laboratory at Wright-Patterson as a Staff Development Engineer. He served as Deputy for Advanced Concepts for the Aeromechanics Division and as Branch Chief of the Aerodynamics and Airframe Branch in the Laboratory. He received a Ph.D. degree in Aerospace Engineering, with a minor in Laser Physics, from the Air

---

Excerpted from: Department of Energy, National Technical Association, and NASA, *Careers in Science and Technology* (Washington: GPO, September 1993), 62, 65–71.

Force Institute of Technology in 1978, and a master's degree in Business Administration from the University of Houston, Clear Lake in 1987.

---

**Dr. Frank Greene**
Electrical Engineer

Dr. Frank Greene is currently a director of Networked Picture Systems, Inc., and has held the position of President from August 1989 until January 1991. Networked Picture Systems sells computer systems for color printing and image retouching. The company's software and hardware emphasize high-quality printed color and creative design.

Dr. Greene founded Technology Development Corporation (TDC) in 1971. TDC specialized in software and services for scientific applications and electronic testing. The name was changed to ZeroOne Systems, Inc., in 1985, when a segment of Technology Development Corporation was spun off and taken public using that name. In 1987, ZeroOne was sold to Sterling Software and Dr. Greene continued as a group president until 1989. Dr. Greene is now a director of the new Technology Development Corporation in Arlington, Texas, as well as a director and president of Networked Picture Systems in Santa Clara, California.

Dr. Greene was a development engineer and project manager with Fairchild Camera and Instrument Co. from 1965 to 1971. His responsibilities included magnetic and semiconductor memory system design and project management for the Illiac IV memory system. From 1961 to 1965 he was an electronics officer in the U.S. Air Force, where he earned the rank of Captain.

Dr. Greene's education and experience in electrical engineering and business has enabled him to corner a unique market in system development. He received his B.S. in electrical engineering from Washington University, his M.S.E.E. from Purdue University, and his Ph.D.E.E. from the University of Santa Clara. He is currently a member of the Board of Trustees of Santa Clara University. He has authored or co-authored ten technical papers and two books and holds a patent on "the uses of faculty circuits in position coding." Dr. Greene entered the computer software business when it was just a budding enterprise.

---

**Dr. Shirley A. Jackson**
Theoretical Physicist

Shirley A. Jackson, a theoretical physicist at AT&T Bell Laboratories in Murray Hill, New Jersey, was the first African American woman to earn a Ph.D. from the Massachusetts Institute of Technology (MIT). Dr. Jackson, a native of Washington, D.C., received the B.S. degree in physics from MIT in 1968 and the Ph.D. in theoretical particle physics from MIT in 1973. During her student years at MIT, she did volunteer work at Boston City Hospital and tutoring at the Roxbury (Boston) YMCA. After earning her doctorate, she was a research associate at the Fermi National Accelerator Laboratory in Batavia, Illinois (1973–74 and 1975–76), and a visiting scientist at the European Center for Nuclear Research (1974–75), where she worked on theories of strongly interacting elementary particles. Since 1976, Dr. Jackson has been at AT&T Bell Laboratories, where she has done research on various subjects including charge density waves in layered compounds, polaronic aspects of electrons on the surface of liquid helium films, and most recently, optical and electronic properties of semiconductor strained layer superlattices. In 1986, she was elected a Fellow of the American Physical Society for her research accomplishments. In 1991, she received an honorary Doctorate of Science from Bloomfield College, New Jersey, and was also elected a Fellow of the American Academy of Arts and Sciences.

Dr. Jackson studied in Colorado, Sicily, and France, taught at MIT, and has lectured at several institutes. She has received numerous scholarships, fellowships, grants, and awards. She is a member of the American Physical Society, a past president of the National Society of Black Physicists, and a member of several other professional societies.

In 1985, New Jersey Governor Thomas Kean appointed Dr. Jackson to the New Jersey Commission on Science and Technology. She was reappointed for a five-year term in 1989. She has also served on committees of the National Academy of Sciences, American Association for the Advancement of Science, and the National Science Foundation, promoting science and research and women's roles in these fields. Dr. Jackson is a trustee of MIT, Rutgers University, Lincoln University (Pennsylvania), and the Barnes Foundation. She has been a vice president of the MIT Alumni Association and is on three MIT visiting committees: Physics, Electrical Engineering and Computer Science, and the Sloan School of Management.

---

**Jerry T. Jones**

Electronics Engineer and Manufacturer

Determination and a dream made the multimillion dollar corporation, Sonicraft, a reality for its president Jerry Jones. He developed a basement operation into one that has its headquarters in Chicago and offices in Boston and in Washington, D.C., and which employs more than 200 people. The makings of that dream were first sparked at the age of four when Jones' father bought him a radio when he was still living in Sledge, Mississippi. He became fascinated with electricity and nurtured that interest until he became his Chicago neighbors' favorite radio repairman by the age of 13.

His technical interests were formally pursued at Illinois Institute of Technology in Chicago, where he earned a degree in physics. Although one of his high school teachers told him that he would never become a successful engineer, he didn't listen. To help support himself through college, he made customized high-fidelity equipment. He conducted research at the Institute for 10 years; however, he always had a business eye and decided to launch his own business interests with $1,000 in capital. The fledgling company started in his basement. His first major contract was with the Navy, making loudspeakers.

Through the years, Sonicraft grew as its contracts with various military branches and Federal Aviation companies increased. After 15 years in business, the reliability and quality of the products produced by Sonicraft led to the signing of a $268 million contract with the Air Force in 1982. Sonicraft agreed to design and manufacture very low frequency receivers that will operate even after exposure to X-rays and gamma radiation from a nuclear attack. The agreement is a multiphase operation that ran through 1989.

Jones—the major shareholder in Sonicraft—set the goal to make it a billion-dollar company by the early 1990s, putting his company in the league with the Fortune 500s. Although the company's basic work involved manufacturing, in the future it will seek to increase its research and development capability and strive to make yet another dream come true.

---

**Emma Littleton**
Electrical Engineer

Fresh out of high school, Emma Littleton had one clear goal in mind—to become a top-notch secretary.

She started as an enthusiastic and inquisitive junior clerk at Indian Hill Laboratory in Naperville, Illinois, and soon was promoted to a secretarial job. The technical work going on around her was intriguing, and after learning about the work in her department—the development of an expanded and improved telephone switching system called No. 4 ESS—she decided that she wanted to become an electrical engineer. Technical staff were impressed by her enthusiasm and obvious aptitude to grasp technical information.

With encouragement, she enrolled in Tennessee State University, a traditionally African American engineering school, where she excelled. She earned an engineering degree with top honors and a cumulative grade point average of 3.8. During her summer vacations she continued to work at Indian Hill, but in a technical capacity. A Bell Labs staff member was her senior project advisor. He was on leave from his regular work assignment and teaching at Tennessee State University. Today Ms. Littleton is a fulltime member of Bell Labs' professional staff. Her career in engineering had been delayed four years because she initially decided to be a secretary (because that was the only thing she knew). In time, she grew restless and a bit dissatisfied. That led her to her real career choice.

Bell Labs was so enthusiastic that they selected her to study for a year under the Bell Labs Graduate Studies Program, which permits promising staff members to work for an advanced degree in their specialty with full tuition and expenses underwritten by the company.

Ms. Littleton's career was successfully launched after receiving the right exposure to career opportunities and the right counseling, which she should have received in high school. Fortunately, engineers at Indian Hill and staff at Tennessee State University were able to help her realize her full potential.

---

**Dr. Walter Massey**

Theoretical Physicist and former Director, National Science Foundation

Dr. Walter Massey was appointed as Director of the National Science Foundation by President George Bush in 1990, a position he held until January 1993. His position prior to that appointment was Vice President of the University of Chicago for Research and for Argonne National Laboratory. He served as President of the American Association for the Advancement of Science (AAAS), the world's largest association of professional scientists, in 1989. He was elected President of the American Physical Society in 1990. He also serves on the Physics Advisory Committee of the National Science Foundation, the National Academy of Sciences Advisory Committee on Eastern Europe and the former USSR, and the Board of Directors of AAAS.

In 1979, at the age of 41, Dr. Massey became Director of Argonne National Laboratory (ANL) in Argonne, Illinois (25 miles southwest of Chicago), which is operated by the University of Chicago. ANL is one of the nation's top energy research laboratories. In addition to its main site near Chicago, there is the Idaho Engineering Laboratory near Idaho Falls. ANL employed 2,000 scientists and engineers among its more than 5,000 employees while Dr. Massey was Director. ANL operates seven experimental nuclear reactors, including the nation's first nuclear breeder reactor which started producing electricity at ANL's site in Idaho Falls. Although nuclear energy is a primary function of the Laboratory, other energy sources researched at ANL include perfecting battery powered automobiles, making synthetic fuels, and better utilizing solar energy.

Dr. Massey has a long association with the University of Chicago. He first joined the faculty in 1968 after his graduate studies. When he was first appointed Director of ANL, he was also appointed Professor of Physics at the University. Prior to becoming Director, Dr. Massey was at Brown University where he was first an associate professor of physics and later appointed Dean of the College.

A native of Hattiesburg, Mississippi, Dr. Massey entered college from the 10th grade. He is a Morehouse College graduate with a B.S. degree in physics and mathematics and received his master's degree and doctorate in physics from Washington University in St. Louis, Missouri. He was a postdoctoral appointee at Argonne from 1966 to 1968 and a consultant there from 1968 to 1975. His research interest was in the theory of strongly interacting systems of many particles, with particular emphasis on low temperature properties of quantum liquids and solids.

Dr. Massey was recently appointed Senior Vice President and provost of the University of California.

## Dr. Warren Miller

Nuclear Engineer

Dr. Warren Miller developed an interest in mathematics and science in grade school and high school. A West Point graduate, he has a B.S. degree in Engineering Sciences with an area of interest in nuclear engineering and nuclear reactor theory. He obtained the master's and Ph.D. degrees from Northwestern University in nuclear engineering.

Dr. Miller currently has a joint appointment as an Associate Director at Los Alamos National Laboratory and as a Dardee Professor (endowed chair) in the Nuclear Engineering Department at the University of California, Berkeley.

Los Alamos National Laboratory is one of the largest research laboratories under contract to the U.S. Department of Energy. Prior to joining the technical management staff at Los Alamos, he was a Group Leader for the Transport and Reactor Theory Group, in the Theoretical Division of the Laboratory. His research interests include neutral and charged particle transport theory, fluid dynamics, radioactive waste management, and radiation shielding. Concurrent with his position at Los Alamos, he serves as a consultant to Sargent and Lundy, and to Argonne National Laboratory.

He has published numerous papers and excelled in the area of computer solutions to partial differential equations, specifically, solutions to the neutron transport equation. He was invited to coauthor a book entitled *Computational Methods in Neutron Transport Theory*, which will be published by Wiley Interscience. His work has led to the development of a general theory that allows a comparison of all methods suggested for accelerating the iterative processes used in neutron transport computer codes. This theory makes it possible for new methods to be developed and compared with existing schemes.

In addition to his research interests, Miller enjoys teaching and has served as a visiting professor of nuclear engineering at Howard University. He also developed and taught a course at Northwestern entitled "Science, Technology, and the Black Experience." One of the stated objectives of the course was to promote thinking in aspiring scientists and engineers toward application of their skills to the development of African American communities at home and abroad.

## Dr. John B. Slaughter

President, Occidental College, and former Director of the National Science Foundation

Dr. John Brooks Slaughter received his academic training in electrical engineering and engineering physics, and his research specialty is in the field of digital control systems theory and applications. Although he began his career as an electronics engineer in industry and later in government, he has always been a part of the academic community. Throughout his career, he has been active in national efforts to involve minorities in engineering and science.

Dr. Slaughter is currently President of Occidental College in Los Angeles, California, a post to which he was appointed in 1988. Prior to that, he served as Chancellor of the University of Maryland from 1982 to 1988.

Dr. Slaughter served as Director of the National Science Foundation from 1980 to 1982, a post he was nominated for by President Jimmy Carter. The National Science Foundation is the major U.S. government agency that supports basic research in the physical and life sciences. NSF also supports science education and applied sciences and is in charge of several major laboratories, such as the National Optical and Radio Astronomy Observatories. Dr. Slaughter served an earlier tour of duty at NSF, also at the request of President Carter, as Assistant Director of Astronomical, Atmospheric, Earth, and Ocean Sciences from 1977 to 1979. In this position, Dr. Slaughter was responsible for five NSF divisions—Earth Sciences, Ocean Sciences, Astronomical Sciences, Atmospheric Sciences, and Polar Programs.

In 1979, between his two appointments at NSF, Dr. Slaughter was Academic Vice President and Provost of Washington State University in Pullman, Washington.

Before going to NSF in 1977, Dr. Slaughter was Director of the Applied Physics Laboratory and Professor in the Department of Electrical Engineering at the University of Washington at Seattle. As Laboratory Director since 1975, he was responsible for the direction and management of a research and development program in ocean and environmental sciences and engineering. He initiated new programs in energy resource conservation and development, geophysics, environmental acoustics, and bioengineering at the Laboratory. Under his direction, the staff of the Laboratory, a leader in the fields of underwater acoustics, underwater vehicles, and physical oceanography, grew from 180 to 240 people.

Prior to going to the University of Washington, Dr. Slaughter was Head of the Information Systems Department of the Naval Electronics Laboratory Center (NELC) in San Diego, California. In that capacity, he was responsible for the management of a 200-person staff engaged in the design and development of Navy command and communication systems. Under his direction there, a major new emphasis in biomedical engineering was started. He was named Scientist of the Year at NELC in 1965, where he served from 1960 to 1975.

Earlier in his career, from 1956 to 1960, Dr. Slaughter was employed by the Convair Division of General Dynamics Corporation in San Diego where he worked on the development of missile flight instrumentation and telemetry equipment.

Dr. Slaughter was born in Topeka, Kansas, in 1934 and earned a Bachelor of Science degree in electrical engineering at Kansas State University in 1956. He earned a Master of Science degree at the University of California at Los Angeles (UCLA) in 1961. He also was awarded a Doctor of Philosophy degree in engineering science at the University of California at San Diego in 1971.

Dr. Slaughter has written widely in his field and is a member of several professional organizations including Eta Kappa Nu, the honorary electrical engineering society; Tau Beta Pi, the honorary engineering society; and was elected to the National Academy of Engineering in 1982. He has been editor of the *International Journal of Com-*

*puters and Electrical Engineering* since 1977. He is also a Fellow of the Institute of Electrical and Electronic Engineers.

He served from 1968 to 1975 as Director and Vice-Chairman of the Board of the San Diego Transit Corporation, specializing in policy matters related to transit planning and technology. He was a member of the affiliate engineering faculty at San Diego State University from 1963 to 1965 and established graduate courses in discrete and optimal control systems. He was appointed to the San Diego Chamber of Commerce Energy Task Force in 1974 to advise on technological approaches to energy conservation.

Dr. Slaughter has been active in efforts to encourage minorities to pursue careers in science and engineering, and he has played a strong role in urging educational institutions, industry, and government to take more aggressive steps to improve opportunities for minorities in those fields. In 1976, he was appointed a member of the National Academy of Engineering Committee on Minorities in Engineering. In 1983 he served as a member of the National Science Board Commission on Pre-College Education in Mathematics, Science, and Technology. In 1987, the *U.S. Black Engineer* magazine named him its first "Black Engineer of the Year."

---

**Howard Smith**
Computer Research Manager

Howard Smith is a computer scientist and President of Clarity Software, Inc. Mr. Smith and a group of computer veterans founded Clarity Software in January 1990 to design, develop, and market personal communications software products for UNIX workstation platforms. These personal communications products have significant value added, both in function and in multimedia document technologies, and are mail-centered. Clarity's initial product is called CLARITY and is composed of a Compound Document Editor, spreadsheet, presentation graphics, and Advanced Mail components. This product will be followed by other personal communications software.

Previously, Mr. Smith was employed by Hewlett-Packard Corporation. Responsible for the development of operating systems and process systems units, he supervised 200 employees for Hewlett-Packard's Computer Systems Division in Cupertino, California. The development of computer equipment is a very competitive and dynamic business, and Howard Smith is one of the managers who helped Hewlett-Packard, a leader in computer design, maintain its competitive edge.

Enthusiastic about his work, Smith considers himself to be very people oriented. Some might find this surprising, since the computer scientist is often stereotyped by others as cold and more oriented to machines than to people. The job as a manager requires long hours that must be divided between the research aspects and the people performing the research. As manager, Mr. Smith must assure that ideas are developed within a given budget and completed within a specified schedule. Any problem that threatens the research project must be resolved by him and the staff working in the division. Some of the problems faced are technical, but many are more related to business and planning.

Mr. Smith is a graduate of Los Angeles State College and San Jose State University. He received a bachelor's and a master's degree in mathematics. He attributes part of his career success to enjoying what he does. He says that it is important to set both short-range and long-term goals so that you can evaluate your progress and redefine your goals as necessary. Reaching your goals creates self-confidence, another important part of succeeding in an organization.

Some of Howard's spare time is spent working with high school and college students helping them prepare for technical careers. He is a member of the advisory board of Project Interface in Oakland, California. This program is community-based and provides tutorial services and career counseling to student participants.

# 17 Accessing the Information Age

# Bridging the Racial Divide on the Internet (1998)

Donna L. Hoffman and Thomas P. Novak

While it is often true that "a rising tide lifts all ships," it is also true that powerful new technologies have the potential to redistribute power. If these technologies are not democratically available, they are more likely to widen than to bridge inequalities of privilege. And, of course, the technology itself is only one element of opportunity. To trade stocks 24-hours a day on the Internet requires not only a computer and a telephone connection but also some money to invest. Worries about the lack of access to the information highway for many Americans has led to a number of studies of who, exactly, is on the Net.

---

The Internet is expected to do no less than transform society (1); its use has been increasing exponentially since 1994 (2). But are all members of our society equally likely to have access to the Internet and thus participate in the rewards of this transformation? Here we present findings both obvious and surprising from a recent survey of Internet access and discuss their implications for social science research and public policy.

Income and education drive several key policy questions surrounding the Internet (3, 4). These variables are the ones most likely to influence access to and use of interactive electronic media by different segments of our society. Looming large is the concern that the Internet may be accessible only to the most affluent and educated members of our society, leading to what Morrisett has called a "digital divide" between the information "haves" and "have-nots (5)."

Given these concerns, we investigated the differences between whites and African Americans in the United States with respect to computer access and Web use. We wished to examine whether observed race differences in access and use can be accounted for by differences in income and education, how access affects use, and when race matters in access.

Our analysis is based on data provided by Nielsen Media Research, from the Spring 1997 CommerceNet/Nielsen Internet Demographic Study (IDS), conducted from December 1996 through January 1997 (6). This nationally projectable survey of Internet use among Americans collected data on race and ethnicity (7).

---

Reprinted from: Donna L. Hoffman and Thomas P. Novak, "Bridging the Racial Divide on the Internet," *Science* 280 (April 17, 1998): 390–391. Reprinted by permission. 

## Computer Access and Web Use

Our survey results (table 17.1, column 1) show that overall whites were significantly more likely than African Americans to have a home computer in their household (8). Whites were also slightly more likely to have access to a PC at work.

Nearly twice as many African Americans as whites stated that they planned to purchase a home computer in the next 6 months. African Americans were also slightly more interested in purchasing a set-top box for Internet television access.

The racial gap in Web use was proportionally larger the more recently the respondent stated that he or she had last used the Web. Proportionally, more than twice as many whites as African Americans had used the Web in the past week. As of January 1997, we estimate that 5.2 million (±1.2 million) African Americans and 40.8 million whites (±2.1 million) have ever used the Web, and that 1.4 million (±0.5 million) African Americans and 20.3 million (±1.6 million) whites used the Web in the past week.

Whites and African Americans also differed in terms of where they had ever used the Web. Whites were significantly more likely to have ever used the Web at home, whereas African Americans were slightly more likely to have ever used the Web at school.

## Possible Causes

Because students behave quite differently from the rest of the respondents with respect to computer access and Internet use, we treat them separately later.

We used the national median household income of $40,000 to divide respondents. For household incomes under $40,000, whites were proportionally twice as likely as African Americans to own a home computer and slightly more likely to have computer access at work (columns 2 and 3 in table 17.1).

However, for household incomes of $40,000 or more, a slightly greater proportion of African Americans owned a home computer, and a significantly greater proportion had computer access at work.

We adjusted race differences in home computer ownership for income and found, as one would expect, that increasing levels of income corresponded to an increased likelihood of owning a home computer, regardless of race. In contrast, adjusting for income did not eliminate the race differences with respect to computer access at work. African Americans were more likely than whites to have access to a computer at work after taking income into account.

What accounts for this result? African Americans with incomes of $40,000 or more in our sample were more likely to have completed college, were younger, and were also more likely to be working in computer-related occupations than whites. These factors led to greater computer access at work.

At lower incomes, the race gap in Web use was proportionally larger the more recently the respondent stated that he or she had last used the Web. Whites were almost six times more likely than their African American counterparts to have used the Web in the past week and also significantly more likely to have used the Web at home and in other locations. Notably, as indicated above, race differences in Web use vanish at household incomes of $40,000 and higher.

Regardless of educational level, whites were significantly more likely to own a home computer than were African Americans and to have used the Web recently (table 17.1, columns 4 and 5). These differences persisted even after statistically adjusting for education. Thus, although income explains race differences in home computer ownership and Web use, education does not: Whites are still more likely to own a home computer than are African Americans and to have used the Web recently, despite controlling for differences in education.

However, greater education corresponded to an increased likelihood of work computer access, regardless of race.

Thus, race matters to the extent that societal biases have either (i) required African Americans to obtain higher levels of education in order to achieve the same income as whites, or (ii) resulted in older African Americans not being able to achieve high incomes.

### Students Are Special: Race Almost Always Matters

Higher education translates into an increased likelihood of Web use. Students were more likely than any other income or educational group to have used the Web (table 17.1, column 6). Students exhibited the highest levels of Web use because, even without home computer ownership or access at work, they presumably had access at school.

The most dramatic difference between whites' and African Americans' home computer ownership was among current students (including both high school and college students). Whereas 73% of white students owned a home computer, only 32% of African American students owned one. This difference persisted when we statistically adjusted for students' reported household income. Thus, in the case of students, household income does not explain race differences in home computer ownership. This is the most disturbing instance yet of when race matters in Internet access.

Our analysis also revealed (table 17.1, column 6) that white students were significantly more likely than African American students to have used the Web, especially in the past week. However, there were no differences in use when students had a computer at home.

White students without a computer in the home (table 17.1, column 8), were more than twice as likely as similar African American students to have used the Web in the past 6 months and more than three times as likely to have used the Web in the past week. Thus, white students lacking a home computer, but not African American students, appear to be accessing the Internet from locations such as homes of friends and relatives, libraries, and community centers.

**Table 17.1**
Percentage (weighted) of individuals in each group responding positively concerning the variable specified in that row. Asterisk indicates that the difference between whites and blacks is statistically significant ($P < 0.05$).

| | (1) Full Sample | | Non-Students (2) <$40,000 Income | | (3) $40,000+ Income | | (4) High School or Less | | (5) Some College | |
|---|---|---|---|---|---|---|---|---|---|---|
| | Whites N=4906 | Blacks N=493 | Whites N=1833 | Blacks N=213 | Whites N=1916 | Blacks N=131 | Whites N=1794 | Blacks N=210 | Whites N=2776 | Blacks N=219 |
| Own home computer | 44.3* | 29.0* | 27.5* | 13.3* | 61.2 | 65.4 | 27.0* | 16.4* | 57.7* | 49.3* |
| PC access at work | 38.5 | 33.8 | 25.9 | 20.7 | 59.1* | 76.7* | 24.2 | 18.4 | 55.0* | 63.9* |
| Buy PC in 6 months | 16.7* | 27.2* | 14.3* | 23.4* | 20.4* | 35.7* | 12.6* | 23.3* | 19.4* | 28.5* |
| Internet TV interest | 11.8 | 14.9 | 9.2 | 9.4 | 15.0* | 23.9* | 8.2 | 12.3 | 13.6 | 16.8 |
| Ever used Web | 26.0 | 22.0 | 13.0* | 7.5* | 36.7 | 38.8 | 10.1 | 11.5 | 36.5 | 29.2 |
| ... in past 6 months | 22.4* | 16.6* | 10.4* | 4.7* | 32.5 | 36.2 | 8.2 | 7.4 | 31.6 | 26.6 |
| ... in past 3 months | 20.6* | 14.9* | 9.5* | 4.3* | 29.9 | 33.8 | 7.6 | 5.9 | 29.2 | 24.7 |
| ... in past month | 17.8* | 9.7* | 8.1* | 2.5* | 26.5 | 24.3 | 6.7* | 3.3* | 25.3* | 16.6* |
| ... in past week | 12.9* | 5.8* | 5.9* | 1.1* | 19.2 | 17.1 | 4.7* | 1.4* | 18.6* | 11.6* |
| ... at home | 14.7* | 9.0* | 6.4* | 2.4* | 22.3 | 22.8 | 5.3 | 3.4 | 21.6 | 16.9 |
| ... at work | 11.1 | 8.4 | 4.9 | 3.7 | 19.8 | 24.5 | 3.6 | 5.0 | 19.2 | 16.8 |
| ... at school | 7.2 | 10.9 | 2.8 | 2.6 | 6.6 | 8.5 | 1.9* | 5.9* | 6.9 | 6.9 |
| ... at other locations | 7.8 | 5.3 | 4.4* | 1.8* | 8.8 | 42.8 | 2.8 | 3.3 | 9.4 | 9.0 |

| Students | | | | | |
|---|---|---|---|---|---|
| (6) | | (7) | | (8) | |
| All Students | | Have Home PC | | No PC at Home | |
| Whites N=336 | Blacks N=64 | Whites N=247 | Blacks N=22 | Whites N=89 | Blacks N=42 |
| 73.0* | 31.9* | 100 | 100 | 0.0 | 0.0 |
| 27.0 | 24.0 | 30.1 | 32.3 | 18.6 | 20.1 |
| 26.3 | 40.3 | 22.3 | 9.3 | 37.1 | 54.8 |
| 23.5 | 21.4 | 26.7 | 26.6 | 14.9 | 19.0 |
| 65.8* | 48.6* | 72.1 | 63.8 | 48.8 | 41.5 |
| 58.9* | 31.1* | 66.7 | 63.8 | 37.8* | 15.9* |
| 51.9* | 28.8* | 58.9 | 56.5 | 32.8 | 15.9 |
| 44.9* | 19.8* | 51.8 | 35.4 | 26.2 | 12.4 |
| 31.9* | 9.9* | 38.0 | 20.8 | 15.5* | 4.8* |
| 33.3* | 13.0* | 43.6 | 36.8 | 5.5 | 1.9 |
| 8.8* | 2.0* | 11.4 | 6.3 | 1.9 | 0.0 |
| 45.5 | 42.8 | 48.3 | 49.9 | 38.1 | 39.5 |
| 23.5* | 4.2* | 24.0* | 5.5* | 22.1* | 3.7* |

## Policy Points

Five million African Americans have used the Web in the United States as of January 1997, considerably more than the popular press estimate of 1 million. This means that African Americans are already online in impressive numbers and that continued efforts to develop online content targeted to African Americans, commercial or otherwise, are likely to be met with success.

Overall, students enjoy the highest levels of Web use. However, white students were proportionally more likely than African Americans to own a home computer, and this disquieting race difference seems to result from factors other than income.

Also, white students who lacked a home computer were more likely to use the Web at places other than home, work, or school than were African Americans. Thus, it is important to create access points for African Americans in libraries, community centers, and other nontraditional places where individuals may access the Internet and to encourage use at these locations.

Overall, increasing levels of education are needed to promote computer access and Web use. Education explains race differences in work computer access, although our findings for African Americans with household incomes above the national median suggest the presence of a powerful bias that could restrict Internet use to a narrow segment of African Americans.

The policy implication is obvious: To ensure the participation of all Americans in the information revolution, it is critical to improve educational opportunities for African Americans.

Finally, access translates into usage. Whites were more likely than African Americans to have used the Web because they were more likely to have access, whereas African Americans in our survey were more likely to want access. This may explain in part the recent commercial success of computers priced below $1000. It follows that programs that encourage home computer ownership and the adoption of inexpensive devices that enable Internet access through the television should be aggressively pursued.

The consequences to U.S. society of a persistent racial divide on the Internet may be severe. If a significant segment of our society is denied equal access to the Internet, U.S. firms will lack the technological skills needed to remain competitive. Employment opportunities and income differences among whites and African Americans may be exacerbated, with further negative consequences to the nation's cities. As Liebling observed regarding the freedom of the press (9), the Internet may provide equal opportunity and democratic communication, but only for those with access.

## References and Notes

1. W. J. Clinton, *State of the Union Address*. Given at the United States Capitol, 4 February 1997 [www.whitehouse.gov/WH/SOU97/].

2. Network Wizards, "Internet Domain Survey," January 1998 [www.nw.com/zone/WWW/report.html].

3. D. L. Hoffman, W. Kalsbeek, T. P. Novak, *Comm. ACM* **39**, 36 (1996).

4. J. Katz and P. Aspden, "Motivations for and Barriers to Internet Usage: Results of a National Public Opinion Survey," Paper presented at the 24th Annual Telecommunications Policy Research Conference, Solomons, MD, 6 October 1996.

5. J. Keller, in *Public Access to the Internet*, B. Kahin and J. Keller, Eds. (MIT Press, Cambridge, MA, 1996), pp. 34–45; T. P. Novak, D. L. Hoffman, A. Venkatesh, paper presented at the Aspen Institute forum on Diversity and the Media, Queenstown, MD, 5 November 1997; P. Burgess, "Study Explores 'Digital Divide,'" *Rocky Mountain News*, 11 March 1997, p. F31A.

6. Nielsen Media Research, "The Spring 1997 CommerceNet/Nielsen Media Internet Demographic Survey," Full Report, Volume I of II (1997).

7. The IDS is based on an unrestricted, random-digit, dial sampling frame and used a computer-assisted telephone interviewing system to obtain 5813 respondents. Weighted, these respondents represent and allow projection to the total population of 199.9 million individuals in the United States aged 16 and over.

8. All significance tests were obtained with Research Triangle Institute's SUDAAN software and incorporate sampling weights provided by Nielsen Media Research (*6*).

9. A. J. Liebling, *The New Yorker* **36**, 105 (14 May 1960).

## Cyberghetto: Blacks Are Falling through the Net (1998)

Frederick L. McKissack, Jr.

One critic, a Web designer and freelance writer in Chicago, assessed the participation of African-Americans in the new cyber culture.

---

I left journalism last year and started working for an Internet development firm because I was scared. While many of my crypto-Luddite friends ("I find e-mail so impersonal") have decided that the Web is the work of the devil and is being monitored by the NSA, CIA, FBI, and the IRS, I began to have horrible dreams that sixteen-year-old punks were going to take over publishing in the next century because they knew how to write good computer code. I'd have to answer to some kid with two earrings, who will make fun of me because I have one earring and didn't study computer science in my spare time.

You laugh, but one of the best web developers in the country is a teenager who has written a very sound book on web design and programming. He's still in his prime learning years, and he's got a staff.

What should worry me more is that I am one of the few African Americans in this country who has a computer at home, uses one at work, and can use a lot of different kinds of software on multiple platforms. According to those in the know, I'm going to remain part of that very small group for quite some time.

The journal *Science* published a study on April 17 which found that, in households with annual incomes below $40,000, whites were six times more likely than blacks to have used the World Wide Web in the last week. Low-income white households were twice as likely to have a home computer as low-income black homes. Even as computers become more central to our society, minorities are falling through the Net.

The situation is actually considerably worse than the editors of *Science* made it seem. Some 18 percent of African American households don't even have phones, as Philip Bereano, a professor of technical communications at the University of Washington, pointed out in a letter to *The New York Times*. Since the researchers who published their study in *Science* relied on a telephone survey to gather their data, Bereano explains, the study was skewed—it only included people who had at least caught up to the Twentieth Century.

---

Reprinted from: Frederick L. McKissack, Jr., "Cyberghetto: Blacks are Falling through the Net," *The Progressive* 62 (June 1998), 20–22. Reprinted by permission from The Progressive, 409 E. Main Street, Madison, WI 53703, www.progressive.org.

About 30 percent of American homes have computers, with the bulk of those users being predominantly white, upper-middle-class households. Minorities are much worse off: Only about 15 percent have a terminal at home.

The gulf between technological haves and have-nots is the difference between living the good life and surviving in what many technologists and social critics term a "cyberghetto." Professor Michio Kaku, a professor of theoretical physics at City University of New York, wrote in his book *Visions: How Science Will Revolutionize the Twenty-first Century*, of the emergence of "information ghettos."

"The fact is, each time society made an abrupt leap to a new level of production, there were losers and winners," Kaku wrote. "It may well be that the computer revolution will exacerbate the existing fault lines of society."

The term "cyberghetto" suggests that minorities have barely passable equipment to participate in tech culture. But most minorities aren't even doing that well.

Before everybody goes "duh," just think what this means down the line. Government officials are using the Web more often to disseminate information. Political parties are holding major on-line events. And companies are using the Web for making job announcements and collecting résumés. Classes, especially continuing-education classes, are being offered more and more on the Web. In politics, commerce, and education, the web is leaving minorities behind.

The disparity between the techno-rich and techno-poor comes to a head with this statistic: A person who is able to use a computer at work earns 15 percent more than someone in the same position who lacks computer skills.

"The equitable distribution of technology has always been the real moral issue about computers," Jon Katz, who writes the "Rants and Raves" column for *Wired* online, wrote in a recent e-mail. "The poor can't afford them. Thus they will be shut out of the booming hi-tech job market and forced to do the culture's menial jobs."

This technological gap, not Internet pornography, should be the public's main concern with the Web.

"Politicians and journalists have suggested frightening parents into limiting children's access to the Internet, but the fact is they have profoundly screwed the poor, who need access to this technology if they are to compete and prosper," Katz said. "I think the culture avoids the complex and expensive issues by focusing on the silly ones. In twenty-five years, when the underclass wakes up to discover it is doing all the muscle jobs while everybody else is in neat, clean offices with high-paying jobs, they'll go berserk. We don't want to spend the money to avoid this problem, so we worry about Johnny going to the *Playboy* web site. It's sick."

In his 1996 State of the Union address, President Clinton challenged Congress to hook up schools to the Internet. "We are working with the telecommunications industry, educators, and parents to connect ... every classroom and every library in the entire United States by the year 2000," Clinton said. "I ask Congress to support this educational technology initiative so that we can make sure the national partnership succeeds."

The national average is approximately ten students for every one computer in the public schools. According to a study by the consulting firm McKinsey & Co., the President's plan—a ratio of one computer to every five students—would cost approximately $11 billion per year over the next ten years.

Some government and business leaders, worried about a technologically illiterate work force in the twenty-first century, recognize the need for increased spending. "AT&T and the Commerce Department have suggested wiring up schools at a 4:1 ratio for $6 or $7 billion," says Katz.

But according to the U.S. Department of Education, only 1.3 percent of elementary and secondary education expenditures are allocated to technology. That figure would have to be increased to 3.9 percent. Given the tightness of urban school district budgets, a tripling of expenditures seems unlikely.

Then there's the question of whether computers in the schools are even desirable. Writer Todd Oppenheimer, in a July 1997 article for *Atlantic Monthly* entitled "The Computer Delusion," argued that there is no hard evidence that computers in the classroom enhance learning. In fact, he took the opposite tack: that computers are partially responsible for the decline of education.

Proponents of computers in the classroom struck back. "On the issue of whether or not technology can benefit education, the good news is that it is not—nor should be—an all-or-nothing proposition," writes Wendy Richard Bollentin, editor of *OnTheInternet* magazine, in an essay for *Educom Review*.

There is an unreal quality about this debate, though, since computer literacy is an indispensable part of the education process for many affluent, white schoolchildren.

Consumers are beginning to see a decline in prices for home computers. Several PC manufacturers have already introduced sub-$1,000 systems, and there is talk of $600 systems appearing on the market by the fall. Oracle has spent a great deal of money on Network Computers, cheap hardware where software and files are located on large networks. The price is in the sub-$300 range. And, of course, there is WebTV, which allows you to browse on a regular home television set with special hardware.

Despite the trend to more "affordable" computers, a Markle Foundation–Bellcore Labs study shows that this may not be enough to help minorities merge onto the Information Superhighway. There is "evidence of a digital divide," the study said, with "Internet users being generally wealthier and more highly educated, and blacks and Hispanics disproportionately unaware of the Internet."

So, what now?

"For every black family to become empowered, they need to have computers," journalist Tony Brown told the *Detroit News*. "There is no way the black community is going to catch up with white society under the current system. But with a computer, you can take any person from poverty to the middle class."

This is the general line for enlightened blacks and community leaders. But having a computer won't bridge the racial and economic divide. Even if there is a 1:1 ratio of students to computers in urban schools, will students' interest be piqued when they don't have access to computers at home? One out of every forty-nine computer-science professors in the United States is black. Will this inhibit black students from learning how to use them? And even if every black student had a computer at home and at school, would that obliterate all racial obstacles to success?

Empowerment is not just a question of being able to find your way around the Web. But depriving minorities of access to the technology won't help matters any. We need to make sure the glass ceiling isn't replaced by a silicon ceiling.

## Troubletown (1998)

**Figure 17.1**
Troubletown cartoon by Lloyd Dangle. Reprinted from the *San Francisco Bay Guardian*, December 30, 1998, available through www.stbg.com.

## The Black Technological Entrepreneur (1999)

Janet Stites

One of the most advertised uses of the Internet is not just how it can keep one informed, but how it can make one rich. This op-ed piece looks at the black technological entrepreneur.

The Internet has given rise to a rush of entrepreneurism as men and women from all walks of life walk away—under the spell of equity—from otherwise lucrative jobs to start their own businesses. New York City's black technologists are no exception. Though comparatively small in number, and nearly invisible in the greater New York Internet and new-media industry, black Internet entrepreneurs are organizing a support network, outlining steps to success and finding their own role models.

On a rainy Friday evening earlier this month at New York University, a group of black executives, professionals and students met to discuss ways to use information technology for "digital freedom," as they put it. The event was sponsored by the Blacks in New Media Association, a student group that was started last fall in the university's interactive telecommunications program.

"The playing field is as level as it's going to get," said B. Keith Fulton, a technologist and lawyer who is now director of technology programs and policy at the National Urban League. "You don't need a $10 million marketing budget to put up a Web site," Mr. Fulton said. "But there is consolidation in the industry, and as the Internet matures, new barriers are going to arise."

Another panelist was Bob Ponce, who wields a great deal of influence in the new-media industry as president of the World Wide Web Artist Consortium (www.wwwac.org), a grass-roots organization that has a far-reaching E-mail discussion list and that is host to numerous events for Web designers. Mr. Ponce, who is African-American, is established in the broader business world, but he believes that catering to the interests of black Americans might provide some of the best opportunities for black entrepreneurs.

"First, we need to help get our community on line," he said, referring to black Americans as a group. "Then you help create your future customers."

Black Americans are definitely coming on line, according to research from Forrester Research of Boston. According to the firm, 2.7 million African-American households (out of about 11 million) have access to the Internet, whether it be from home, work or school. That is up from two million households at the beginning of 1998.

What's important from a business standpoint, Mr. Ponce said, is that people *know* that blacks are alive and well in the wired world. "We have to let people know we are technology savvy," he said.

Another panelist, Auriea Harvey, who is a Web site designer and founder of Entropy8 (www.entropy8.com), is doing just that. She has a digital camera pointed at her desk so that visitors to her Web site can watch her work. "I want to make a point that I'm here, that I'm a woman and that I'm black," Ms. Harvey said.

Ms. Harvey, who has designed the Web sites of Virgin Records America and PBS, among others, wasn't always so eager to emphasize her race. "At first, the anonymity was important to me," she said. "I wanted people to see my work and judge me solely for that."

But Ms. Harvey said she had come to realize that working in the virtual world offered her advantages as a black woman over working in the real world. "On the same day you'll be working with someone on the West Coast and then someone in Switzerland," she said. "They may not even know your color, and if they find out later, it's just trivia."

When it comes to raising money for a new business, though, the interpersonal may play a role, one panelist, Omar Wasow said. "The primary barriers are more just about being a start-up, not being a *black* start-up," said Mr. Wasow, an African-American who is the Internet correspondent for the cable channel MSNBC. "But this is a business where social networks count for a lot," he said, in an interview after the panel.

Mr. Wasow, a 28-year-old New York native and Stanford graduate, is a Silicon Alley pioneer, having started New York Online (www.nyo.com), originally a subscription-supported dial-up on-line service, in 1993. He initially sought to create an on-line community that would reflect the diverse personalities of the city. But as on-line services moved to the Web and as community tended to take a back seat to commerce, Mr. Wasow shifted New York Online's business to concentrate on Web site design.

Mr. Wasow, who divides his time between MSNBC and his company, says that the bulk of the challenges he faces as a businessman have nothing to do with race. "Because this industry is so young and new and there's so much turmoil, it's still a wide open game with plenty of opportunity," he said.

The black entrepreneurs on the panel hesitated to say much specific about the trials of raising venture capital, and there seems to be scant industry data about the amount of capital that goes to black-owned new-media firms. But the consensus of that Friday night discussion seemed to be that while the venture capital door is open, it may not always be open quite wide enough.

Locally, the New York City Investment Corporation, which is backed by 10 banks, has a mandate to finance minority ventures. The fund, started in 1996, acts more like a bank than a venture firm, giving out loans of $50,000 to $1 million in return for equity or royalties. Though the $25 million fund does not necessarily focus on technology companies, it has lent close to $1 million to two New York-based black-owned technology firms—Data Industries, a computer personnel staffing company, and Axicom Communications Group, an international telecommunications firm. Howard Sommer, who manages the fund, says it is now focusing more on Internet and new-media companies.

Nationally, a smattering of venture funds have been started to take advantage of underserved, or overlooked, minority businesses. None are yet catering to the New York Internet market, though.

But organizations, like Blacks in New Media, are sprouting up to help give African-American businesses a foothold in New York's Internet economy. And the New York new-media industry as a whole, which prides itself on diversity, is seeking ways to attract more minority entrepreneurs.

"We look at the attendees at our events," said Alice Rodd O'Rourke, executive director of the New York New Media Association. "And we ask, 'Where are the people of color?' We want everyone to know this industry exists and what the opportunities are."

# 18 Technological Troubles

## A Place at the Table: A *Sierra* Roundtable on Race, Justice, and the Environment, and The Letter That Shook a Movement (1993)

Technology is usually discussed in terms of what it can do *for* people rather than *to* people. Since the 1960s however, it has been the signal task of the country's environmental movement to point out that all technologies use up (often nonrenewable) resources, and that all technologies create waste—some of it appallingly toxic. Just as all sectors of American society do not share equally in technology's benefits, neither do they bear the same burden of its inevitable costs. For every successful black engineer there is an entire African-American community forced to live in an equally intimate, but negative, relationship with modern technology. The Sierra Club, one of the oldest and largest of the nation's environmental organizations, held a roundtable discussion with leading environmental justice activists. "The Letter That Shook a Movement" follows.

### A Place at the Table

The era of an American environmental movement dominated by the interests of white people is over. The beginning of the end came in September 1982, in Warren County, North Carolina, when more than 500 predominantly African-American residents were arrested for blocking the path of trucks carrying toxic PCBS to a newly designated hazardous-waste landfill. Among those taken into custody was the Reverend Benjamin Chavis, executive director of the United Church of Christ Commission for Racial Justice (CRJ). His suspicions as to why North Carolina would choose a black community as a dump for its poison were confirmed in a milestone report by the CRJ in 1987, which demonstrated that the single most significant factor in the siting of hazardous-waste facilities nationwide was race. A subsequent report, by the *National Law Journal*, found that the EPA took 20 percent longer to identify Superfund sites in minority communities, and that polluters of those neighborhoods were fined only half as much as polluters of white ones.

Armed with proof of what has become known as "environmental racism," a loose alliance of church, labor, civil rights, and community groups led by people of color arose to demand environmental justice. Part of doing so meant confronting the so-called "Group of Ten," the nation's largest—and largely white—environmental groups, and bluntly accusing them of racism.

The charges came in early 1990 in a jolting series of letters from Louisiana's Gulf Coast Tenant Leadership Development Project and the Southwest Organizing

Reprinted from: "A Place at the Table" and "The Letter That Shook a Movement," *Sierra* 78 (May/June 1993): 51–58, 90–91. Reprinted with permission from *Sierra* magazine.

Project in Albuquerque (see "The Letter That Shook a Movement"). Abashed, many of the mainstream groups vowed reformation, if not transformation. Last May, in a speech celebrating the Sierra Club's centennial at Harpers Ferry, West Virginia, then—Executive Director Michael Fischer called for "a friendly takeover of the Sierra Club by people of color." The alternative, he said, was for the Club to "remain a middle-class group of backpackers, overwhelmingly white in membership, program, and agenda—and thus condemn[ed] to losing influence in an increasingly multicultural country.... The struggle for environmental justice in this country and around the globe must be the primary goal of the Sierra Club during its second century."

Recently *Sierra* invited some of the leading proponents of the environmental justice movement to San Francisco to explore how we might arrive at that multicultural future. Attending were the Reverend Chavis; Richard Moore, co-chair of the Southwest Network for Environmental and Economic Justice in Albuquerque and longtime community organizer; Vivien Li, chair of the Sierra Club's Ethnic Diversity Task Force and director of the Boston Harbor Association; Scott Douglas, then community organizer with the Sierra Club's Southeast Office (he has since been named director of Greater Birmingham Ministries); and Winona LaDuke, director of the White Earth Recovery Project in Minnesota. The discussion was moderated by Carl Anthony, president of Earth Island Institute in San Francisco and founder of its Urban Habitat Program.

Carl Anthony: We've all been involved in the struggle for environmental justice for a very long time, even if we didn't always call it that. Reverend Chavis, you invented the term "environmental racism," didn't you?

Benjamin Chavis: We coined it, but the reality was out there—we just gave language to it. This movement for environmental justice is a definitive movement: we're redefining our realities. Our guiding principle is that our work must be done from a grassroots perspective, and it must be multiracial and multicultural. We are learning how to do that. There's no blueprint, but there are guiding principles that emerge, and that's what we want to share with you.

The good news is, we're being inclusive, not exclusive. We're not saying to take the incinerators and the toxic-waste dumps out of our communities and put them in white communities—we're saying they should not be in *anybody's* community. When the movement first got going, I think some whites actually became afraid, because they thought it was a movement of retribution. It is not a movement of retribution—it is a movement for justice. You can't get justice by doing an injustice on somebody else. When you have lived through suffering and hardship, you want to remove them, not only from your own people but from all peoples.

By the way, I'm a history buff, and I would like to note that the Sierra Club is the first one of the Group of Ten to actually invite members of the environmental justice movement to its national headquarters. I've come to the environmental justice movement from a 30-year-long career in the civil rights movement, in the fight for racial justice. I find that the environmental justice movement is emerging as the bridge movement, bringing diverse racial and ethnic communities together in a profound way, a way no other issue even has the potential to do.

Richard Moore: It's certainly the first meeting that I've attended where we've been asked to come together and have this kind of dialogue. We appreciate that—that's what the letters and all that was about in the first place.

Anthony: Richard, you've written quite a few letters, and they all made a lot of waves. Could you tell us about the one that you wrote to the big environmental groups?

Moore: Back in 1990 we had classes at the Southwest Network in how to write. A lot of our people don't know how, and we figured if we were going to offer literacy classes, we should at least do something productive. It was very difficult for us; we spent a lot of time talking about it, because we knew that there would be ramifications. It meant a lot to people because it was the first time that we've had the opportunity, as poor people, as working-class people, as people of color, to sit down and talk about how we feel about things and then transfer that to paper.

Basically, we raised three issues. One, the issue of some—not all—of the mainstream environmental organizations accepting money from the same corporations that are killing our people.

Secondly, we had concerns about the staffing of mainstream environmental organizations and the composition of their boards. If you put one black person or brown person or red person or yellow person on the board and think, "OK, everything's going to be cool," that's not what we're talking about. We have not seen the kind of forward movement that we would have hoped to see. We knew nothing was going to change overnight, but at the same time we have not seen that movement.

The third one is more basic. Who is it you are advocating for? In New Mexico, where I've lived 25 years now, there has been a history of problems and conflict between our communities and environmental organizations. We're talking about land issues, water issues, regulations that environmental organizations have been pushing forward—for the protection of who? For what? If it's for the protection of us, then how come we haven't been involved in it? Why do we have to hear something third down the line, sixth down the line, or never down the line? If it is to protect our interests, then bring us to the table, because we do very fine at protecting our own interests.

Anthony: A good example of the attitudes of the more established groups toward communities of color is *Blueprint for the Environment*, which was submitted to George Bush when he took office. It contained 750 detailed recommendations from 18 established environmental groups, including proposals for every Cabinet department except two: the Department of Housing and Urban Development and the Department of Labor. The groups could have made recommendations about lead poisoning, energy conservation in public housing, siting of affordable housing near transportation corridors, occupational health-and-safety issues in the workplace. But they didn't. In retrospect, it is clear that ignoring these two departments clearly reflected the movement's racial and class biases.

Your letters made public what a lot of us in social justice movements were feeling. At first, many of the environmental groups denied charges of racism. But then, gradually, they realized that the charges were serious. It was a kind of wake-up call. Vivien, how did the Sierra Club respond?

Vivien Li: The Sierra Club started looking at the issue of environmental justice in the 1970s. We cosponsored, with two other environmental groups and the National Urban League, a conference in 1978 in Detroit that brought together 700 people from around the country, people of color as well as more traditional environmentalists, to look at how the civil rights movement could work together with the environmental movement.

Unfortunately, once the Reagan/Bush administration came in, people went back to protecting their own turf. Had there been a continuation of that kind of coalition effort at the grassroots level, I think we would be a lot further ahead today. I think your letters were important, because they focused people's attention on working together again.

As to the funding issue, I've served for three years on a three-person committee that reviews every corporate gift to the Sierra Club over $1,000. We do not take money from the oil industry, from the paper industry, from chemical companies, tobacco companies, or major polluters. If there is ever any doubt about a corporation's environmental record, we don't take money from them.

The diversity of our staff and board is an issue that concerns the Club. We don't have an easy fix for it, but we're committed to trying to change, starting at the grassroots and working all the way up. It should never be a question of a token minority on the Board of Directors. One of the things that we're trying to do is to ensure that diversity is an issue throughout the Sierra Club, for both staff and volunteers.

This past year, we funded 12 grassroots organizing projects proposed by Sierra Club activists. These ranged from a lead-poisoning prevention effort in San Francisco's Chinatown to a campaign against toxic dumping on the Rosebud Reservation in South Dakota. We're trying to connect traditional Sierra Club work with community-based efforts.

The type of change we're talking about is very fundamental, and it will not happen in one or two or even five years. The Sierra Club is a hundred years old, and some of the baggage that we carry, frankly, goes back a long time. I think we've made significant progress, but clearly we're not totally there yet.

Chavis: I was very happy when I heard that the Sierra Club had hired Scott Douglas; I think that was an indication of the Club's seriousness.

Scott Douglas: You know, I think my position as grassroots organizer owes its existence to those letters. The writing project worked, Richard, at least that aspect of it.

My commitment, coming out of the civil rights movement and the peace movement, is to renegotiate the relationship between peoples, and between peoples and the earth. Some of the best models for that process are from Native American traditions. They teach us that part of that renegotiation is mutuality and respect for people's cultures, respect for the lessons that they have learned. We're erasing peoples at a hellish pace, and with them goes their body of knowledge. We erase indigenous peoples, and then give some university a $10-million grant to discover one-tenth of what those people had accumulated over eons. That's not very efficient.

Anthony: We often forget that there are 60 million people of color in the United States; soon we aren't going to be "minorities" any more. Already more than a third of the actual physical territory of the North American continent has indigenous people as its majority population. Winona LaDuke, are Native Americans ready to be part of the "mainstream"?

Winona LaDuke: We are not part of and do not wish to be part of the mainstream of America. We are different. America has to come to terms with our difference, and to recognize our need for territorial integrity and self-determination.

My reservation is in northern Minnesota, 36 miles by 36 miles, located between Bemidji and Fargo, one of seven Ojibwa reservations in the North. We were ceded a huge area under a treaty of 1867, a land we call the White Earth. It's a wealthy land full of lakes, pinelands, farmlands, prairie, and most of the medicines our people have used for centuries. That's why we don't have it today. By 1920, 90 percent of our reservation was in non-Indian hands, seized by a bunch of illegal land transactions. Most of our people were forced off the reservation and into poverty. Three-quarters of them are refugees. Most Native Americans, in fact, are refugees.

Our work is about trying to reclaim our land base. We've tried every legal recourse, but have had no success in the court system. So we are trying to reclaim that land through other processes, including negotiations with large absentee landholders like the Potlatch timber company or the Boy Scouts of America, who own thousands of acres there.

We've also had federal legislation introduced for the return of 50,000 acres—the Tamarac National Wildlife Refuge. Right now the refuge is pretty much used as a hunting ground for sport hunters from Minneapolis; nine times as many deer are taken by non-Indians as by Indians there, and only 40 percent of the lakes are closed to duck hunting. We hunt, but only for subsistence. We hunt because we're poor, because we need that food, because that's what we've always done.

I have to say that historically, environmental organizations, including the Sierra Club, have sometimes opposed land transfers back to Indians. We're hoping that that doesn't occur in our case. So far it's been pretty good.

Chavis: I would like to pay deference here to our Native American sisters and brothers, who have been trying to focus the attention of the environmental justice movement on the sacredness of the air, the water, the land, the sacredness of the Creation. Social justice movements often leave spirituality out, but the environmental justice movement holds spirituality as a very key element.

LaDuke: In my language, most nouns are animate. A rock, *asin*, is animate, and *mandamin*, corn, is animate. They have standing on their own, they have spirit. They are not recognized as objects or resources; they are instead recognized as vital living things that we have to respect and have a relationship to.

Native people consider themselves a part of nature. There's no separation, like the one that necessarily exists in the industrial mind. Unfortunately, most environmentalism comes out of that mind, not out of the indigenous mind. The challenge faced by environmentalists is to decolonize their industrial minds.

Anthony: It's interesting the way language and culture work. I would say most African-Americans and Latinos have long been aware that our neighborhoods were dumping grounds for locally unwanted land uses, but not until Ben Chavis invented the term "environmental racism" did we have a name for it. When the United Church of Christ Commission on Racial Justice published *Toxic Waste and Race*, which documented the disproportionate siting of hazardous-waste facilities in our communities, everyone knew right away what they were talking about.

Li: For Native Americans subjected to toxic- and hazardous-waste facilities, we're talking about very detrimental health effects. For example, Native American infants suffer from the highest rates of sudden-infant-death syndrome—also known as crib death—in the country. Doctors don't know why Native Americans are afflicted more than any other group, but exposure to environmental poisons can't help.

LaDuke: There have been a hundred separate proposals to dump toxic wastes on Native communities. Fifteen of the 18 federal research grants for Monitored Retrievable Storage Facilities [for nuclear wastes] went to Indian reservations. One-third of all low-sulfur coal and two-thirds of the uranium mined in this country are on Indian reservations. We have nuclear radiation all over our land, but no major environmental group in this country has a uranium campaign. No major environmental group in this country has dealt consciously with the issues of Native people. Our communities are bearing the brunt of America's energy policy, yet no one has seen fit to address our concerns in their policy-making.

Anthony: Few people realize how much communities of color suffer from bad energy policies, from inappropriate hydroelectric dams and nuclear power projects to over-reliance on fossil fuels. Navajo teenagers still suffer from radiation exposure from uranium mining. Several years back, the Center for Third World Organizing in Oakland found that these exposures were causing cancer at a rate 17 times the national average.

It's not just nuclear energy; dependence on fossil fuels also places burdens on our communities. From the extraction of fossil fuels to their distribution, use, and waste, our communities get fewer benefits and pay a greater price. Stripmining in the Four Corners region for energy users hundreds of miles away creates a national sacrifice zone on sacred Native lands, without even providing for local energy needs. Poor people of color in the cities use up to 35 percent of their income to purchase energy; renters get none of the incentives to weatherize their homes, but they are stuck with big heating bills. And even though people of color drive fewer vehicle miles per year than other city dwellers, freeways often cut their communities in half, destroying their economic and social lives in addition to exposing them to a disproportionate amount of air pollution.

My own work in the Urban Habitat Program focuses on these issues, trying to build a multicultural urban environmental leadership. I'm sorry that we haven't heard much yet today about the cities. Historically, people of color have been concentrated in barrios and ghettos without adequate neighborhood services, schools, or open space. This concentration is the result of a long history of discrimination.

Li: Think back to how Chinatowns were started. They were the only place that Chinese people could work and congregate and have an identity. Up until the early 1960s, Chinese immigration was greatly restricted, and Chinese tended to settle along the two coasts where they had gained entry. Since they were considered inferior to Caucasians, no one wanted to work with them, so they developed their laundries and their restaurants, frankly for economic survival.

It's been different since the 1960s and '70s. Now there are Vietnamese, Cambodians, and Laotians who have come not just to seek a better future but because they have been subjected to political persecution in their native lands. Many of them tend to be poor and less well-educated; they have more difficulty identifying a cultural niche for themselves. Here in San Francisco, the Chinese are very well-organized politically, as is the Japanese community, but amongst Laotians and Cambodians it has been much more difficult. In order for them to get attention paid to urban environmental issues like rodent control and lead poisoning, they must first develop that political power.

Anthony: I think it's very important for us to understand and connect to our history. When we talk about history, of course, we're really talking about peeling an onion, because we begin to see connections that we didn't see before. I'm thinking about Thoreau, who was imprisoned for protesting the Mexican-American War, which relates to the Treaty of Guadalupe Hidalgo; or Martin Luther King, who was shot in Memphis protecting garbage workers. The connections between the civil rights movement and the environmental movement are really quite rich.

LaDuke: In our case, unfortunately, the trouble is that environmental groups have, historically, come from a Eurocentric perspective. This is not an inclusive perspective, and it's not something we can relate to. Many times, in fact, environmental groups make decisions that affect other communities without the input of those communities. One of them even purchased land on our reservation without ever talking to us about it, and restricted our use of an area that had medicinal plants.

Douglas: In Alabama, we have the largest hazardous-waste dump in the country, at a place called Emelle. It got put in Alabama because white environmentalists negotiated the site with Chemical Waste Management without talking to the black community, which was getting the message that there was going to be a brick factory there. These guys had the audacity to negotiate that county's future—and still think they were doing it in our interest!

Moore: After our letter went out, we got calls from communities all over northern New Mexico thanking us. "For the first time," they said, "legislators called us and said, 'We're developing a bill. Do you want to come to the meeting and help us?'" Because what usually happens is that after the bill is already developed, someone says, "This bill affects some Indian people—we better find an Indian person and see how they feel about it." You know how it goes—it comes on down through the fax machine, and you have 15 minutes to decide: "We'd like to give you more time but we're in a hurry, 'cuz it's got to go to Senator So-and-So ..."

We are talking about providing grassroots organizations the possibility of bringing issues to the table. That's what it's all about. It's about us speaking for ourselves. It's not about what's taking place under the table, it's about what's taking place on top.

LaDuke: I've been working for 16 years on Native environmental issues: with groups that have five members in the middle of the Navajo Reservation to Innu from northern Labrador who have been fighting the siting of a military base and a bunch of dams up there. Those groups are frontline environmental groups, but they are seen instead as *Native* groups, because the big environmental movement wants to position itself clearly as *the* environmental movement. We need to broaden the definition and to recognize these grassroots groups that have been struggling over these issues for all these years. What we need is a place at the table.

Anthony: With its focus on wilderness, the traditional environmental movement on the one hand pretends there were no indigenous people in the North American plains and forests. On the other, it distances itself from the cities, denying that they are part of the environment. It's interesting what we talk about and what we avoid talking about. For example, in this roundtable at the headquarters of the Sierra Club, it occurs to me that we, as people of color, have had very little to say about nature. Some people may think that people of color are insensitive to plants and animals, that we don't care about the biosphere ...

Chavis: I agree it's important to be able to see through a lens that doesn't filter nature out of this discussion. One of the things that has emerged is the bifurcation between wanting to protect animal species and protecting the human species. That is a false bifurcation. The notion that you can protect the ozone layer from further deterioration without seeing the degradation in the neighborhoods, the barrios, and the ghettos is a false notion.

Douglas: My job assignment in Birmingham is to assist grassroots groups working against the imposition of injustice in the environmental field. As I went to those communities, I noticed that they didn't separate the hazardous-waste incinerator from the fact that lead poisoning is not being dealt with in their schools, from the fact that their schools have been underfunded, that they have no day care, no jobs, no access to jobs. They don't separate it, because their quality of life as a whole is going down. So who was I when I came to visit their community organizations? Was I Scott Douglas the environmentalist, or Scott Douglas the get-out-the-vote guy? These problems are all wrapped up together, and it teaches us a very important lesson: oppressed people do not have compartmentalized problems. Very seldom do you go to a low-income black community, and the only problem is the incinerator: "Man, if we got rid of the incinerator, we'd be fine!" The incinerator is merely the external reflection of a whole host of problems.

Anthony: The people that I work with, young people, don't make artificial divisions between homicide in the inner city, gang violence, toxics that come from incinerators, and the slaughter of dolphins. They don't make all these distinctions that other people do when they have budgets to submit to foundations.

Moore: Sometimes the big environmental groups seem to think that we're imagining things. If you choose, you *can* come to the conclusion that we're just a bunch of crazy people who are trying to raise some hell and get our names in the paper. And you would be making a very serious mistake.

Li: Richard, I've yet to hear anybody here in the Sierra Club differ with you on the issues. I think you could probably cite some examples, but overall the kinds of things that you're talking about are exactly the same types of things that Sierra Club activists at all levels have been talking about. No one wants more lead paint. We're not pushing for more landfill dumps. We're not pushing for incineration. We were the ones who got $10 million for job retraining into the Clean Air Act . . .

Moore: Let's not act as though what we're saying is not a reality. The Sierra Club has been responsible, has been a co-conspirator in attempting to take away resources from our communities. Like we said in our letter: your organizations are supporting policies that emphasize the cleanup and preservation of the environment on the backs of working people, and people of color in particular. When you come into our communities talking about closing down the plant, who's working in the plant? We've had to close down plants, let me tell you that. In the final analysis that plant may have to go: it's killed people inside, and has also poisoned our groundwater and our air and our children outside. But we went through a process first, attempting to bring workers into the decision.

Li: I think we have to be careful we do not allow ourselves to be pitted against each other—people of color should not be divided and conquered. All of us care about jobs, be it Sierra Club people, people in the community, whoever. But what is happening, frankly, is that you believe that there is a split between jobs and the environment. And if you believe that, then mainstream America has suceeded in dividing the civil rights movement and the environmental movement.

Douglas: This is important: people, if given a choice, will choose safe, clean water, land, and air over "economics"—but only if they're given a choice. They're not given a choice in places like Sumter County, Alabama, where you can lose your livelihood—and sometimes your life—if you speak out against the biggest polluters. People aren't choosing between jobs and the environment; they're choosing between death—their jobs are killing them—and unemployment. It's a sick choice. The workers are choosing early death so their families can eat. They know they're going to die.

This is the sickness we're up against. This is the same sickness as racism. Until we're able to address it, we won't be able to protect ourselves against it. Unless we can immunize ourselves against it, there's always a tool that can be used against us.

Racism is what makes foreigners of people in their own lands. We have a society that makes foreigners out of Native Americans. We're very selective about who we make foreigners of now, with our immigration policy. Until we begin to address the use of that fear to get people to act like lemmings, they will stampede off the cliff, killing themselves, their families, their inheritance, and their legacy, because they've been successfully panicked.

Chavis: The denial of racism in this country perpetuates it. One of the things that we are demanding in the environmental justice movement is a coming to grips with the phenomenon of environmental racism, and coming to grips with the broader phenomenon of racism in general. You have to understand that racism is not natural. There's a purpose for racism: it serves the economic interests of those who would exploit. That is the history of racism in the world, not only in the United States. Apartheid in South Africa exists not just because some whites in South Africa don't like black people; it's because some whites in South Africa want to live in a privileged position, and take the diamonds and the gold and the natural resources of the people. Racism has always been used to justify the rape of the environment and the rape of people, and to deprive them of economic rights.

One of the things I've come to appreciate in dealing with the environmental justice movement is patience. We should be impatient with injustice; we have to confront it, we have to challenge it, but we have to be patient with the victims of injustice. There will not be an overnight cure.

Li: It's very important that the environmental agenda we develop is one that is developed *with* communities of color. Not imposed on them, but rather forged together in the spirit of mutual respect and trust. That's a very hard thing to do, be it in this room here or anyplace else. It's hard not to try to dominate, hard not to tell people "This is how it should be done."

Douglas: One of the problems I face in trying to increase the troops working on environmental justice issues within the Sierra Club is that we don't have enough activists to go around. When I got the Sierra Club job, two friends of mine, both of them European-Americans, walked up to me in Birmingham and said, "I'm glad you got the job. I'm a member of the Sierra Club." I said ,"You are? I never knew that." The first one was a woman who is very active in low-income housing; she's a technical assistant, she teaches people to read through HUD regulations so they can access some of these crazy grants. The other one is a church-related person who works with families about to be evicted, providing emergency food, housing, and utilities. Unless this is a very rare coincidence, I bet there are some other folks in the present membership who have daily connections with people in the struggle. They don't bring it to the Sierra Club because that's not the agenda.

When was the last time the Sierra Club did a survey to determine what the members are active in when they're not doing the traditional conservation issues? If you could find that out, you'd also find out your points of connection with the rest of the community. We have the skills in-house, but they just haven't been pulled together yet.

LaDuke: We have totally common issues, but environmental groups have to embrace a broader position. Last year the Greenpeace board of directors adopted a position in support of the sovereignty of Native people. The Sierra Club should adopt a similar policy. Environmental groups need to not feel threatened by the taking of land out of the so-called public domain and returning it to Native people; instead, they need to recognize that our traditional stewardship of land has been very sound to the extent that we are able to restore our traditional values and continue our traditional spiritual practices. Sure we've got problems, like tribal councils that are trying to site toxic-waste dumps. But what environmental groups need to do is shore up their relations with traditional people, because traditional people don't subscribe to the ethics of pollution.

Chavis: I said to The Nature Conservancy a little while ago: if you really want to conserve the earth, then join the environmental justice movement, because this is the movement that is going to constrain the destroyers of the earth, because the destroyers of the environment are the destroyers of our neighborhoods and our communities.

I'm very optimistic about the extent to which we can continue to build the environmental justice movement from the grassroots up. This is not a fad, this is not a momentary blip on the social-justice graph of this nation, but an effort that will have very long-term implications for the future of our nation and the future of our world.

## The Letter That Shook a Movement

*The environmental justice movement, a loose coalition of hundreds of grassroots groups led by people of color, was born out of a challenge to the country's largest environmental organizations. Below are excerpts from a letter sent by the Southwest Organizing Project on March 15, 1990, to what's known as the "Group of Ten": the Sierra Club, Sierra Club Legal Defense Fund, Friends of the Earth, The Wilderness Society, National Audubon Society, Natural Resources Defense Council, Environmental Defense Fund, National Wildlife Federation, Izaak Walton League, and National Parks and Conservation Association.*

We are writing this letter in the belief that through dialogue and mutual strategizing we can create a global environmental movement that protects us all. . . .

For centuries, people of color in our region have been subjected to racist and genocidal practices, including the theft of lands and water, the murder of innocent people, and the degradation of our environment. Mining companies extract minerals, leaving economically depressed communities and poisoned soil and water. The U.S. military takes lands for weapons production, testing, and storage, contaminating surrounding communities and placing minority workers in the most highly radioactive and toxic work sites. Industrial and municipal dumps are intentionally placed in communities of color, disrupting our cultural lifestyle and threatening our communities' futures. Workers in the fields are dying and babies are born disfigured as a result of pesticide spraying.

Although environmental organizations calling themselves the "Group of Ten" often claim to represent our interests, in observing your activities it has become clear to us that your organizations play an equal role in the disruption of our communities. *There is a clear lack of accountability by the Group of Ten environmental organizations towards Third World communities in the Southwest, in the United States as a whole, and internationally.*

Your organizations continue to support and promote policies that emphasize the cleanup and preservation of the environment on the backs of working people in general and people of color in particular. In the name of eliminating environmental hazards at any cost, across the country industrial and other economic activities which employ us are being shut down, curtailed, or prevented while our survival needs and cultures are ignored. We suffer from the end results of these actions, but are never full participants in the decision-making which leads to them....

We ... call upon you to cease operation in communities of color within 60 days, until you have hired leaders from those communities to the extent that they make up between 35 and 40 percent of your entire staff. We are asking that Third World leaders be hired at all levels of your operations.... Also provide a list of communities of color to whom you furnish services, or Third World communities in which you have organizing drives or campaigns, and contacts in those communities....

It is our sincere hope that we can have a frank and open dialogue with your organization and other national environmental organizations. It is our opinion that people of color in the United States and throughout the world are clearly an endangered species. Issues of environmental destruction are issues of our immediate and long-term survival. We hope that we can soon work with your organization in helping to assure the safety and well-being of all peoples.

# Neighbors Rally to Fight Proposed Waste-Burner (1992)

The movement for environmental equity, begun in 1982 by the Reverend Benjamin Chavis, spread across the country to other communities threatened by the concentration of pollution in their neighborhoods.

## Cleveland

A busload of residents from Cleveland's Lee-Seville neighborhood packed a courtroom yesterday to hear arguments over a "trash-to-energy" plant proposed for the East Side.

About 50 Clevelanders and environmental activists attended a brief rally outside the old Lakeside Courthouse before moving into the Ohio Court of Appeals (8th District) courtroom.

For the residents, the issue is a simple one, reflected by the buttons they wore: "Don't Waste Our Neighborhood."

"We're here to let them know we don't need garbage in our community," said Esther M. Jones, president of the Lee-Seville-Miles Citizens Council. "Lee-Seville is just saying, 'Let us keep our community a residential community and not a testing ground or a laboratory for unproven technology.'"

No plants in the United States use the technology planned for the proposed plant. Thus the arguments heard by a three-judge panel dealt as much with chemistry as with law.

A city ordinance prohibits "incineration or reduction" of garbage except in private incinerators in areas zoned for general industry. The ordinance also bars "trade, industry or use that will be injurious, hazardous, noxious or offensive to an extent equal to or greater than" other operations prohibited by the ordinance.

The city's Board of Zoning Appeal rejected Cleveland Distillation Energy Corp.'s request for a permit to build a plant at 16007 Seville, saying it would violate the zoning ordinance.

The company appealed to Common Pleas Court and on Feb. 12, 1991, Judge John E. Corrigan overturned the decision and ordered the city to issue Cleveland Distillation a permit. Corrigan, in a one-paragraph ruling, said the board's decision was "arbitrary and unreasonable."

Yesterday's hearing before the appellate judges was Cleveland's appeal of Corrigan's order.

---

Figure 18.1
Esther M. Jones, president of the Lee-Seville-Miles Citizen Council, speaks at a rally outside the old Lakeside courthouse.

The proposed plant would take in at least 35 truckloads of garbage a day. The process involved in turning that garbage into useable energy is what is at issue.

Garbage would be heated in a flameless oven, reducing it to a char that would be hauled away. Gases from the heated trash would be burned in boilers to make steam and generate electricity.

Does that burning of the gas constitute incineration? Deborah J. Nicastro, lawyer for Cleveland Distillation, says it does not. City lawyers say it does.

The appellate judges—David T. Matia, Leo M. Spellacy and John F. Corrigan—did not indicate when a ruling might be issued.

Spellacy hinted at another issue raised in the case. He questioned the lawyers about Cleveland Distillation's arguments that the ordinance is vague and overbroad and, thus, unconstitutional.

Spellacy noted that the issue of the law's constitutionality was not addressed by Corrigan in his Common Pleas Court decision.

City lawyers William Ondrey Gruber and Christopher A. Holecek have argued that the plant would be in the heart of a residential community "which has worked together over the years to keep its neighborhood clean and its homes well-groomed."

Cleveland Distillation is one of three companies trying to build trash-to-energy plants in Greater Cleveland. While the companies are legally separate they share common officers and investors. One major investor in each company is Paul Voinovich, brother of Gov. George V. Voinovich.

## United States to Weigh Blacks' Complaints about Pollution (1993)

*The New York Times* reported the federal government's response to the environmental justice issues raised by African-Americans.

**Washington, Nov. 18**

The Clinton Administration has for the first time agreed to investigate complaints that states are violating the civil rights of blacks by permitting industrial pollution in their neighborhoods.

In a step that opens new avenues for legal challenges to the placement of hazardous waste sites and other pollution sources, the Environmental Protection Agency's office of civil rights notified Louisiana and Mississippi last month that it had opened investigations under the 1964 Civil Rights Act, which bars racial discrimination in federally supported programs.

Blacks in those states contend that state decisions involving hazardous waste treatment plants have the effect of unfairly exposing them to more toxic pollution than whites. They say that the states' permit procedures, which are supported with Federal money, are partly to blame.

In recent years, a great deal of data has been collected suggesting that racial minorities are disproportionately exposed to air pollution, hazardous wastes, pesticides and the like. But Federal environmental laws do not directly address racial disparities, so efforts to block pollution permits by claiming discrimination have proved to be difficult.

For example, when blacks from the communities of Genesee and Flint, Mich., asked the E.P.A.'s board of appeals to block construction of an incinerator in a predominantly black neighborhood, the board rejected the appeal. An administrative law judge ruled in September that the Clean Air Act did not provide grounds for racial distinctions to affect decisions about the location of such operations.

The new tactic of using the Civil Rights Act, while still untested in court, would allow appeals without requiring changes to existing environmental laws like the Clean Air Act.

Even when decisions are made on the basis of factors like the cost of land, population density, geological conditions or for other reasons not directly connected to race, the Civil Rights Act leaves room to challenge actions that have the effect of discriminating, according to lawyers involved in the cases.

Environmental and civil rights groups have been urging the Administration to use the civil rights law in environmental cases. The Mississippi case was called to the environmental agency's attention by the United States Commission on Civil Rights, which urged Carol M. Browner, the agency's administrator, to begin a high-level, priority review of the state's hazardous waste permit program.

**E.P.A. Point of View**

"I don't think that there is any doubt that low-income and minority communities have borne the brunt of our industrial life style," Ms. Browner said in an interview today. "The cases that have been filed demonstrate the real frustrations that these communities feel."

The E.P.A.'s decision to open the civil rights investigations is the latest sign of the Administration's broad effort to make racial equality a keystone of its environmental policies.

Within a few weeks, the White House is expected to issue an executive order instructing all Federal agencies to make equal treatment of racial groups an important consideration in any environmental decisions.

Congress is also considering several bills intended to redress environmental inequality. Some call for a list of "environmental high impact areas" where a moratorium would be imposed on new toxic chemical sites.

In addition, the Administration plans to incorporate the theme of equal rights in legislation it will soon propose to extend the Superfund law, which provides for the cleanup of hazardous waste dumps, as well as in the Clean Water Act and other environmental laws that are due to be revised.

But for now, the Civil Rights Act appears to be the weapon of choice.

State officials in Louisiana and Mississippi said they were not deliberately steering pollution toward black populations and said they would cooperate fully with the E.P.A.

Kai Midboe, the secretary of Louisiana Department of Environmental Quality, said he expected that the agency would try to satisfy the complaints by local citizens that they have not had adequate opportunity to comment on the permit in question.

In both cases, the plaintiffs have said that the states have not considered potential discrimination from state decisions.

The plaintiffs in the two cases have contended that the civil rights law does not require them to prove that states intended to discriminate but that the practical effects on racial minorities must be considered by themselves.

A study of toxic emissions along the lower Mississippi River, published in May by the E.P.A., showed that many sites in Louisiana emitting large amounts of chemicals were in predominantly minority areas.

In Louisiana, where blacks are 34 percent of the population, about 105 pounds of toxic material is released into the environment on a per-capita basis, the data show. But in Iberville Parish, which is 46 percent black, 168 pounds of toxic material per capita is released, and 9 out of 10 of major sources of industrial pollution are in predominantly black areas. In Carville, near the proposed site of a hazardous waste plant, the population is 70 percent black and 353 pounds of toxic material per capita is released.

In the Mississippi case, the complaint seeks to block permits to build a cement kiln just outside Noxubee County, for a hazardous waste dump in the county and for a landfill in the county.

## Presidential Executive Order 12898—Environmental Justice

In 1994 the Clinton administration issued Executive Order 12898, committing the federal government to addressing issues of environmental equity. Because of its areas of responsibility, the Environmental Protection Agency (EPA) had a particularly important role to play in this effort for justice.

### Overview

Environmental justice is a movement promoting the fair treatment of people of all races, income, and culture with respect to the development, implementation, and enforcement of environmental laws, regulations, and policies. Fair treatment implies that no person or group of people should shoulder a disproportionate share of the negative environmental impacts resulting from the execution of this country's domestic and foreign policy programs. (The environmental justice movement is also occasionally referred to as Environmental Equity—which EPA defines as the equal treatment of all individuals, groups or communities regardless of race, ethnicity, or economic status, from environmental hazards).

The environmental justice movement is generally acknowledged to have emerged in the early 1980's in response to large demonstration opposing the siting of a PCB-landfill in a predominantly black community in Warren County, North Carolina. Subsequent studies and public attention raised concerns of the fairness and protection afforded under existing environmental programs—concerns that are now receiving increased attention at all levels of government as well as within the private community.

Today, environmental justice is a priority both within the White House and EPA. The Administration has documented its concern over this issue through issuing Executive Order 12898, Federal Actions To Address Environmental Justice in Minority Populations and Low-Income Populations (February 11, 1994). This Order requires that federal agencies make achieving environmental justice part of their mission (a summary of the requirements imposed under E.O. 12898 is provided below.... Similarly, the EPA has identified environmental justice a key priority under the Browner Administration. EPA created an Office of Environmental Justice (originally the Office of Environmental Equity) in 1992, commissioned a task force to address environmental justice issues, oversees a Federal Advisory Committee addressing environmental justice issues

Excerpted from: "Presidential Executive Order 12898—Environmental Justice," 1994, available at http://es.epa.gov/program/exec/eo-12898.html.

(the National Environmental Justice Advisory Council), and has developed an implementation strategy as required under E.O. 12898.

The environmental justice movement has both direct and indirect links with pollution prevention. At the most basic level, pollution prevention provides a means of achieving or improving environmental justice through reducing the environmental and health impacts that must be borne by any element of society. In recognition of this, the environmental justice grants program makes achieving pollution prevention an express objective of projects eligible for grants. In addition, environmental education efforts focused on promoting fair treatment emphasize the importance of pollution prevention in achieving programmatic objectives. Many environmental justice leaders also participate in the Common Sense Initiative, EPA's program to work with segments of industry to promote waste reduction, compliance, and the streamlining of environmental regulations. Finally, EPA is examining integrating considerations of environmental justice in the use of Supplemental Environmental Projects to promote compliance and program objectives....

In November, 1992, EPA created an Office of Environmental Justice (originally named the Office of Environmental Equity) to examine and integrate environmental justice concerns into EPA's existing environmental programs. The Office of Environmental Justice (OEJ) serves as the focal point for environmental justice concerns within EPA and provides coordination and oversight regarding these concerns to all parts of the Agency. The OEJ also coordinates communication and public outreach activities, provides technical and financial assistance to outside groups investigating environmental justice issues, and serves as a central environmental justice information clearinghouse. The OEJ provides technical support to environmental justice research and demonstration projects examining whether EPA programs contribute to disproportionate risks faced by some low-income and minority populations, as well as responding to inquiries from Congress and other interested parties....

The OEJ coordinates EPA's Minority Academic Institutions (MAI) Program. This program attempts to increase the number of minority students receiving science and engineering degrees and to improve the quality of minority student education in the sciences....

In June 1993, EPA's OEJ was delegated authority to solicit and select environmental justice projects, issue grants for such projects, supervise and evaluate these projects, and disseminate information on the effectiveness of the projects and the feasibility of the practices, methods, techniques, and processes examined as applied to environmental justice issues. In FY 1994 the environmental justice grants program was initiated. EPA divided approximately $500,000 among it 10 Regional Offices, and directed these offices to make grants awards in amounts not to exceed $10,000 per grant. Seventy-one (71) grants totalling $507,000 were awarded in FY 1994....

The grant program is intended to provide financial assistance and promote the public interest by supporting projects undertaken by affected community groups to address environmental justice issues. Eligible groups include any affected community group (i.e., community-based/grassroots organizations, schools, educational agencies, colleges or universities, and non-profits organizations) as well as Tribal governments. . . .

Under the auspices of the Federal Advisory Committee Act (5 U.S.C., App. II) EPA has established the National Environmental Justice Advisory Council (NEJAC) to advise, consult with, and make recommendations to the Administrator of EPA on matters relating environmental justice. The NEJAC holds meetings, analyzes issues, conducts reviews, performs studies, produces reports, makes recommendations and conducts other activities as appropriate given its mission and the objectives of EPA's environmental justice program. NEJAC is composed of a parent Council and four subcommittees (Public Participation and Accountability, Enforcement, Waste and Facility Siting, and Health and Research). NEJAC's members include representatives of academia, industry, community groups, non-governmental organizations, state, tribal and local governments, and environmental organizations. . . .

## Study Attacks "Environmental Justice" (1994)

Not everyone saw the justice in "environmental justice."

### Washington

In February, President Clinton ordered federal agencies to take steps to prevent low-income neighborhoods from suffering further toxic threats.

"All Americans have a right to be protected from pollution, not just those who can afford to live in the cleanest, safest communities," Clinton said while signing an executive order.

In Congress and state legislatures, bills aim to block dumping and waste incineration in neighborhoods inhabited mainly by minorities. Internationally, 66 nations teamed up to forbid the U.S. and industrial nations to export hazardous materials to developing nations. In short, "environmental justice" has become a catch phrase from county boards to the United Nations, with policymakers working to prevent exploitative dumping and to correct problems from the past.

But the idea that minorities and disadvantaged people suffer disproportionate risk from pollution is coming under attack; the latest salvo has been fired from the Center for the Study of American Business at Washington University in St. Louis.

A new study by two research fellows, Christopher Boerner and Thomas Lambert, asserts that studies that gave birth to the environmental justice movement are flawed and that governments are misdirected in seeking remedies. The two conclude in the study—"Environmental Justice?"—that depressed areas should be encouraged to negotiate payments for becoming waste repositories and thereby improve people's lives.

"Many may argue that it is immoral to pay individuals to expose themselves to health risks," the study asserts. "(But) as long as environmental regulations guarantee minimal risk, there should be no moral difficulties with compensating individuals for voluntarily accepting the nuisances associated with waste and polluting facilities."

Copies of their study have been sent to the White House and distributed in Congress, and the authors hope that their arguments become part of the debates.

The study is the second in two weeks to try to debunk the concept of environmental justice and the efforts by governments to tackle problems that are often considered widespread. Last month, researchers at the University of Massachusetts published

Reprinted from: "Study Attacks 'Environmental Justice,'" *Plain Dealer* (Cleveland), April 7, 1994.

a study concluding that white working-class people, and not blacks, are more likely to live by waste facilities in the 25 largest metropolitan areas. That study, which analyzed census tracts, was partly funded by the waste-disposal industry.

In a release accompanying the study, the author, Douglas L. Anderton, asserted that "we found some relationships that run counter to the general wisdom of the day, and further research is needed on the question of environmental racism." The words environmental racism further inflame people who feel their rights and health are threatened by waste disposal.

Nationally, the NAACP has a 2-month-old environmental justice program that promotes legislation in Congress and programs around the nation. John Rosenthall, who heads the program, contended that "environmental justice is very valid, just like our other rights" under the Constitution.

Rosenthall said he had returned recently from visiting a public housing complex in Portsmouth, Va., where residents claim that children suffer unusually high levels of developmental disabilities as a result of lead pollution from a plant that operated next door.

"It's hard for me to see how you can pay people enough money for them to feel comfortable living with some of these waste sites," Rosenthall said.

## New York Seminarian Promotes Environmental Justice in Africa (1996)

Paulette V. Walker

Once again, conditions for African-Americans were compared to those of blacks still in Africa. Exposure to waste in South Africa helped highlight the injustice of waste exposure in the United States, and the recent American experience with "environmental justice" served as a helpful and hopeful guide for Africa.

### New York

The similarities between Cape Town and New York City disturb Larry L. Rasmussen, a professor of social ethics at Union Theological Seminary here.

"The poorest neighborhoods in both Cape Town and New York City are the most toxic, the most environmentally damaged," he says. "Look at West Harlem, in New York. City officials have put the sewage-treatment plant *and* the bus barns there—which exacerbates the pollution in the area. The city garbage trucks unload their trash" there. "And all of this next to a grade school."

"In Cape Town," he says, "the townships are located on the worst land. It's impossible for the residents to exist on what this land produces."

Mr. Rasmussen spent most of January in South Africa. While there, he met with community organizers and like-minded idealists at the University of Cape Town and the University of the Western Cape to plan a bi-national coalition on "eco-social justice."

This coalition, he says, will exchange ideas to help cities like Cape Town and New York deal with poverty, environmental racism, and ethnic divisiveness—conditions that undermine eco-social justice.

### Society and the Environment

Eco-social justice does not have a precise definition. Mr. Rasmussen says the movement is about more than the environment: It's about the equitable distribution of natural resources, and the way people from different backgrounds and economic classes treat one another.

"There's a connection between the health and well-being of human society and the health and well-being of nature," Mr. Rasmussen says. "When resources are exhausted, people suffer. When the society is a very unjust one, with major gaps between the rich and the poor, the world of nature is also negatively affected."

Such degradation, he says, has been going on since humans made the switch to agriculture from hunting and gathering 10,000 years ago. But continuing this habit of sacrificing nature for the sake of society, he says, will eventually destroy the planet's life support system.

"The present course is unsustainable," he says. "How do we get from the unsustainable present to the sustainable future?" He offers an answer to that question in his book *Earth Community, Earth Ethic,* to be published this fall by Orbis Books. In the book, he says that the agricultural, industrial, and informational revolutions "reorganized society so as to produce more effectively," but also "reconfigured nature for the sake of society."

To reverse the damage, he says there must be a fourth revolution, one that will teach society how to be more like nature, where the waste from one organism becomes food for the next.

For example, in Karundborg, a small city in Denmark, the power plant's excess steam is given to the oil refinery and to the pharmaceutical company, where it's used for heat and as a source of power for equipment.

The oil refinery uses the water a second time as a coolant for its refining process. The surplus gas produced from the refining process is stripped of sulfur and sold to the sheet-rock factory and power plant. The sulfur is sold to the chemical company; the factory also uses some of the sulfur as a substitute for the mineral, gypsum. The ash generated from the burned coal is used to build roads.

### Work in Harlem

Big cities also can be scenes of ecological and social justice, he says. He offers, as an example, Bernadette Kosar, a Harlem community activist who works with teen-agers to clear vacant lots and then plant gardens. She also visits grade schools to plant gardens on small plots in the play areas.

"Then she teaches the kids about soil, gardening, and protecting the neighborhood," Mr. Rasmussen says. "Pretty soon, it's not just the kids who care about the garden, but the people in the neighborhood. They become protective of it, volunteer their time to keep it up. So in the end she has not only provided a green space, but she's provided a community-building project."

Kalundborg's changes were on a larger scale, Mr. Rasmussen says. But the idea—there, in New York, and in Cape Town—is to make the necessary incremental changes in a community "from the bottom up."

Union Theological could have tried this approach with any number of countries, but Mr. Rasmussen says Union has had a relationship with South Africa since the 1960s, when Union students—in protest of apartheid—initiated the first boycott against American banks that were lending money to the South African government.

### Sharing Information

As part of the new effort, the University of Cape Town will share with Union the results of a study to explore what the churches in South Africa can do to end the ecological degradation in rural and urban parts of the country.

A representative from Harlem Initiatives Together, a community group of which Union is a member, will travel this spring to South Africa to share the group's programs with community activists there.

Barney Pityana, South Africa's Human Rights Commissioner, and Stanley Mahoba, a South African Methodist minister, will visit Union this month to discuss urban social and environmental issues.

Officials at Union are leading a project to incorporate the shared information into curriculum changes at Union and at the two universities in South Africa.

"So what we've got going," he says, "is an ongoing exchange of research, and an exchange of information with a view toward what is the proper action to be taken to address the social and environmental degradation in both places."

## Further Readings


Adas, Michael. *Machines as the Measure of Men: Science, Technology, and Ideologies of Western Dominance*. Ithaca, N.Y.: Cornell University Press, 1989.

Baker, Henry E. *The Colored Inventor: A Record of Fifty Years*. New York: The Crisis Publishing Co., 1913.

Bedini, Silvio. *The Life of Benjamin Banneker*. New York: Charles Scribner's Sons, 1972.

Bragg, Janet Harmon. *Soaring above Setbacks: The Autobiography of Janet Harmon Bragg, African American Aviator*. Washington, D.C.: Smithsonian Institution Press, 1996.

Brooks, Sara. *You May Plow Here: The Narrative of Sara Brooks*, ed. Thordis Simonsen. New York: W. W. Norton & Co., 1986.

Carney, Judith Ann. *Black Rice: The African Origins of Rice Cultivation in the Americas*. Cambridge, Mass.: Harvard University Press, 2001.

Dew, Charles B. *Bond of Iron: Master and Slave at Buffalo Forge*. New York: W. W. Norton & Co., 1994.

Dinerstein, Joel. *Swinging the Machine: Modernity, Technology, and African American Culture between the Wars*. Amherst: University of Massachusetts Press, 2003.

Dorson, Richard M., comp. *American Negro Folktales*. Greenfield, Conn.: Fawcett Publications, 1967.

Douglass, Frederick. *Life and Times of Frederick Douglass*. Hartford, Conn.: Park Publishing Co., 1882.

Fouche, Rayvon. *Black Inventors in the Age of Segregation: Granville T. Woods, Lewis H. Latimer & Shelby J. Davidson*. Baltimore, Md.: Johns Hopkins University Press, 2003.

Hindle, Brooke. *Technology in Early America: Needs and Opportunities for Study*. Chapel Hill: University of North Carolina Press, 1966.

Green, Venus. *Race on the Line: Gender, Labor, and Technology in the Bell System, 1880–1980*. Durham, N.C.: Duke University Press, 2001.

James, Portia P. *The Real McCoy: African-American Invention and Innovation, 1619–1930*. Washington, D.C.: Smithsonian Institution Press, 1989.

Loving, Neal V. *Loving's Love: A Black American's Experience in Aviation*. Washington, D.C.: Smithsonian Institution Press, 1994.

Meaders, Daniel. *Advertisements for Runaway Slave in Virginia, 1801–1820*. New York: Garland Publishing, 1997.

Northrup, Herbert R. *The Negro in the Aerospace Industry,* Report No. 2, The Racial Policies of American Industry. Philadelphia, Penn.: University of Pennsylvania Press, 1968.

Pennington, James W. C. *The Fugitive Blacksmith; or, Events in the History of James W. C. Pennington.* 3d ed., 1850.

Pursell, Carroll. *The Machine in America: A Social History of Technology.* Baltimore, Md.: Johns Hopkins University Press, 1995.

Sinclair, Bruce, ed. *Technology and the African-American Experience: Needs and Opportunities for Study.* Cambridge, MA: MIT Press, 2004.

Sprague, Stuart Seely, ed. *His Promised Land: The Autobiography of John P. Parker, Former Slave and Conductor on the Underground Railroad.* New York: W. W. Norton, 1996.

Trotter, Joe William, Jr. *The African American Experience.* Boston: Houghton Mifflin Co., 2001.

United States, Department of Energy, with the National Technical Association and NASA. *Careers in Science and Technology.* Washington, D.C.: GPO, September 1993.

———, Department of Labor, Women's Bureau. *Women at Work: A Century of Industrial Change,* Bulletin No. 115. Washington, D.C.: GPO, 1934.

———, National Science Foundation. *Women and Minorities in Science and Engineering.* Washington, D.C.: NSF, January 1982.

Washington, Booker T. *Working With the Hands.* New York: Doubleday. Page & Co., 1904.

Windley, Lathan Algerna. *A Profile of Runaway Slaves in Virginia and South Carolina from 1730 through 1787.* New York: Garland Publishing, 1995.

Windley, Lathan A., comp. *Runaway Slave Advertisements: A Documentray History From 1730s to 1790. Volume I, Virginia and North Carolina.* Westport, Conn.: Greenwood Press, 1983.

# Index

Printed in the United States
by Baker & Taylor Publisher Services